Fishers' Knowledge in Fisheries Science and Management

Coastal Management Sourcebooks series

Coping with Beach Erosion

Remote Sensing Handbook for Tropical Coastal Management

*Underwater Archaeology and Coastal Management:
Focus on Alexandria*

Fishers' Knowledge in Fisheries Science and Management

Edited by
Nigel Haggan, Barbara Neis and Ian G. Baird

Coastal Management Sourcebooks 4

UNESCO Publishing

The ideas and opinions expressed in this publication are those of the authors and are not necessarily those of UNESCO and do not commit the Organization.

The designations employed and the presentation of material throughout this publication do not imply the expression of any opinion whatsoever on the part of UNESCO concerning the legal status of any country, territory, city or area or of its authorities or concerning the delimitation of its frontiers or boundaries.

Published by the United Nations Educational, Scientific and Cultural Organization
7, place de Fontenoy, 75352 Paris 07 SP, France

Cover: *Halibut* © Lyle Wilson. Acrylic on rag paper, 11.5 × 17 inches
Typeset by Desk
Printed by Barnéoud, Bonchamp-lès-Laval

ISBN 978-92-3-104029-0

© UNESCO 2007

All rights reserved

Printed in France

This book is dedicated to the memory of Bob Johannes for his groundbreaking research on traditional marine resource management and knowledge.

Contents

Contributors	11
Foreword	21
Preface	23
Acknowledgements	25
List of figures	27
List of tables	31
List of acronyms	33
Introduction: Putting fishers' knowledge to work	35
Nigel Haggan, Barbara Neis and Ian G. Baird	
1. The value of anecdote	41
R.E. Johannes and Barbara Neis	
I. Indigenous practitioners and researchers	59
2. Life supports life	61
Klah-Kist-Ki-Is; Chief Simon Lucas	
3. My grandfather's knowledge: First Nations' fishing methodologies on the Fraser River	71
Arnie Narcisse	
4. Indigenous technical knowledge of Malawian artisanal fishers	83
Edward Nsiku	
5. Application of Haida oral history to Pacific herring management	103
Russ Jones	
6. The use of traditional knowledge in the contemporary management of a Hawaiian community's marine resources	119
Kelson K. Poepoe, Paul K. Bartram and Alan M. Friedlander	
II. Indigenous and artisanal fisheries	145
7. Traditional marine resource management in Vanuatu: Worldviews in transformation	147
Francis R. Hickey	

8. Tropical fish aggregations in an indigenous environment in northern Australia: Successful outcomes through collaborative research 169
Michael J. Phelan
9. Sustaining a small-boat fishery: Recent developments and future prospects for Torres Strait Islanders, Northern Australia 183
Monica E. Mulrennan
10. Sawen: Institution, local knowledge and myth in fisheries management in North Lombok, Indonesia 199
Arif Satria
11. Fishers' perceptions of the seahorse fishery in the central Philippines: Interactive approaches and an evaluation of results 221
Jessica Meeuwig, Melita Samoilys, Joel Erediano and Heather Koldewey
12. Local ecological knowledge and small-scale freshwater fisheries management in the Mekong River in Southern Laos 247
Ian G. Baird
13. Use of fishers' knowledge in community management of fisheries in Bangladesh 267
Parvin Sultana and Paul Thompson
14. The role of fishers' knowledge in the co-management of small-scale fisheries in the estuary of Patos Lagoon, Southern Brazil 289
Daniela C. Kalikoski and Marcelo Vasconcellos
15. The value of local knowledge in sea turtle conservation: A case from Baja California, Mexico 313
Kristin E. Küyük, Wallace J. Nichols and Charles R. Tambiah
16. Can historical names and fishers' knowledge help to reconstruct the distribution of fish populations in lakes? 329
Johan Spens

III. Commercial fisheries 351
17. Putting fishers' knowledge to work: Reconstructing the Gulf of Maine cod spawning grounds on the basis of local ecological knowledge 353
Ted Ames
18. Integrating fishers' knowledge with survey data to understand the structure, ecology and use of a seascape off south-eastern Australia 365
Alan Williams and Nicholas Bax
19. Using fishers' knowledge goes beyond filling gaps in scientific knowledge: Analysis of Australian experiences 381
Pascale Baelde

20. Fishers' knowledge? Why not add their scientific skills while you're at it? 401
    Richard D. Stanley and Jake Rice
21. The changing face of fisheries science and management 421
    Nigel Haggan and Barbara Neis
22. The last anecdote 433
    Ian G. Baird

Contributors

Ted Ames
Penobscot East Resource Center
PO Box 27
Stonington, ME 04681
Tel: +1 207 367- 2708
amest@verizon.net
Ted Ames is a lifelong commercial fisherman from Stonington, Maine and has been active in fisheries research throughout his career. He holds an MS in Biochemistry and has numerous publications in fisheries-related topics. Ted is an active member of the Stonington Fisheries Association and is current president of the Penobscot East Resource Center, an organization supporting community-based fisheries management.

Pascale Baelde
School of Anthropology, Geography and Environmental Studies
University of Melbourne
24 Ashmore Street, Brunswick East, 3057, Australia
pbaelde@unimelb.edu.au
Pascale Baelde worked as a government fisheries scientist for ten years before becoming an independent researcher focusing on how to integrate fishers' and scientists' knowledge and explore the meaning and expectations of partnership approaches between industry, government and NGOs. She assists industry groups in developing environmental management systems to promote industry stewardship and self-governance.

Ian G. Baird
1235 Basil Ave
Victoria, B.C.,
V8T 2G1
Canada
ianbaird@shaw.ca
Ian Baird has worked with local fishers in the Mekong River Basin for fourteen years. He is a PhD candidate in the Geography Department of the University of British Columbia, Vancouver, Canada, and executive director of the Global Association for People and the Environment (GAPE), a Canadian NGO mainly active in mainland South East Asia.

Paul K. Bartram
Hui Malama o Mo'omomi
P.O. Box 173,
Kualapu'u,
Hawaii 96757
US
pbartram@tripleb.com
Paul K. Bartram, of part Hawaiian ancestry, augmented his formal education in marine biology with the study and application of traditional Hawaiian ecological thinking under Mr Kelson Poepoe. Bartram integrates both traditional and contemporary information to further community resource management along the northwest coast of Moloka'i.

Nicholas Bax
CSIRO Marine and Atmospheric Research
PO Box 1538
Hobart
Tasmania
Australia, 7001
nic.bax@csiro.au
Dr Bax studies the impact, management and mitigation of biodiversity loss on ecosystem services. He works with Alan Williams in the interpretation of individual fishers' knowledge and is particularly interested in determining approaches that would manage habitat as a productive asset for biodiversity and fisheries.

Joel A. Erediano
429 Balico Cor. Balite Streets
Poblacion, Talibon 6325, Bohol
Philippines
joelerediano@hotmail.com
Joel Erediano has a political science degree from the University of the Philippines. He worked for ten years as a community organizer in the Philippines and was Social Research Officer for Project Seahorse from 1999 to 2003. His work focuses on empowering fishing communities to manage their own resources, form their own cooperatives, and thereby take initiative and control of the resources on which they depend.

Alan M. Friedlander
NOAA, NOS, National Centers for Coastal Ocean Science
Biogeography Team, and Oceanic Institute
Makapu'u Point/41-202 Kalanianaole Highway

Waimanalo
Hawaii 96795, US
afriedlander@oceanicinstitute.org
Alan Friedlander is Pacific Coral Reef Science Coordinator for NOAA's National Ocean Service-Biogeography Team and holds a PhD from the University of Hawaii. He has been involved in the research and management of marine resources for over twenty years in Hawaii, the Caribbean, and throughout the Indo-Pacific region.

Heather Koldewey
Senior Curator, Aquarium
Zoological Society of London
Regent's Park
London NW1 4RY
heather.koldewey@zsl.org
Heather Koldewey is Senior Aquarium Curator for the Zoological Society of London and is actively engaged in many aspects of aquatic conservation. Previously she conducted research in salmonid genetics. She co-founded and is Associate Director of Project Seahorse and a board member of the Project Seahorse Foundation for Marine Conservation (Philippines).

Nigel Haggan
Nigel Haggan and Associates
1777 East 7th Ave.
Vancouver, BC
V5N 1S1
Canada
n.haggan@fisheries.ubc.ca
Nigel Haggan is a consultant and research associate at the University of British Columbia Fisheries Centre. His research focus is on how different systems of knowledge can contribute to collective understanding of past and present aquatic ecosystems, the full range of values they provide and how to predict the ecological and social as well as economic consequences of our actions.

Francis R. Hickey
Vanuatu Cultural Centre
P.O. Box 184, Port Vila,
Vanuatu, SW Pacific
francishi@vanuatu.com.vu
Francis R. Hickey has a degree in marine biology from the University of Victoria, BC. He moved to Vanuatu in 1991 to take up a five-year management post with the

Vanuatu Department of Fisheries. Since 1996, he has been a researcher with the Vanuatu Cultural Centre, documenting and supporting traditional fisheries and management systems.

R.E. (Bob) Johannes
Deceased
Robert E. (Bob) Johannes earned his MSc in Fisheries at the University of British Columbia, Canada, and his PhD in Marine Biology from the University of Hawaii. His work with traditional small-scale fishers in the tropical Indo-Pacific led to a rapid appreciation of the enormous value of their knowledge for science, conservation and management. He dedicated his life to seeking out this knowledge and exploring how it could go where 'classic' fisheries science cannot. He started a new field of inquiry, inspired a generation of researchers and challenged us to find even more effective ways for communities, natural and social scientists to increase our knowledge of the ocean and how to live with it. He was the recipient of several prestigious awards, including a Guggenheim Foundation Fellowship, the CSIRO Medal and a Pew Fellowship.

Russ Jones
PO Box 98
Queen Charlotte City, Haida Gwaii
Canada
V0T 1S0
rjones@island.net
Russ Jones, a Haida from the Raven-Wolf clan of Tanu, comes from a fishing family and spent a decade as a fisher. He has an MSc in fisheries and for the past sixteen years has been a consultant to First Nation clients in British Columbia on fisheries matters.

Daniela C. Kalikoski
Department of Geosciences
Fundação Universidade Federal do Rio Grande
CEP: 96201-900
Rio Grande, RS
Brazil
danielak@furg.br
Daniela C. Kalikoski is a professor at the Federal University of Rio Grande (FURG, Brazil). She did her PhD at the Institute for Resources, Environment and Sustainability (University of British Columbia, Canada), in resource management and environmental studies. Her research focuses on fisheries co-management, community-based management, and traditional ecological knowledge.

Kristin E. Küyük
c/o 23 Filomena Drive
Pittsfield, Massachusetts 01201
US
Kristin_Kuyuk@yahoo.com
Kristin Küyük is an applied anthropologist with an interest in community-based conservation of both cultural and natural resources. Her academic background is in Wildlife Management, Anthropology and Environmental Ethics; she holds degrees from the University of New Hampshire and Oregon State University.

Chief Simon Lucas, Klah-Kist-Ki-Is
Hesquiaht Nation Council
Box 2000, Tofino, BC
Canada
V0R 2Z0
Tel: (250) 670 1100
Chief Simon Lucas, Klah-kisht-ki-is, is tenth-ranking hereditary chief of Hesquiat on the west coast of Vancouver Island. His entire life has been dedicated to serving his people, and their rights to manage and access salmon and other aquatic resources for sustenance, livelihood, and mental, emotional, physical and spiritual health. He received an honorary degree in 2002 for his local, national and international work connecting clean water, healthy ecosystems, environmental and individual wellness.

Jessica Meeuwig
Oceans Liaison Officer
National Oceans Office,
Australian Department of Environment and Heritage
10/16 Phillimore St.
Fremantle, WA 6160 Australia
Jessica.Meeuwig@oceans.gov.au
Jessica Meeuwig holds an interdisciplinary master's degree and a PhD in Marine Ecology. She has spent twenty years working in community-focused marine resource management in the Caribbean, South East Asia, Canada and Australia.

Monica E. Mulrennan
Dept. of Geography, Planning and Environment
Concordia University
1455 de Maisonneuve Blvd. W
Montreal, QC
Canada, H3G 1M8
monica.mulrennan@concordia.ca

Monica Mulrennan is an associate professor in the Department of Geography, Planning and Environment at Concordia University, Montreal. Her research interests include community-based resource management, local knowledge, protected areas management and environmental change in the context of the traditional lands and seas of Torres Strait Islanders (northern Queensland) and James Bay Crees (northern Quebec). She was born, raised and educated in Ireland.

Arnie Narcisse
Chair, BC Aboriginal Fisheries Commission
Suite 707 – 100 Park Royal
West Vancouver, BC
Canada. V7T 1A2
ANarcisse@bcafc.org
Arnie Narcisse is a Stlatlimx-Blackfeet Indian and Chair of the BC Aboriginal Fisheries Commission. He is dedicated to advancing conservation and a major role for aboriginal people in all aspects of aquatic ecosystem and fisheries research, management and use, consistent with aboriginal rights and title. He was appointed to the Pacific Salmon Commission in 2004.

Barbara Neis
Dept. of Sociology, Memorial University,
St. John's, NL
Canada A1C 5S7
bneis@mail.mun.ca
Barbara Neis is Co-Director of SafetyNet, a community research alliance on health and safety in marine and coastal work at Memorial University, Newfoundland and Labrador, Canada. Her main research focus is on finding ways to link the knowledge of fishers, their families and processing industry workers with science to help us better understand the interactive effects of environmental, industrial and policy restructuring on the health of people and the environment.

Wallace J. Nichols
Post Office Box 324
Davenport, California 95017
US
j@oceanrevolution.org
Wallace J. Nichols is a research associate at the California Academy of Sciences. His academic background is in wildlife ecology and evolutionary biology, natural resources economics and policy. He collaborates with numerous non-profit organizations and government agencies on ocean and coastal research, conservation and education programmes. He holds graduate degrees from Arizona and Duke University.

Edward Nsiku
25-13262 72nd Avenue
Surrey BC
V3W 2N6
Canada
emnsiku@aol.com
Edward Nsiku is a Malawian fisheries researcher with an MSc in resource management and environmental studies from UBC, Canada. He worked for the Malawi Fisheries Department between 1986 and 1997 and participated in a course on human and social perspectives in natural resources management by IUCN/ROSA and University of Zimbabwe in 1994.

Michael J. Phelan
Department of Business, Industry and Resource Development
Berrimah Research Farm, Darwin, Australia.
GPO Box 3000, Darwin NT 0801, Australia
Tel: 08 8999 2144
michael.phelan@nt.gov.au
Michael J. Phelan is a research scientist with a background in natural and social science. During the present study he was employed by Balkanu and the Queensland Department of Primary Industries and Fisheries. He has worked on several projects focusing on indigenous fishers and is presently employed with the Northern Territory's Fisheries Research Group.

Kelson K. Poepoe
Hui Malama o Mo'omomi
P.O. Box 173,
Kualapu'u,
Hawaii 96757
US
Kelson K. Poepoe is a lifelong resident of a native Hawaiian community engaged in subsistence fishing along the north-west coast of Moloka'i, Hawaii. During forty-five years of local fishing experience, he has verified the teachings of his ancestors and expanded on this traditional knowledge to become the community's resource manager.

Jake Rice
Canadian Science Advisory Secretariat
Fisheries and Oceans Canada
200 Kent St.
Ottawa ON

Canada, K1A OE6
Ricej@dfo-mpo.gc.ca
Jake Rice is Director of Advice and Assessment, Fisheries and Oceans Canada. He provides scientific advice on ocean and fisheries policy and management in Canada and internationally. He has also conducted research on marine ecosystem dynamics and has managed marine research divisions on the Atlantic and Pacific coasts.

Melita Samoilys
Regional Coordinator — Marine and Coastal Ecosystems
IUCN Eastern Africa Regional Office
PO Box 68200, Nairobi, Kenya
melita.samoilys@iucn.org
Melita Samoilys spent twenty years as a research scientist specializing in artisanal coral reef fisheries and fish reproduction in the Pacific and Indian Oceans. She now works more holistically in marine conservation in Eastern Africa. She is a founding Board member of the Project Seahorse Foundation for Marine Conservation in the Philippines.

Arif Satria
Center for Coastal and Marine Resources Studies
Bogor Agricultural University
Kampus IPB Darmaga, Lingkar Kampus 1
Bogor 16680, Indonesia
arifsatria@ipb.ac.id
Arif Satria is a PhD candidate at Kagoshima University. His research interests include fisheries sociology, political ecology, marine policy and community-based fisheries management. His recent position is Head of Coastal Communities and Institutional Empowerment at the Center for Coastal and Marine Resources Studies, Bogor Agricultural University, Indonesia.

Johan Spens
Department of Aquaculture, SLU,
S-901 83 Umeå
Sweden
johan.spens@vabr.slu.se
Johan Spens is an environmental manager in the municipality of Örnsköldsvik, Sweden. During the last twenty years, chiefly in this position, he has worked in the field of limnology and conservation. Currently he is finishing his graduate studies at the Department of Aquaculture, Swedish University of Agricultural Sciences in Umeå.

Richard D. Stanley
Pacific Biological Station
Fisheries and Oceans, Canada
Nanaimo, BC
Canada, V9T 6N7
Stanleyr@pac.dfo-mpo.gc.ca
Rick Stanley has been a Research Biologist for twenty-five years with Fisheries and Oceans Canada in the Groundfish Section at the Pacific Biological Station in Nanaimo. His work has focused primarily on the biology and stock assessment of rockfishes (*Sebastes* spp.).

Parvin Sultana
Flood Hazard Research Center
Middlesex University
Queensway
Enfield
EN3 4SA
UK
parvin@dhaka.agni.com
Parvin Sultana is a researcher, originally a zoologist, with a PhD in natural resource management; she has worked in agriculture, water resources and fisheries, mostly in Bangladesh, for more than twenty-five years. Recently she has focused on community-based fisheries management, participatory processes and institutional arrangements.

Charles Tambiah
ANU Institute for Environment
GPO Box 2527
Canberra ACT 2601
Australia
turtlecommunity@yahoo.com
Charles Tambiah is a research and conservation facilitator with a background in biological and social sciences. He focuses on multidisciplinary and participatory approaches to project management. His research interests include community-based conservation of sea turtles and integrated conflict resolution, with collaborations in over twenty countries, primarily in the developing world.

Paul Thompson
Flood Hazard Research Center
Middlesex University
Queensway

Enfield
EN3 4SA
UK
paul@agni.com
Paul Thompson is a social science researcher, with a PhD in Development Studies, who has worked mainly on floodplain resource management in Bangladesh. He has been involved in projects developing and assessing community management of floodplain fisheries for the last nine years.

Marcelo Vasconcellos
Marine Resources Service
Food and Agriculture Organization (FAO)
Rome, Italy 00100
marcelo.vasconcellos@fao.org
Marcelo Vasconcellos is a fishery resources officer at the Food and Agriculture Organization (FAO). He holds a PhD from the University of British Columbia in Resource Management and Environmental Studies. One of the central themes of his research has been the development of ecosystem approaches to fisheries assessment and management.

Alan Williams
CSIRO Marine and Atmospheric Research
PO Box 1538
Hobart
Tasmania
Australia, 7001
alan.williams@csiro.au
Dr Alan Williams specializes in the mapping of deep-sea marine habitat and biodiversity toward its conservation and maintenance of fisheries productivity. He works closely with individual fishers to include their detailed knowledge in management decisions, and to aggregate their individual knowledge into an industry-wide perspective.

Foreword

In 1985, UNESCO published the groundbreaking volume edited by Ken Ruddle and Richard (Bob) Johannes on *The Traditional Knowledge and Management of Coastal Systems in Asia and the Pacific*. In those early days, environmental knowledge possessed by local and indigenous communities was beginning to gain recognition, but only in isolated circles and at disparate localities. This changed with the 1992 Convention on Biological Diversity and its explicit requirement in Article 8(j) that contracting Parties must *'respect, preserve and maintain the knowledge, innovations and practices of indigenous and local communities'*. Today, local and indigenous knowledge is widely recognized as a key element in biodiversity conservation, even though its role continues to stir considerable controversy and debate.

The present volume caps more than twenty years of partnership between UNESCO and Bob Johannes, and his tireless efforts to bridge the persistent gap between scientists and fishers. On UNESCO's side, this partnership was initiated in the early 1980s by its Coastal Marine programme. This programme has evolved into the Coastal Regions and Small Islands Platform (CSI), but its focus remains on interdisciplinary work that crosses the boundaries between the natural and social sciences, and between ecological and cultural systems. UNESCO's Local and Indigenous Knowledge Systems (LINKS) programme, established in 2002, expands this work with an explicit focus on traditional knowledge and customary management. LINKS embraces several of the goals espoused by Bob Johannes and exemplified in this volume. It focuses on empowering local knowledge holders in biodiversity governance by strengthening collaboration among local communities, scientists and decision-makers. It also contributes to the safeguarding of traditional knowledge and practices within local communities by enhancing their transmission to the next generation.

As the fourth contribution to the UNESCO series entitled *Coastal Management Sourcebooks*, this volume provides analyses and case studies that convincingly support Bob Johannes' contention that, throughout the world, the knowledge of local fisherfolk must become an integral part of decision-making on renewable resource management.

Walter Erdelen
Assistant Director-General for Natural Sciences
UNESCO

Preface

This book had its origin in a 2001 conference called 'Putting Fishers' Knowledge to Work' inspired by the late Bob Johannes. For Bob, this conference was one more step in a lifetime of work concerned with bridging the gap between fishers and natural and social scientists. It was also part of his larger vision to create an active, global community of researchers, fishers and fishing communities working together to deepen our collective understanding of marine resources and management (see Ruddle, 2003). At the conference, Bob brought to our attention that although there are many international centres devoted to terrestrial indigenous knowledge, no such centre exists for fisheries knowledge. He urged conference participants to seek support to create one or more such centres dedicated to the collection, synthesis, archiving and dissemination of fishers' ecological knowledge.

This first major international conference of its kind explored the nature of indigenous, small-scale and industrial fishers' knowledge and its relevance for improved science and management in fisheries. The conference focused on examples of how fishers' knowledge was already being used to expand and strengthen fisheries science and management. In Johannes' words from the conference brochure:

> 'Small scale traditional fisheries are often set in environments where scientific knowledge is poor and conventional remedies are prohibitively costly. Yet local fishers often know much about where and when marine animals migrate or aggregate, how they behave and how fishing and marine environmental conditions have changed over time. Understanding this knowledge, and how fishers act on it, can contribute very substantially to marine resource management, environmental impact assessment and the location and size of marine protected areas. In developed commercial fisheries, local knowledge includes elements of the above, but other factors also come into play. Market constraints and technology changes, for example, can have major influences on fishing behaviour.'

The Fishers' Knowledge conference brought together researchers and practitioners from twenty-five countries and thirty-five aboriginal and indigenous nations and organizations. Chapters in the book were selected from some forty-eight contributions to the *Proceedings* (Haggan et al., 2003), and subjected to peer review and further development. The book brings together examples from indigenous, small-scale industrial and recreational fisheries in marine and freshwater environments across the globe. These examples of collaboration between traditional and 'modern' fisheries science and management are particularly refreshing after years of frustration and misunderstanding. The emerging synergy is reflected in the title, *Fishers' Knowledge in Fisheries Science and Management*.

The cover picture 'Halibut', seen through the eyes of Haisla Nation artist Lyle Wilson, conveys the continuing cultural and economic significance of what is now a major commercial species.

'The Haisla Beaver Clan were originally from a state now known as Alaska. During their migration, a series of supernatural events was observed by Clan members. The appearance of a giant halibut was one of these events. When the Beaver Clan eventually settled at their present location of Kitamaat, the Giant Halibut was taken as a crest.'

(Lyle Wilson, 2006)

We are a long way philosophically, culturally and economically, from a time when fish were spiritual beings, holding the power of life and death. Today, BC Halibut (*Hippoglossus stenclepis*) are managed as an individual transfer quota (ITQ) fishery, with license values as high as $US 1 million (CCPFH 2005), making them inaccessible to small-scale fishers and posing a problem for settlement of modern day treaties which require transfer of fish to Aboriginal peoples. If economic systems like this do not make it irrelevant, fishers' knowledge is a tremendous resource.

No one book can do justice to the wealth of fishers' knowledge that exists around the world. We hope that we have at least opened the door a bit wider.

Acknowledgements

Bob Johannes had the original idea of convening a conference at the University of British Columbia (UBC) and was a source of many ideas and inspiration in planning. Tony Pitcher, then Director of UBC, enthusiastically agreed to host the conference and provided strong ongoing support. We thank Claire Brignall and Janice Doyle for conference coordination and administrative assistance. We are grateful to the conference co-hosts: the UBC Fisheries Centre, the BC Aboriginal Fisheries Commission and the UBC First Nations House of Learning. We gratefully acknowledge funding from the government of Canada through the Departments of Fisheries and Oceans and Indian and Northern Affairs; BC Hydro and Power Authority; the British Columbia Ministry of Environment, Lands and Parks and Fisheries Renewal BC; UNESCO-Coastal Regions and Small Islands programme; and the David Suzuki Foundation. We are particularly grateful to our referees whose insightful comments strengthen and inform the contributions and to our authors for their unfailingly cheerful and positive response to our editorial suggestions.

We thank Lyle Wilson of the Haisla Nation for letting us use the painting *Halibut* for the cover art that gives the book its unique character. We are also grateful to our publishers for their support and patience over a long gestation period. We acknowledge the help of many others and our families, without their support this book would not exist.

Nigel Haggan, Vancouver, BC, Canada
Barbara Neis, St John's, Newfoundland, Canada
Ian G. Baird, Victoria, British Columbia, Canada

REFERENCES

CANADIAN COUNCIL OF PROFESSIONAL FISH HARVESTERS. 2005. A crisis of sustainability in the fish harvest labour force. Press Release, URL: http://www.ccpfh-ccpp.org/cgi-bin%5cfiles%5c050824-Press-Release-and-Backgrounder-E-Final.pdf (Last accessed 14 June 2006).

HAGGAN, N.; BRIGNALL, C.; WOOD, L. (eds). 2003. Putting fishers' knowledge to work. Fisheries Centre Research Reports, Vol. 11, No. 1, 504 pp. Available online at: www.fisheries.ubc.ca/publications/reports/report11_1.php (accessed on 30 May 2006).

Ruddle, K. (ed.). 2003. *Traditional marine resource management and knowledge.* Information Bulletin Special Edition, March 2004. Secretariat of the Pacific Community. Available online at: http://www.spc.int/coastfish/News/Trad/Sp1/Trad-Johannes.pdf (accessed on 4 September 2004).

List of figures

2.1 First Nations of British Columbia 63
2.2 Map showing landmarks and fishing areas described with inset showing location of Nuu-chah-nulth Tribes and La Perouse Bank. 65
3.1 Map of Stl'atl'imx territory, courtesy of Lillooet Tribal Council. The Stl'atl'imx people fished throughout the entire area delineated. The fishing sites described here are located between Bridge River to the north to just south of the Cayoosh Creek/Fraser confluence 72
3.2 Arnie and grandfather. West bank of the Fraser River, immediately south of the Bridge River confluence, ca. 1959 73
3.3 Arnie Narcisse with grandmother. West bank of the Fraser River, immediately south of the Bridge River confluence, ca. 1959 74
3.4 Drying racks 75
3.5 Arnie with gillnet rigged to gin pole 76
3.6 Arnie wants to use a dipnet. West bank of the Fraser River, immediately south of the Bridge River confluence, ca. 1959 77
3.7 Dipnetting. West bank of the Fraser River, immediately south of the Bridge River confluence, ca. 1959. 78
3.8 Intergenerational equity. My grandparents, my immediate family and my grandson 80
4.1 Map of Malawi showing main water bodies, fishing district towns and some cities 85
5.1 Map of Haida Gwaii showing herring locations 105
5.2 Herring catch (a) spawn length (b) and number of spawn records (c) at Skidegate Inlet, 1930–2001 107
5.3 Herring catch (a) and spawn length (b) in Haida Gwaii major stock area (Cumshewa-Louscoone), 1930–2001 108
6.1 (A) Location of the main Hawaiian Islands, (B) island of Moloka'i (Landsat 7 ETM/1G Satellite Imagery) and, (C) Mo'omomi and Kawa'aloa Bays located on the north shore of Moloka'i 121
6.2 Hawaiian moon calendar showing months, seasons and moon phases that are used to guide fishing activities. Names used for months in this calendar are specific to the island of Moloka'i 128

6.3 Hawaiian names for nights of the rising (*ho'onui*), full (*poepoe*) and falling (*emi*) moon phases and prohibition (*kapu*) periods 130
6.4 Comparison of fish biomass (t/ha) at Mo'omomi Bay and similar exposed north shore locations around the main Hawaiian Islands. Error bars are standard error of the mean 137
6.5 (a) Fork length (cm) of *moi* (*Polydactylus sexfilis*) harvest along windward O'ahu and in Mo'omomi Bay in 1999. (b) Length frequency distributions for *aholehole* (*Kuhlia sandvicensis*) caught at Hilo Bay after gillnet ban and at Mo'omomi Bay in 1999 138
7.1 Map of Vanuatu 149
8.1 Map of the north end of Cape York Peninsula 170
8.2 An adult *Protonibea diacanthus* 172
8.3 Composition of the size classes of *P. diacanthus* harvested in the Northern Peninsula Area in 1999 and 2000 175
8.4 Location of the area within the Northern Peninsula Area closed to the harvest of *P. diacanthus* under the regional agreement 177
9.1 Torres Strait, Northern Australia 185
10.1 Map of North Lombok, Nusa Tenggara Province, Indonesia 201
10.2 Scope of *sawen* 202
10.3 Two domains of management in *paer* philosophy underlying *sawen* 203
10.4 Linkages of ecology and authority triad 204
10.5 Organization structure of *Kelompok Nelayan Pantura Penyawen Teluk Sedayu*, Kayangan 212
11.1 Map of the Philippines showing the study area of Danajon Bank in northern Bohol, central Visayas 223
11.2 Structure of (a) feedback sessions to validate personal and fishing effort data and repeat scoping survey for catch and effort data, (b) focus group discussions on fishing ground habitat type and quality 226
11.3 Focus group discussion methods using graphic symbols to solicit information from seahorse fishers 228
11.4 The number of grounds fished per year on Danajon Bank, Bohol 231
11.5 Correlations of effort by (a) fisher and (b) fishing ground in the group of overlapping fishers (n = 71) and grounds (n = 25) for the Scoping (S) and Feedback (F) studies 233
11.6 Mean fisher's relative ranking (FRR) of habitat quality by fishing ground with standard errors demonstrating general consistency of response among fishers for each ground 234

LIST OF FIGURES

11.7	Correlation between percentage of fishers indicating a site is 'good' and percentage of rubble cover measured on ecological surveys	235
11.8	Trends in status of (a) fishing ground condition, (b) seahorse populations and (c) fishers' livelihood assessed by fishers from past (1990), present (2000) to future (2010)	236
12.1	Map of Laos showing Siphandone Wetlands area in Khong District	249
12.2	Khong district showing sixty-three villages with regulations to manage and conserve fish and other aquatic resources in the Mekong River, streams, wetlands and rice paddies	255
13.1	Map of Bangladesh showing locations mentioned in text and main river system	270
13.2	Estimated total number ('00) of gear units operated in Ashurar Beel, 1997–2002	281
13.3	Estimated volume of fish (tons) caught in Ashurar Beel, 1997–2002 (excluding sanctuary harvest in 1999)	281
14.1	Location of the Patos Lagoon estuary in Southern Brazil	291
14.2	Small-scale fisheries landings in the estuary of Patos Lagoon	294
14.3	Four-phase model of estuarine and coastal fisheries resource dynamics	295
14.4	Fishing calendars for small-scale fisheries in the estuary of Patos Lagoon and coastal waters during the 1960s and the early 1990s	300
15.1	Survey results showing occasions for turtle consumption in Magdalena Bay	314
15.2	Map of the Baja California Peninsula	316
15.3	Map of Estero Banderitas, an estuary in the northern reach of Magdalena Bay identified by fishers as a productive area for sea turtles. This map shows movements of turtles tracked over several years	320
16.1	The 1,509 lakes within the study area, the northern boreal region of Sweden	332
16.2	Coverage of sixty-three fishery management organizations (FMOs) within the study area	333
16.3	Brown trout and non-brown trout lakes within the study area	337
16.4	Scale bar (A.D.) illustrating temporal range of methods to reconstruct brown trout distribution in lakes within the current study	340
16.5	*Rö*-named lakes with brown trout populations	341

17.1 Historical cod spawning grounds and recent distribution patterns of cod eggs in the northern Gulf of Maine 361
18.1 A coarse-scale map of habitats – the 'fisher map'– made for the 'Twofold Shelf Bioregion' an area of the continental shelf off SE Australia. The map is a mix of fisher-delineated geomorphological features (mostly sediment plains and rocky banks) ground-truthed with physical samples and photographs from surveys 371
19.1 SE Australian fishing zone 383
20.1 Location of widow rockfish study area (inner box) and silvergray rockfish assessment regions 405
20.2 Location of widow rockfish study site off the north-west coast of Vancouver Island 406
20.3 Location of transects relative to longitudinal axis of the widow rockfish shoal 407
20.4 Locations of bottom-trawl tows which captured at least 200 kg of silvergray rockfish 411
20.5 Percent composition by age of silvergray rockfish samples taken during February 2001 observer trip 412
20.6 Location of silvergray rockfish samples used in 2000 silvergray rockfish assessment for area 5E 413

List of tables

6.1	Important seafood food resources and methods of harvest for the Hoʻolehua homesteaders	124
6.2	Moʻomomi Bay fish spawning calendar for the year 2000 for key resource species. Black boxes indicate months of peak spawning. Grey boxes indicate other months when spawning was observed	131
6.3	Season movement patterns of *aholehole* (*Kuhlia xenura*) in relation to changes in habitat at Moʻomomi Bay	132
6.4	Seasonal movement and aggregation of *moi* around the island of MolokaʻI	133
6.5	Observations of the edible seaweed *limu kohu* (*Asparagopsis taxiformis*) at the major shallow-water (0–1 m) harvest site, January 2000–January 2001	135
6.6	Environmental factors affecting the distribution and growth of the edible seaweed *limu kohu* (*Asparagopsis taxiformis*) in and around Moʻomomi Bay	136
10.1	Roles of all *mangkus* in traditional resources management practices	205
10.2	Some original *sawen* practices and their underlying scientific rationales	206
10.3	Goals of revitalized *sawen* rules and sanctions for violation	211
10.4	Comparison of original *sawen* and revitalized *sawen*	213
11.1	List of villages participating in the scoping and community meetings	227
11.2	Annual lantern fishing effort on Danajon Bank as reported by fishers from the scoping and feedback surveys	232
11.3	Ranking of the responses from the marine resource discussions on the destruction of fishing grounds from most important to least important	237
11.4	Ranking of the responses from the marine resource discussions on declines in seahorse populations from most important to least important	238
11.5	Ranking of the responses from marine resource discussions on the status of fishers' livelihoods	238

13.1 Changes in natural resource status, use and access
in Shuluar Beel 274
13.2 Changes in fisher's ranking of fish species (importance in catch)
in Shuluar Beel 275
13.3 Main problems identified in a Participatory Action Plan
Development Workshop in 2002 in Shuluar Beel 277
13.4 Local knowledge of reasons, impacts and potential solutions for
main natural resource problems in Shuluar Beel (summarized
across stakeholder groups) 278
13.5 Species composition of catch (per cent by weight) from sampling
on 2–4 days/fortnight in Ashurar Beel 282
13.6 Changes in fishing income, house construction and sanitation
(1996–2001) in Ashurar Beel 283
13.7 Mean difference in change in perceived indicator scores
(1997 compared with 2001) between NGO participants
and non-NGO respondents in Ashurar Beel 284
14.1 Summary of biology and lifecycle of main small-scale fisheries
resources in the estuary of Patos lagoon 294
14.2 Comparison between selected principles of the Code of Conduct
for Responsible Fisheries (FAO, 1997) and adjustments to local
fisheries management suggested by small-scale fishers during
interviews and Forum of Patos Lagoon meetings 305
16.1 *Rö*-named lakes and methods elucidating past and present brown
trout populations 338
16.2 Estimation of maximum (E_{MAX}) permanent extinctions, 1672–1920 . 341
16.3 Factors associated with the extinction of brown trout populations
in *Rö*-named lakes 342
18.1 Sources and types of information used to describe the continental
shelf seascape in the south-eastern South East Fishery during
the 'ecosystem project' 369

List of acronyms

AFLP	Amplified fragment length polymorphisms
AFMA	Australian Fisheries Management Authority
ANZECC	Australian and New Zealand Environment and Conservation Council
BC	British Columbia
BMC	Beel management committee
BWDB	Bangladesh Water Development Board
CAS	Catch Assessment Survey (Malawi)
CBFCM	Community-based fisheries co-management
CBFM	Community-based fisheries management
CBNRM	Community-based natural resource management
CESVI	Cooperazione Sviluppo (Cooperation and Development)
CGRCS	Canadian Groundfish Research and Conservation Society
CM	Co-management
CO	Community organizer
CPR	Common-property resources
CPUE	Catch per unit effort
CSIRO	Australian Commonwealth Scientific and Research Organization
DFO	Canadian Department of Fisheries and Oceans
DLNR	(Hawaiian) State Department of Land and Natural Resources
DPA	Department of Fisheries and Aquaculture (Brazil)
EEZ	Exclusive economic zones
EIA	Environmental impact assessment
EMC	Environmental Management Committee
ENSO	El-Niño Southern Oscillation
EPCDSWP	Environmental Protection and Community Development in Siphandone Wetland Project
FAO	Food and Agriculture Organization of the United Nations
FCZ	Fish conservation zone
FD	Fisheries Department (Malawi)
FK	Fishers' knowledge
FMO	Fishery management organization
FRDC	Fisheries Research and Development Corporation (Australia)
FRR	Fisher's relative ranking

FS	Frame survey
GIS	Geographic information system
IBAMA	Federal Institute for the Environment (Brazil)
ICES	International Council for the Exploration of the Sea
IIRR	International Institute of Rural Reconstruction
IPHC	International Pacific Halibut Commission
IRN	International Rivers Network
ITK	Indigenous technical knowledge
ITQ	Individual transferable quota
IUCN	World Conservation Union
IVQ	Individual vessel quota
LCFDPP	Lao Community Fisheries and Dolphin Protection Project
LEK	Local ecological knowledge
LMNLU	Lembaga Musyawarah Nelayan Lombok Utara, or Representative Council of North Lombok Fishers
MCS	Monitoring, control and surveillance
MPA	Marine protected area
MTF	Malawi Traditional Fisheries survey system
NEFMC	New England Fisheries Management Council
NGO	Non-governmental organization
NMFS	National Marine Fisheries Service (USA)
NOAA	National Oceanic and Atmospheric Administration (USA)
NPA	Northern Peninsula Area of Cape York, Australia
NRAES	Natural Resource, Agriculture, and Engineering Service (USA)
PAPD	Participatory action plan development
PNG	Papua New Guinea
PZJA	Torres Strait Protected Zone Joint Authority
SEAP	Special Secretariat for Fisheries and Aquaculture (Brazil)
STCNC	Sea Turtle Conservation Network of the Californias
SUDEPE	Federal Sub-Secretary for Fisheries Development (Brazil)
TEK	Traditional ecological knowledge
TMT	Traditional marine tenure
TSFMC	Torres Strait Fisheries Management Committee
TSFSAC	Torres Strait Fisheries Scientific Advisory Committee
TSFT	Torres Strait Fisheries Taskforce
TSRA	Torres Strait Regional Authority

Introduction
Putting fishers' knowledge to work

Nigel Haggan, Barbara Neis and Ian G. Baird

THE FISHERS' KNOWLEDGE CONFERENCE

This book grew out of a 2001 conference inspired by the late Bob Johannes, called 'Putting Fishers' Knowledge to Work'. Over 200 people from twenty-four countries and thirty-four aboriginal and indigenous organizations came to present results and discuss situations, problems and solutions (Haggan et al. 2003).

Johannes' strong opinion, and he was not one for timid views, was that we had heard more than enough of the 'conflicting dogmas of the omniscience of science and fishers' knowledge' (Stanley and Rice, this volume). He felt there had been enough reporting of the frustration of indigenous, artisanal and industrial fishers that their information was ignored. There had also been enough acknowledgements by scientists that there is indeed valuable information in their knowledge, but not in a form they can readily use.

Bob Johannes believed that the conference and this book should focus on where and how fishers' knowledge is *already* being used in collaboration with scientists and government managers, by indigenous peoples and artisanal fishers in their own unique contexts, and by large and small-scale commercial fishers. To this end, this book presents practical examples of how fishers' knowledge is being applied in fisheries science and management. In practice, as many chapters attest, either co-management, customary tenure or both are essential preconditions for the successful application of fishers' knowledge. Fishers rely on their knowledge for their livelihood, so it has always been 'put to work' in the most practical sense. Fishers' knowledge is not just of academic interest. It is a way of life that evolves continuously to address changes in fisheries and fishers. It is attracting increasing interest from non-fishers with an interest in fisheries.

DEVELOPMENT OF THE BOOK

The editors are trained in the social and natural sciences and have a history of involvement in interdisciplinary research. Two have considerable experience working with indigenous peoples. Social scientists have done a great deal of work on issues related to different knowledge systems but, with the notable exception of Bob Johannes and a few like-minded individuals, the topic has received less serious attention by natural scientists, although this is certainly changing. Fair and effective recognition of fishers' knowedge in fisheries science and management requires more collaboration and increased mutual understanding between fishers and natural and social scientists. For this reason, we chose to have most of the chapters peer reviewed by one natural and one social scientist. This made for some interesting comments, pointing to important gaps between social and natural scientists with regard to issues of knowledge creation, the value of local knowledge, and how local knowledge can be best and most appropriately used by fishers and outsiders alike. Our hope is that this book helps to narrow these gaps and broaden the scope for future dialogue and collaboration.

Contributions were selected on the basis of the strength of the research and in order to capture a broad range of situations and practical examples. We sought a balance between indigenous and artisanal, freshwater and marine, as well as small-scale and industrial commercial fisheries and broad, international representation. We do not suggest that one or two chapters can do justice to a particular country, never mind a continent. We know the collection provides a more comprehensive set of examples from Canadian and Australian fisheries than from most other parts of the world. We hope, however, that it will encourage the development of future collaboration between fishers, social and natural scientists, and managers, and will lead to further collections.

The term 'fishers' knowledge' (FK) was deliberately chosen as inclusive of the men and women who 'fish' in the broadest possible sense: those who depend on marine and freshwater species and ecosystems for their physical and cultural survival. The chapters contain a rich variety of terms for FK, including *fishers' knowledge* (Johannes and Neis, this volume) *indigenous technical knowledge* (Nsiku, this volume), *traditional ecological knowledge* (five authors), *local ecological knowledge* (eight authors) and ***-*knowledge* (Stanley and Rice, this volume). We left these terms as the authors wanted to present them. We have, however, used the term 'fisher' throughout as inclusive of the women and men whose contributions are described. We know that this will not please everybody. Most men involved in Canada's west coast fishery dislike the term intensely, and quite a few women in the fishery describe themselves as 'fishermen', as they believe it confers not just equality, but group membership. While some raise a concern that using 'fishers' to describe exclusively male groups actually masks the gender of those involved in fishing, we note that people fish from communities where the involvement of both sexes is crucial to the success of the fishing enterprise, regardless of who does the actual catching.

The diverse fisheries situations, knowledge and belief systems and geographic scope of the chapters almost defy organization. Where do indigenous fisheries end and artisanal and industrial fisheries begin? Some indigenous authors and co-authors describe indigenous fisheries in the developed world (Australia, Canada and the US/

Hawaii). Others describe indigenous fisheries in post-colonial contexts (Hickey, this volume) or after major political change (Satria, this volume). Still others identify the scope of indigenous knowledge in the non-industrialized world (Nsiku, this volume).

Chapter 1, 'The value of anecdote' sets the stage for exploration of the full scope of FK by beginning with what 'classic' fisheries science and management perceive as its weakest point. The chapter draws on the August 2001 conference keynote address by Bob Johannes. Before his untimely death in 2003, Bob asked Barb Neis to contribute to his chapter for this book. In his original draft, Bob described fishers' knowledge as anecdotal information that can be gathered by the astute scientist in order to inform scientific work. Barb thinks of fishers' knowledge and science as different knowledge systems, both of which need to be scrutinized for their underlying assumptions.

INDIGENOUS PRACTITIONERS AND RESEARCHERS

The main body of the book is divided between *Indigenous and Artisanal fisheries* and *Commercial fisheries*, imperfect as such a classification system may be. We begin the first section with works by indigenous people with practical experiences based on their own knowledge systems. In Chapters 2–6, indigenous knowledge holders and researchers take us on a journey that starts on the west coast of Canada with thoughts from a hereditary chief of the Hesquiaht Nation on the antiquity, scope and continuing capacity of traditional ecological knowledge (TEK) to expand and link concepts of biodiversity and environment (Lucas, this volume). Still in British Columbia (BC), on the Fraser River, we get an aboriginal perspective on the importance of fine-scale knowledge in the struggle to protect wild salmon and how this knowledge is passed down the generations (Narcisse, this volume). From there, we travel halfway around the world to explore the rich indigenous knowledge of fish species and behaviour, and the enormous range of ethno-botanical knowledge used in the manufacture of fishing gear and boats in Malawi (Nsiku, this volume). We return to Haida Gwaii on the west coast of Canada to explore how traditional knowledge of the Haida Nation could be used to improve the management of industrial herring fisheries (Jones, this volume). Heading west, we stop in mid-Pacific, where native Hawaiians have successfully applied traditional management principles to restore fish stocks to higher levels than surrounding areas (Poepoe et al., this volume).

INDIGENOUS AND ARTISANAL FISHERIES

In Chapters 7 to 17, researchers working with indigenous and artisanal fishing communities guide us on a journey that starts in the Pacific with the Republic of

Vanuatu. Since gaining independence in 1980, the islanders are rediscovering the resource management value of traditional beliefs and practices suppressed by colonial government and missionaries (Hickey, this volume). From there, it is a relatively short distance to the Aboriginal community of Injinoo in Northern Australia, where tribal leaders concerned about the depletion of an important foodfish species, were successful in negotiating a two-year closure supported by government and commercial and sport fishers (Phelan, this volume). Due North of Injinoo, on the Torres Strait Islands, indigenous people are struggling with government and industry to re-establish a seasonal, multi-species fishery that would sustain access to resources essential for their cultural and physical survival (Mulrennan, this volume).

Traditional management in North Lombok, Indonesia, was suppressed under the 'New Order' regime of ex-President Suharto. Empowered by decentralization since 1998, villagers in North Lombok are re-establishing their traditional integrated management system with benefits to fisheries resources, lifestyle and identity. Satria's account makes an interesting contrast to examples where indigenous people have been unable to achieve legislated management authority and access to more than a very modest amount of local resources (see also Hickey, Mulrennan, Poepoe and others, this volume).

From Indonesia, we travel north to the Philippines where researchers from Project Seahorse conducted surveys and workshops with fishers to explore the extent of resource depletion, identify the primary causes and develop workable solutions to conserve and rebuild resources, as well as exploring alternative livelihood options (Meeuwig et al., this volume). Continuing north, we come to the landlocked country of Laos, where local fishers' knowledge has been validated and incorporated into the management of local and migratory resources in the face of substantial scepticism from outside 'experts'. Unlike Lombok, where success is largely attributable to bottom-up efforts with no outside help, a small, flexible non-government organization (NGO) was of assistance to Lao fishers in facilitating a government-recognized but village-centred process for establishing socially and ecologically sound management regulations (Baird, this volume).

From Laos we travel to Bangladesh and another freshwater fisheries environment where local people depend on freshwater resources and demand often exceeds supply. In an interesting parallel with the Lao fishers, local people have established conservation zones in deep-water areas where fish can survive seasonal droughts and high temperatures. In partial contrast to the Lao example, success in developing local ecological knowledge (LEK)-based management is threatened by insecurity of tenure and the absence of a legal framework (Sultana and Thompson, this volume).

From Bangladesh, we go halfway round the world to the estuary of Patos Lagoon in Brazil, where small-scale fishers are collaborating to protect the environment and resources they depend on from overfishing by the industrial sector. Some success has been achieved inside the lagoon, but outside fisheries continue to threaten stocks and livelihoods, pointing to the need for a systemic approach that links LEK to large-scale fisheries management and legal recognition (Kalikoski and Vasconcellos, this volume). From Brazil, we travel north-west across the equator to Mexico, where researchers and artisanal fishers are developing a sea turtle monitoring programme with the objective of providing year-round coverage while respecting the harvest of turtles as an important cultural practice (Küyük et al., this volume).

In a different kind of study, the knowledge of long-dead fishers, perpetuated in the names of Swedish lakes, was used to identify lakes that had once supported populations of brown trout (*Salmo trutta*). A map-based survey identifying lake names with the root *Rö* (an archaic term for brown trout) was ground-truthed in a sub-sample, indicating that the technique could identify in two months what it would have taken two people with gillnets five years to accomplish. The research also indicated the presence of a fish-classification system around 1,900 years old (Spens, this volume).

COMMERCIAL FISHERIES

THE knowledge of commercial fishers has been used to map the seabed in the United States of America and Australia. Decades of overfishing in the Gulf of Maine had extirpated stocks of cod (*Gadus morhua*) and haddock (*Melanogrammus aeglefinus*). Ted Ames, a fisher and marine biologist, applied the knowledge of retired commercial fishers to identify spawning grounds for these stocks, with a view to using it for the purposes of restocking. The maps created were later validated using sidescan sonar. The project generated very useful guidelines for collecting such knowledge, for example that fishers had to be convinced it would be not be used to threaten their livelihood, but would be put to what they considered to be 'good use' – in this case, restocking (Ames, this volume).

In an Australian case, Williams and Bax (this volume) provide a concrete example of the point made by Johannes and Neis in Chapter 1: that there will never be enough time, money or trained people. The knowledge of commercial fishers is used to create detailed seabed maps at a scale that is useful to fishers.

Another example from the same area in Southeast Australia explores the usefulness of FK in three very different management situations. Excluding fishers from decisions such as establishing a marine protected area can pose severe problems for management and enforcement (Baelde, this volume). In British Columbia, Canada, collaboration between commercial fishers and government scientists in stock assessment for rockfish (*Sebastes* spp.) – from hypothesis formulation, through survey design, implementation and data analysis – added to the sum total of knowledge on both sides. The 'Swapping' of vessels and gear was also found to be an innovative and productive element (Stanley and Rice, this volume).

Chapter 21 presents a synthesis of some of the important ideas presented in the book, assesses progress made towards Johannes' goals, and renews his challenge to establish an international facility for the ethical collection, preservation, dissemination and application of FK (Haggan and Neis, this volume). The book ends with a chapter, appropriately named 'The last anecdote' (Baird, Chapter 22 this volume), which shows that even the most experienced and diligent researchers sometimes fail to get it right. There's hope for us all!

REFERENCES

HAGGAN N.; BRIGNALL, C.; WOOD, L. (eds.). 2003. *Putting fishers' knowledge to work*. UBC Vancouver, Fisheries Centre Research Reports, Vol. 11, No. 1, 504 pp.

CHAPTER 1   The value of anecdote

R.E. Johannes and Barbara Neis

ABSTRACT

THE knowledge that indigenous, artisanal and commercial fishers and marine hunters accumulate over the course of their fishing careers can be invaluable to marine researchers despite its low scientific repute among methodological purists. Over the past several decades, and in tropical, temperate and Arctic fisheries, it has cast considerable light on important subjects such as stock structure, interannual variability in stock abundance, migrations, the behaviour of larval/post-larval fish, currents and the nature of island wakes, nesting site fidelity in sea turtles, spawning aggregations and locations, local trends in abundance and local extinctions. It has also cast light on the dynamics of fisheries and their relationship to scientific understanding. This chapter draws on a series of examples from indigenous, artisanal and commercial fisheries to explore ways in which the knowledge of fishers and fisheries scientists can complement each other and, in the process, drive forward not only our knowledge about fisheries' resources but also our capacity to manage our degraded marine ecosystems to recovery.

INTRODUCTION

PROVIDING more relevant and timely scientific data and gaining better understanding of interactions between human societies and their marine environments are very high priorities in tackling our planet's marine environmental problems.[1] But the quest for more funds for these purposes tends to inhibit researchers from making an important admission: these problems have become far too big, too many and too complex for there ever to be enough time, money and qualified people to address them all effectively (e.g. Johannes, 1998a). Related to this is the mounting evidence

---

1. As a fisheries biologist, Bob Johannes worked primarily with artisanal fishers and fisheries science in tropical contexts (Johannes, 1981, 1993, 1998a, b). Barb Neis is a social scientist who has done similar work with commercial fishers in Newfoundland and Labrador, Canada, and with temperate fisheries researchers (Neis and Felt, 2000). Before his untimely death in 2003, Bob asked Barb to contribute to his 'The value of anecdote' paper. Unfortunately, she did not get a chance to work on the paper while he was alive. She hopes he would approve of her contributions to the work.

that fisheries scientists are often just following fisheries around, providing advice that is more likely to document resource decline than to inform sustainable management (Haedrich et al., 2001; Neis and Kean, 2003).[2] Support for these claims comes from indications of continuing decline in many of the world's marine seafood stocks and ecosystems (Pauly and Maclean, 2003), coupled with ongoing increases in the efficiency of fishing fleets.

Fisheries scientists and managers, but perhaps particularly fisheries-dependent communities, are confronting major challenges. The changes happening all over the world are so numerous, dynamic and multifaceted that the physical–chemical environment, the estuaries and benthic environments, the population and species diversity and, more generally, the marine ecosystems we see today are not the same as those that existed even in the recent past. If we are to stop the degradation, understand the productive capacity of these environments and begin the long, hard process of achieving recovery, we need to understand what was there in the past (Jackson et al., 2001; Pitcher et al., in press), the interactive social–ecological processes that are driving the decline (Frank et al., 2005), what is left, and how these altered ecosystems work.

It is now critical that we do everything possible to improve our marine environmental information base and share our expanded knowledge with those interacting with marine ecosystems to increase our collective capacity for stewardship and enhancement. Fishers' knowledge may often be the only source of information on the history of changes in local ecosystems and on their contemporary state that is of sufficiently fine scale to help us design ways to protect stock remnants and critical habitats.

'Anecdote-gathering' from fishers is one approach to broadening the information available to science; another is to treat fishers' knowledge, fisheries natural science and fisheries social science as different knowledge systems that have interacted over time and space to influence the history of fish and fisheries (Murray et al., in press). A third approach is to involve fishers, natural scientists and social scientists in the design, conduct and review of research that seeks to collect their knowledge in a systematic fashion (Davis and Wagner, 2003) and in a form that makes it commensurate with other knowledge forms (Neis et al., 1999a, b).

The word 'anecdote' comes from the medieval Latin *anecdota*: unpublished items or narratives. However, its more common meaning today is 'a short, usually amusing account of an incident, especially a personal or biographical one', or 'a particular or detached incident or fact of an interesting nature … a single passage of private life'.[3] Certainly, fishers' knowledge is largely unpublished, a feature that

---

2. In a 1992 meeting with the Canada's Department of Fisheries and Oceans (DFO), Percy Walkus, an elder of the Wuikinuxv Nation on the Central Coast of British Columbia, said, 'All DFO are doing is managing the rate of decline' (cited in Haggan, 1998).
3. www.wordreference.com Dictionary.

distinguishes it from science. Fishers often convey information in the form of storytelling about a particular biographical event, another feature of their knowledge that tends to distinguish it from scientific knowledge. However, it is often the 'detached', 'amusing' and 'biographical' features that those inclined to devalue fishers' knowledge are referring to when they think of it as anecdotal. This approach ignores the potential for systematic research involving fishers and their knowledge that pays attention to its social, ecological and historical context (Neis and Felt, 2000; Neis et al., 1999a). It also conveniently overlooks the frequency with which scientific research is 'detached' from highly significant historical and local contexts (Neis, 1998) and the extent to which science is 'biographical' in that it reflects the training, experience and prejudices of scientists as well as the institutional structures within which they are embedded (Finlayson, 1994).

Information obtained from natural resource users has been variously described as 'pre-scientific' or 'natural history', or as 'inductive' if one is being especially polite (Fuller, 1997). But if we can lay aside our graduate school prejudices long enough to examine the facts, it is impossible not to acknowledge that fishers' and marine hunters' knowledge about the sea has sometimes proven a fast and inexpensive shortcut to information essential to our scientific understanding of the marine environment, even when that knowledge is from the distant past (Spens, this volume). Juxtaposing their observations and interpretations with the results of scientific work can provide important insights for scientists and managers, as well as for fishers themselves. As with science, concerns that fishers' interpretations of observations may be mistaken should not preclude paying attention to the observations themselves.

Some of the information possessed by fishers in developing and developed countries may well never become available to science if we depend solely on conventional research to obtain it. Conversely, if natural and social scientists and fishers do not begin working together more effectively, we are unlikely to protect the fish that remain, let alone enhance the potential for recovery. The remainder of this chapter presents some established examples of the practical scientific value of the 'anecdotal' information of indigenous and artisanal fishers and marine hunters, and the different purposes it can serve.

FISHERS AND 'EXCUSES' FOR POOR CATCHES

How many times have we heard fishers explain poor catches by saying 'they've just gone somewhere else; they'll be back'? And how many times have scientists thought to themselves 'Yeah sure; you just don't want to admit that catch reductions are needed because of overfishing'? Although such suspicions are sometimes warranted, they are not always correct. In the Torres Strait, hunters caught roughly a fifth as many dugong (*Dugong dugon*) with essentially the same hunting effort during a survey made in 1983–4 as they had during the previous survey in 1976–8 (Johannes and MacFarlane, 1991). Many hunters expressed no concern about this change, saying that the animals had simply gone somewhere else and that sooner or later they would return. We and some other biologists were sceptical (Johannes and MacFarlane, 1991). However, aerial surveys of dugong in the Strait subsequently revealed great interannual variation in numbers and dugong catch per unit effort did subsequently rebound. Hunters predicted both outcomes (Marsh et al., 2002).

Research also supports a related claim made by the hunters, that great interannual variation occurs in the extent and location of the seagrass upon which the dugong feed, a pattern that may well influence their distribution significantly (Marsh et al., 2002). Whether or not overhunting is occurring in the Strait is still unknown because of difficulties associated with getting reliable stock estimates. However, the hunters' claims about the scale of natural interannual variability in both the number of dugong and their seagrass food were undoubtedly correct (Marsh et al., 2002).

As scientific awareness of the natural interannual variation in stock sizes has grown in the past fifteen years, so has the realization that stocks can sometimes recover in the absence of more stringent management. Awareness has also grown that social and technological dynamisms, as well as persistent problems with bycatch, discarding and under-reporting, are all contributing to sustained catch rates in the context of resource decline and to problems in estimating fish mortality. Not surprisingly, precautionary concerns may motivate scientists to support conservation initiatives on the basis of less than adequate data – as when they think stocks are overfished but cannot prove it. Conflict is the inevitable consequence of management actions based on scientific assumptions that run counter to biological information held by fishers. But fishers too need to consider the spatial, temporal and ecological dynamics of their fisheries and the ways these could be influencing their observations and related perceptions about stock health (Murray et al., 2006). Respect, collaboration and the successful application of knowledge require both fishers and scientists to understand not only the strengths but also the weaknesses of their *own* knowledge systems.

We need to study fishers' knowledge of their resources as a matter of high priority so as to be able to understand it in all its complexity, test it (preferably in collaboration with fishers) as soon as possible, and thereby reduce the likelihood of such conflicts (many chapters, this volume). Similarly, fishers need help to understand that there are different paradigms within science and that science, like fishers' knowledge, is to some degree a socio-ecological product. Paying particular attention to areas of agreement and disagreement between scientific and fishers' knowledge can contribute significantly to improved understanding and to advancing the knowledge of both groups.

FISHERS' KNOWLEDGE AND STOCK ASSESSMENT SCIENCE

Personal interviews with fishers can elicit large amounts of information pertaining to the past and the present for both commercial and noncommercial species. This information can be very useful in scientific stock assessments. Local knowledge of the time and place fish are caught can indicate seasonal and directional fish movements. Fishers can also provide information on stock structure, spawning grounds and juvenile habitat. They can provide catch-rate data which may reflect local changes in abundance. In addition, they can provide information on spatial and other changes in effort and fishing practices that are critical for interpreting catch-rate data (Hutchings, 1996; Neis et al., 1999b).

On Canada's east coast and elsewhere in the North Atlantic, stock-assessment scientific data were largely derived from offshore areas from research vessel surveys and commercial trawlers. Cod (*Gadus morhua*) and other groundfish stock collapses in the 1990s have made it critically important to understand the stock structure and related behaviour so as to ensure that remnant populations are not overfished. In Newfoundland and Labrador, most of the remaining cod live in the coastal bays. In both Newfoundland and Norway, fishers' knowledge has been used to help identify actual and potential local stocks of cod in fjords and bays (Maurstad and Sundet, 1998; Wroblewski, 2000; Wroblewski et al.). In the Gulf of Maine, it has been used to identify coastal spawning areas for cod and haddock (*Melanogrammus aeglefinus*) (Ames et al., 2000; Ames, 2004; Ames, this volume). Careful management of remnant coastal cod and haddock stocks may be critical to the long-term recolonization of offshore areas (Wroblewski et al., in press).

In Newfoundland, scientists had a history of collecting data from commercial capelin (*Mallotus villosus*) fishers by means of a logbook programme and, since the 1990s, an annual phone survey. Interviews with some of these fishers indicated, however, that the index of relative capelin abundance based on the logbooks of inshore capelin trap fishers may have been positively biased. These fishers described significant changes in the design and size of their capelin traps, as well as other efficiency-related changes that should have been taken into account in interpreting data from the relative index of abundance that was based on the capelin trap catch rate (Neis and Morris, 2002). Similarly, detailed interviews with lobster (*Homarus americanus*) fishers in the Magdalen Islands of Quebec, Canada, helped document changes in fishing equipment, strategies and efficiency that were crucial to interpreting catch-rate data in their fishery (Gendron et al., 2000).

LARVAL BIOLOGY OF REEF FISH

Coral reef fish larvae spend several weeks to several months in the oceanic plankton. Research has shown that once they are sufficiently well developed to take up demersal existence, they can detect reefs from distances of more than a kilometre and swim toward them (e.g. Leis et al., 1996; Stobutzki and Bellwood, 1998). Tobian fishers described this phenomenon clearly to Johannes in the mid-1970s. They named at least five species that they had commonly observed abandoning drifting logs with which they had been associated, and heading over deep water directly towards reefs 'many hundreds of yards away'. A Palauan master fisher later reiterated these observations (Johannes, 1981). Until biologists got around to confirming that such behaviour occurred more than fifteen years later, they routinely assumed that reef fish larvae were entirely at the mercy of the currents (e.g. Roberts, 1997). This incorrect assumption led to major errors in reef fish stock modelling and management (e.g. Leis and Carson-Ewart, 2000).

Such swimming control implies that reef fish larvae have some ability to determine where and when they settle out of the plankton. The way this ability is employed needs to be better understood if efforts to collect settling reef fish larvae for the purpose of aquaculture research are to be refined (e.g. Hair et al., 2002). The two collection devices most commonly used by researchers are light traps and specially designed plankton nets. Both are expensive. Neither has proven very effective in collecting groupers, the most important species among cultured reef food fish (e.g. Hair et al., 2002). Thousands of South East Asian fishers capture settling groupers for sale to aquaculturists. They use more than a dozen different methods and demonstrate considerable knowledge about precisely where and when to deploy them productively (Johannes and Ogburn, 1999; Sadovy, 2001).

Some of these methods are environmentally destructive and should be discouraged (Johannes and Ogburn, 1999; Sadovy, 2001). Others, which focus on pre-settlement larvae, appear not to be destructive and are very inexpensive. In some cases, the gear consists of nothing more than small clumps of old netting, or of particular species of algae or terrestrial vegetation suspended from sticks or ropes at times and places that fishers have learned through trial and error are good for catching grouper larvae or post larvae. Learning from and, indeed, teaming up with some of these fishers offers researchers opportunities to develop the science needed to help increase catches while reducing costs and environmental damage.

SEA TURTLES

For many generations, tropical sea turtle hunters asserted that green turtles (*Chelonia mydas*) usually returned to the same beach to nest over many years.

They believed this because they recognized individual turtles by distinguishing marks or wounds – things like chunks bitten out of their flippers or shells by sharks. If these fishers' claims are true, they are obviously of great importance in the design of useful scientific studies of turtle movements and population and reproductive trends – studies that are absolutely essential to enable researchers to understand turtle biology well enough to detect overharvesting and design useful conservation measures. However, biologists ignored or dismissed this knowledge for a long time before one of them finally decided to take it seriously and test it. The result was biologist Archie Carr's famous turtle-tagging experiments that demonstrated the truth of the turtle hunters' claims in the 1950s (see for example Carr, 1972). Research on sea turtles took a great leap forward after this discovery as scientists realized how much they could learn from turtle-tagging studies where they could usually rely on the turtles to return to the tagging site. Tens of thousand of turtles have since been tagged. A great deal has been learned about growth rates, longevity, reproductive rates and nesting frequency as a direct result of one scientist deciding to take turtle fishers' knowledge seriously (see also Küyük, this volume, for practical application of turtle fishers' knowledge).

BEING IN THE RIGHT PLACE AT THE RIGHT TIME

Fishers and hunters everywhere focus on where and when to find ample prey. This means that marine fishers and hunters often know a lot about how the distribution and abundance of marine animals vary from year to year with type of habitat, season, weather, time of day, stage of tidal cycles, lunar phase and other factors. They can also often relate important observations about behaviours of marine animals that contribute to these changing distributions and abundances. Such subjects are key areas of focus in fisheries research, but scientific observations of them tend to be temporally and spatially limited (Fischer, 2000).

For example, at higher latitudes most scientific research is done in the summer months because, first, the university-based researchers can more easily get away from teaching, and second, summer is often the most comfortable time to be in the field. The Inuit,[4] in contrast, traditionally hunted on sea ice throughout the winter. As a result, they learned much about the biology of their prey that can only be learned in winter when most biologists are snug in their offices. Milton Freeman (in Johannes et al., 2000) describes what Inuit know about the winter biology of the whales they hunt. They were able to tell him where and how they move, and how they can navigate and migrate under the ice. Until recently, biologists had never seen this

---

4. The correct name for Arctic aboriginal people formerly and incorrectly referred to as 'Eskimos'.

ice-related behaviour and initially doubted the accuracy of these claims. Subsequent research showed that Inuit information was not only correct but also essential for developing better estimates of whale population sizes (Johannes et al., 2000).

Similarly, Nakashima's (1993) studies have shown the scientific value of Inuit knowledge about eider duck (*Somateria mollissima*) behaviour during the winter, a time when ice cover forces them to change their behaviour and distribution greatly from that of warmer months when most scientific field research is done. Nakashima states that, 'for many species of Arctic wildlife, [traditional ecological knowledge] far outstrips current scientific knowledge', and that natural resource managers 'make decisions and take actions based upon deficient scientific data, declaring that for the time being it is the only information available. In so doing they choose to ignore the traditional ecological knowledge of Native peoples' (Nakashima, 1993, pp.108, 103). By tapping Inuit knowledge, Nakashima was able to show that the eider duck population of Hudson Bay was almost twice the size estimated by biologists, and reveal a host of interesting biological facts such as how substantial numbers of eiders can shelter on the water under ice domes of their own making. More recently, Labrador hunters' ecological knowledge about the history of interactions between eider duck (*Somateria* spp.) populations and hunting in St Peter's Bay, Labrador, has been combined with scientific research to enhance our understanding of long-term trends in eider abundance and the relationship between industrial and regulatory changes and shifts in mortality among eider populations (Chaffey et al., 2003).

FISHERS' KNOWLEDGE, MARINE PROTECTED AREAS AND ENVIRONMENTAL IMPACT ASSESSMENTS

THE value of fishers' knowledge extends beyond stock assessment science and management. For example, recording the spatial and temporal distribution of coastal marine plants and animals is fundamental both to environmental impact assessments (EIAs) (Johannes, 1993) and to the design of marine protected areas (MPAs). Here, the relevant knowledge possessed by fishers can be invaluable. The locations of rare or endangered species are more likely to be pointed out by local resource users than they are to be identified by outside researchers doing site inventories on their own over limited time periods. The same is true of the timing and location of animal migrations and aggregations (see, for example, Wroblewski, 2000).

An important body of fishers' knowledge in tropical nearshore waters that relates to the siting of MPAs concerns reef fish spawning aggregations. Groupers, snappers, jacks, emperors, mullets, bonefish, rabbitfish, surgeonfish and other species of coral reef food fish aggregate to spawn at the same location, season and moon phase each year. More than thirty researchers or research groups have acknowledged in their publications that it was fishers who first led them to the spawning aggregations that they subsequently studied (for a list of twenty-three of these see Johannes et al., 1999). There are, in fact, very few published examples of biologists locating important spawning aggregations of reef food fish without such aid. Although this subject has often been discussed in print, it deserves continuing emphasis because fisheries managers in many tropical regions have proven incomprehensibly resistant to obtaining fishers' knowledge on spawning aggregations and using it for better reef fish management. Like other small, local stocks, many of these aggregations are highly vulnerable to rapid depletion or complete elimination (reviewed by Johannes et al., 1999). Although supporters of MPAs routinely assert that their most important function is to protect spawning stock biomass, disappointingly few MPA planners make the effort to locate spawning aggregations or to incorporate them into MPAs.

The trade in live reef food fish, in which cyanide is often used to stun the fish, is depleting grouper spawning aggregations and the stocks they represent at unprecedented rates in South-East Asia (e.g. Johannes and Riepen, 1995; Pet-Soede and Erdmann, 1998; Bentley, 1999). Yet there is nothing in the scientific literature in the region on the timing or location of these aggregations. The fishers who are depleting them clearly know more about them than fisheries biologists, illustrating the fact that knowledgeable fishers are by no means always environmentally sound fishers. Just as tomb-robbers make outstanding guides for archaeologists, so too live reef fish operators might prove useful in helping reef fish researchers locate spawning aggregations. So devastated are grouper stocks in South East Asia that such drastic measures may be called for.

CURRENTS

The inhabitants of the tiny isolated oceanic island of Tobi south of Palau were intimately familiar with a form of island wake that strongly influenced where and for what they fished (Johannes, 1981). Johannes was unable to find any such form of wake described in the oceanographic literature, but eventually stumbled across it in the experimental hydrographic literature where it was known as a stable eddy pair (Johannes, 1981). At that time, this basic oceanographic feature had never been observed by oceanographers, yet Tobians had known about it for generations if not centuries.

Trochus (Trochidae) are large, commercially valuable, tropical gastropods found in the western and central tropical Pacific. Although they have an unusually short pelagic larval life, local villagers can reap the benefits of local trochus reproduction when protective breeding sites are located within their waters and the currents retain the larvae in the area until they settle. Surveying the currents on the fishing grounds of many small fishing villages for the essential information to site the trochus reserves would be prohibitively expensive. Local canoe fishers, however, are usually very familiar with these currents, and government fisheries personnel in Vanuatu therefore use this local knowledge to help villagers determine where best to locate their trochus reserves (M. Amos quoted in Johannes, 1998b).

HABITAT ISN'T EVERYTHING

THERE are countless examples of shallow-water spawners that characteristically choose particular types of bottom habitat on or over which to spawn. For example, many seahorses give birth in seagrass beds, where rabbitfish (Siganidae) also lay their eggs, and grunion (*Leuresthes tenuis*) spawn in the sand on the beach on high spring tides. Johannes had always assumed that nearshore species consistently chose a particular spawning habitat, until he learned a lesson from the aboriginal fishers of northern Australia about the spawning habits of barramundi (*Lates calcarifer*). Barramundi are an important food and sportfish in Australia, but in the 1970s little was known about their biology. There had been a long-running argument among marine biologists about where barramundi spawned. Some insisted that they migrated from the ocean into rivers to spawn, while others argued just the opposite – that they migrated from rivers into coastal waters at spawning time – but there were no scientific data available to settle the issue.

The question appeared to have been solved conclusively when research demonstrated that barramundi in the Fly River system in Papua New Guinea migrate out of the river and as much as 100 kilometres along the coast to spawn (Moore, 1982). However, shortly after this discovery and several hundred miles away in northern Australia, Johannes interviewed aboriginal fishers who claimed that barramundi in their waters migrated from the ocean into rivers, sometimes tens of kilometres upstream before spawning. Johannes told some of them about Moore's work and they all responded in the same way saying, in essence, 'With all due respect to your friend Mr Moore, where we live, barramundi migrate into some rivers to spawn.' Eventually, their assertions that barrramundi move upstream to spawn in this region were confirmed (Davis, 1985).

How can the same fish have such conflicting spawning habits in different parts of its range? Further research revealed that fertilized eggs of barramundi survive best in water with a salinity of around 30 parts per thousand. The Fly River, where Moore did his research, has a very large discharge of freshwater that dilutes the seawater for many tens of kilometres out to sea. For this reason, Fly River barramundi have to migrate around 100 kilometres along the coast from the Fly River mouth in order to find water of high enough salinity to spawn. In northern Australia, however, most rivers are relatively small and slope very gently to the sea. The very big tides in this area push seawater many tens of kilometres upstream into some of the rivers, forcing the barramundi to leave the sea and swim inland to find water of low enough salinity. We biologists had assumed that barramundi were choosing a special bottom type in order to spawn, just like most other shallow-water spawners we were familiar with. It had simply never occurred to us that some other environmental factor could be more important. For centuries, the Aborigines knew the barramundi spawning movements in their waters that were determined by this need, even if they didn't know the reason for it.

FISHERS' MISTAKES

Like other resource users and like scientists, fishers sometimes draw false conclusions from accurate observations (Gunn et al., 1988 and similar researchers). We should think twice before discounting a fisher's conclusion even when we are positive it is wrong, because the observation on which it is based may be right – and valuable. For example, fishers in Belize told a group of marine biologists they had periodically seen a large group of whale sharks (*Rhincodon typus*) swimming through milky clouds of water. They said, 'We figured they must be spawning.' As biologists, we knew this was wrong because whale sharks do not release eggs; they are viviparous. However, these were good fishers whose knowledge had proven reliable and valuable in the past. What was going on here? Some days later we discovered the answer. In the same area, we saw a group of whale sharks swimming through a large cloud of very milky water. Its milkiness, we discovered, was due to spawn produced by a large aggregation of snappers. The whale sharks were there because they were feeding on this spawn. The fishers' 'mistake' had helped lead us to an important discovery: nobody had ever recorded this event before, nor had biologists known that the largest shark in the world could feed on eggs smaller than the head of a pin (Heyman et al., 2001).

FISHERS' KNOWLEDGE AS LONG-TERM DATABASES

Fishers' knowledge was recently used to document inshore spawning areas for cod and haddock in the Gulf of Maine. In some cases, these spawning areas were fished out decades ago, making fishers' knowledge the only potential source of information on their location (Ames et al., 2000 and this volume). Similarly, Dulvy and Polunin (2004) mention a large food fish, the bumphead parrotfish (*Bolbometapon muricatus*) that was once common in waters around some Fijian villages but which some village men under 25 have never seen because of overfishing. This species, they discovered, can be considered locally extinct at six islands where the last date of capture was prior to the 1990s, and stocks are severely depleted around other islands in the group.

These scenarios of depletion would never have been discovered if researchers had not canvassed the knowledge of older fishers. Their findings demonstrate that interviews with older fishers can be used to identify declining species and confirm the disappearance of exploited fishes, potentially in time for conservation action. They also suggest that fishers' knowledge, like the knowledge of scientists, is vulnerable to the 'shifting baseline syndrome' in that their sense of an ecosystem and its potential tends to be based on what they encounter when they enter the system and, as a system degrades, successive generations may come to expect less and less in terms of abundance and diversity (Pauly, 1995; Neis and Kean, 2003). In the case of Dulvy and Polunin (2004), their concurrent underwater censuses and subsequent statistical analyses suggest that, compared with surveying fishers' knowledge, conventional ecological censusing is far more expensive and has relatively little power to detect extinctions of large, vulnerable fishes. It took two to three weeks to determine the status of the fish stocks with the aid of fishers' knowledge, whereas it took between a year and eighteen months to do the same by means of ecological censusing (see also Spens, this volume). Dulvy and Polunin's (2004) findings dramatize the fact that in remote areas where few written records are kept, important knowledge about natural resources may die with each generation unless someone records it. Later, conventional biological field research is unlikely to recover this information. The only long-term databases that exist for such areas may reside in the heads of the elders (See Johannes and Yeeting, 2001, for another example).

FISHERS' KNOWLEDGE: A PRICELESS BUT FRAGILE OPPORTUNITY

VAST areas of marine habitat have never been studied scientifically in any detail. Most will remain unstudied because there simply are not enough dollars and scientific personnel to do the job (Johannes, 1998a; Prince, 2003). In addition, our marine ecosystems are changing rapidly in response to the effects of overfishing, climate change and other anthropogenic and natural forces. Vital knowledge about local areas and about the history of fish and fisheries in these areas — knowledge that is critical to the recovery of our marine ecosystems and the communities that depend upon them — resides in the heads of indigenous, artisanal and commercial fishers and hunters around the world. When given the opportunity, fishing experts from these groups have made researchers aware not only of ecological processes but also of customary tenure and local management systems that have been eroded through the interactive effects of external management interventions and resource degradation. In some areas, these insights have fuelled the development of innovative, community-based management initiatives that have helped local fishers and their communities bring about the recovery of marine ecosystems. The 'renaissance of community-based marine resource management in Oceania' described in one of Bob Johannes' last publications contains important examples of success stories that have global relevance (Johannes, 2002). He argued at the Putting Fishers' Knowledge to Work Conference that:

> 'Biologists are much better trained to ask useful questions about local ecological knowledge, put the answers into broader biological context and help restrain social scientists from framing management recommendations that ignore critical biological realities. Social scientists are better skilled in achieving good collaboration and rapport with local people, in interviewing, and in restraining biologists from drawing management conclusions that ignore equally critical cultural realities. The two types of researchers should be working in teams (Johannes, 2003).'

His reflections on the presentations based on indigenous, artisanal and commercial fisheries led him to argue that social and natural scientists working in both venues could learn from each other, but great care needed to be taken with generalizations from temperate to tropical and from commercial to artisanal and indigenous fisheries. In his concluding remarks to the conference, he emphasized that there were some thirty-eight institutions in the world dedicated to the study of indigenous knowledge of the terrestrial environment, but none for aquatic ecosystems. He challenged both natural and social scientists to set one up as a matter of urgency.

> 'Fishers have knowledge that often exists nowhere else. They need to be involved in the careful and systematic collection and evaluation of that knowledge, as well as in decisions about where, when and how it is put to use. The planet loses something precious every time one of these people dies without having had an opportunity to have this knowledge recorded. One Palauan fishing expert who taught Bob Johannes his knowledge recognized this. He told him, "Through you I can leave my footprints in this world before I move on to the next" (Johannes, 1981).'

REFERENCES

AMES, E.P. 2004. Atlantic cod stock structure in the Gulf of Maine. *Fisheries*, Vol. 29, No. 1, pp. 10–28.

AMES, E.P.; WATSON, S.; WILSON, J. 2000. Rethinking overfishing: Insights from oral histories of retired groundfishermen. In: B. Neis and L. Felt (eds.), *Finding our sea legs: Linking fishery people and their knowledge with science and management.* St John's, Newfoundland, ISER Books, pp. 153–64.

BENTLEY, N. 1999. Fishing for solutions: Can the live trade in wild groupers and wrasses from Southeast Asia be managed? *TRAFFIC Southeast Asia Report*, Petaling, Jaya, Malaysia, 100 pp.

CARR, A. 1972. Great reptiles, great enigmas. *Audubon*, Vol. 74, pp. 24–34.

CHAFFEY, H.L.; MONTEVECCHI, W.L.; NEIS, B. 2003. Integrating scientific and local ecological knowledge (LEK): Studies of common eiders in southern Labrador, Canada. In: N. Haggan, C, Brignall and L. Wood (eds.), *Putting fishers' knowledge to work*. Fisheries Centre Research Reports, Vol. 11, No. 1. Vancouver, University of British Columbia, pp. 426–32.

DAVIS, A.; WAGNER, J.R. 2003. Who knows? On the importance of identifying 'experts' when researching local ecological knowledge (LEK). *Human Ecology*, Vol. 31, No. 3, pp. 463.

DAVIS, T.L.O. 1985. Seasonal changes in gonad maturity and abundance of larvae and early juveniles of barramundi, *Lates calcarifer* (Bloch), in Van Diemen Gulf and the Gulf of Carpentaria. *Australian Journal of Marine and Freshwater Research*, Vol. 36, pp. 177–90.

DULVY, N.K.; POLUNIN, N.V.C. 2004. Using informal knowledge to infer human-induced rarity of a conspicuous reef fish. *Animal Conservation*, Vol. 7, pp. 365–74.

FINLAYSON, A.C. 1994. *Fishing for truth: A sociological analysis of northern cod stock assessments from 1977–1990*. St John's, Newfoundland, ISER Books, 176 pp.

FISCHER, J. 2000. Participatory research in ecological fieldwork: A Nicaraguan study. In: B. Neis and L. Felt (eds.), *Finding our sea legs: Linking fishery people and their knowledge with science and management*. St John's, Newfoundland, ISER Books, pp. 41–54.

FRANK, K.T.; PETRIE, B.; CHOI, J.S.; LEGGETT, W.C. 2005. Trophic cascades in a formerly cod-dominated ecosystem. *Science*, Vol. 308, pp. 1621–3.

FULLER, S. 1997. *Science*. Minneapolis, University of Minnesota Press, 159 pp.

GENDRON, L.; CAMIRAND, R.; ARCHAMBAULT, J. 2000. Knowledge-sharing between fishers and scientists: Towards a better understanding of the status of lobster stocks in the Magdalen Islands. In: B. Neis and L. Felt (eds.), *Finding our sea legs: Linking fishery people and their knowledge with science and management*. St John's, Newfoundland, ISER Books, pp. 56–71.

GUNN, A.; ARLOOKTOO, G.; KAOMAYOK, D. 1988. The contribution of the ecological knowledge of Inuit to wildlife management in the Northwest Territories. In: M. Freeman and L. Carbyn (eds.), *Traditional knowledge and renewable resource management in northern regions*. Edmonton, Alberta, Boreal Institute for Northern Studies, pp. 22–30.

HAEDRICH, R.L.; MERRETT, N.L.; O'DEA, N.R. 2001. Can ecological knowledge catch up with deep-water fishing? A North Atlantic perspective. *Fisheries Research*, Vol. 51, No. 3, pp. 113–22.

HAGGAN, N. 1998. Reinventing the tree: Reflections on the organic growth and creative pruning of fisheries management structures. In: T.J. Pitcher, P.J.B. Hart and D. Pauly (eds.), *Reinventing fisheries management*. London, Chapman and Hall, pp 19–30.

HAIR, C.A.; DOHERTY, P.J.; BELL, J.D.; LAM, M. 2002. Capture and culture of pre-settlement coral reef fishes in the Solomon Islands. In: Kasim Moosa, M.K., Soemodihardjo, S., Nontji, A., Soegiarto, A., Romimohtarto, K., Sukarno and Suharsono (eds.), *Proceedings of the 9th International Coral Reef Symposium*. Jakarta, Ministry of Environment, the Indonesian Institute of Sciences and the International Society for Reef Studies. pp. 819–29.

HEYMAN W.D.; GRAHAM, R.T.; KJERFVE, B.; JOHANNES, R.E. 2001. Whale sharks, *Rhincodon typus*, aggregate to feed on fish spawn in Belize. *Marine Ecology Progress Series*, Vol. 215, pp. 275–82.

HUTCHINGS, J.A. 1996. Spatial and temporal variation in the density of northern cod and a review of hypotheses for the stock's collapse. *Canadian Journal of Fisheries and. Aquatic Science*, Vol. 53, pp. 943–62.

JACKSON; J.B.C.; KIRBY; M.X.; BERGER; W.H.; BJORNDAL; K.A.; BOTSFORD. L.W.; BOURQUE; B.J.; BRADBURY; R.H.; COOKE; R.; ERLANDSON; J.; ESTES. J.A.; HUGHES. T.P.; KIDWELL. S.; LANGE. C.B.; LENIHAN; H.S.; PANDOLFI; J.M.; PETERSON. C.H.; STENECK. R.S.; TEGNER. M.J.; WARNER; R.R. 2001. Historical overfishing and the recent collapse of coastal ecosystems. *Science*, Vol. 293, pp. 629–37.

JOHANNES, R.E. 1981. *Words of the lagoon: Fishing and marine lore in the Palau district of Micronesia*. Berkeley, University of California Press, 245 pp.

——. 1993. Integrating traditional ecological knowledge and management with environmental impact assessment. In: J.T. Inglis (ed.), T*raditional ecological knowledge: Concepts and cases*. Ottawa, International Program on Traditional Ecological Knowledge and International Development Research Centre, pp. 33–40.

——. 1998a. The case for data-less marine resource management: Examples from tropical nearshore fisheries. *Trends in Ecology and Evolution*, Vol. 13, pp. 243–6.

——. 1998b. Government-supported, village-based management of marine resources in Vanuatu. *Ocean and Coastal Management Journal*, Vol. 40, pp. 165–86.

——. 2002. The renaissance of community-based marine resource management in Oceania. *Annual Review of Ecological Systems*, Vol. 33, pp. 317–40.

——. 2003. Fishers' knowledge and management: Differing fundamentals in artisanal and industrial fisheries. In: N. Haggan, C. Brignall and L. Wood (eds.), *Putting fishers' knowledge to work*. Fisheries Centre Research Reports Vol.11, No. 1. Vancouver, University of British Columbia, pp. 15–19.

JOHANNES, R.E.; FREEMAN, M.M.R.; HAMILTON, R. 2000. Ignore fishers' knowledge and miss the boat. *Fish and Fisheries*, Vol. 1, pp. 257–71.

JOHANNES, R.E.; MACFARLANE, W. 1991. *Traditional fishing in the Torres Strait islands*. Hobart, CSIRO, 210 pp.

JOHANNES, R.E.; OGBURN, N.J. 1999. Collecting grouper seed for aquaculture in the Philippines. *SPC Live Reef Fish Information Bulletin*, Vol. 6, pp. 35–48.

JOHANNES, R.E.; RIEPEN, M. 1995. *Environmental, economic and social implications of the live reef fish trade in Asia and the Western Pacific*. Honolulu, Hawaii, The Nature Conservancy, 87 pp.

JOHANNES, R.E.; SQUIRE, L.; GRAHAM, T.; SADOVY, Y.; RENGUUL, H. 1999. Spawning Aggregations of Groupers (*Serranidae*) in Palau. *The Nature Conservancy Marine Conservation Research Report No. 1*, The Nature Conservancy and the Forum Fisheries Agency, 144 pp. Available online at http://www.conserveonline.org/2001/06/g/Grouper;internal&action=buildframes.action (last accessed 24 June 2005).

JOHANNES, R.E.; YEETING, B. 2001. *I-Kiribati* knowledge and management of Tarawa's lagoon resources. *Atoll Research Bulletin*, Vol. 489.

LEIS J.M.; CARSON-EWART, R.M. 2000. Behaviour of pelagic larvae of four coral-reef fish species in the ocean and an atoll lagoon. *Coral Reefs*, Vol. 19, pp. 247–57.

LEIS, J.K.M.; SWEATMAN, H.P.A.; READER, S.E. 1996. What the pelagic states of coral reef fishes are doing out in blue water: Daytime field observations of larval behaviour capabilities. *Marine and Freshwater Research*, Vol. 47, pp. 401–11.

MARSH, H.; EROS, C.; PENROSE, H.; HUGHES, J. 2002. The dugong (*Dugong dugon*): status reports and action plans for countries and territories in its range. *UNEP, SSC, IUCN, WCMC and CRC Reef*, 172 p. Available online at http://www.unep.org/dewa/reports/dugongreport.asp (last accessed 24 April 2006).

MAURSTAD, A.; SUNDET, J. 1998. The invisible cod: Fishermen's and scientist's knowledge. In: S. Jentoft (ed.), *Commons in a cold climate: Reindeer pastoralism and coastal fisheries*. Paris, Casterton Hall, Parthenon Publishing, pp. 167–85.

MOORE, R. 1982. Spawning and early life history of barramundi (*Lates calcarifer* Bloch), in Papua New Guinea. *Australian Journal of Marine and Freshwater Research*, Vol. 33, pp. 647–61.

MURRAY, G.; NEIS, B.; JOHNSEN, J.P. 2006. Lessons learned from reconstructing interactions between local ecological knowledge, fisheries science and fisheries management in the commercial fisheries of Newfoundland and Labrador, Canada. *Human Ecology*.

NAKASHIMA, D.J. 1993. Astute observers on the sea ice edge: Inuit knowledge as a basis for Arctic co-management. In: J.T. Inglis (ed.), *Traditional ecological knowledge: Concepts and cases*. Ottawa, International Program on Traditional Ecological Knowledge and International Development Research Centre, pp. 99–110.

NEIS, B. 1998. Fishers' ecological knowledge and stock assessment in Newfoundland. In: J.E. Candow and C. Corbin (eds.), *How deep is the ocean? Historical essays on Canada's Atlantic fishery*. Sydney, N.S., University College of Cape Breton Press, pp. 243–60.

NEIS, B.; FELT, L. (eds.) 2000. *Finding our sea legs: Linking fishery people and their knowledge with science and management*. St John's, Newfoundland, ISER Books, 318 pp.

NEIS, B.; FELT, L.; HAEDRICH, R.; SCHNEIDER, D. 1999a. An interdisciplinary method for collecting and integrating fishers' ecological knowledge into resource management. In: D. Newell; R. Ommer (eds.), *Fishing places, fishing people: Traditions and issues in Canadian small-scale fisheries*. Toronto, University of Toronto Press, pp. 217–38.

NEIS, B.; KEAN, R. 2003. Why fish stocks collapse: An interdisciplinary approach to the problem of 'fishing up'. In: R. Byron (ed.), *Retrenchment and regeneration in rural Newfoundland*. Toronto, University of Toronto Press, pp. 65–102.

NEIS, B.; MORRIS, M. 2002. Fishers' ecological knowledge and fisheries science. In: R. Ommer (ed.), *The resilient outport: Ecology, economy, and society in Rural Newfoundland*. St John's, Newfoundland, ISER Books, pp. 205–40.

NEIS, B.; SCHNEIDER, D.; FELT, L.; HAEDRICH, R.; HUTCHINGS, J.; FISCHER, J. 1999b. Northern cod stock assessment: What can be learned from interviewing resource users? *Canadian Journal of Fisheries and Aquatic Science*, Vol. 56, 1949–63.

PAULY, D. 1995. Anecdotes and the shifting baseline of fisheries. *Trends in Ecology and Evolution*, Vol. 10, No. 10, p. 430.

PAULY, D.; MACLEAN, J. 2003. *In a perfect ocean: The state of fisheries and ecosystems in the North Atlantic Ocean*. Washington, D.C., Island Press, 175 pp.

PET-SOEDE, C.; ERDMANN, M.V.E. 1998. An overview and comparison of destructive fishing practices in Indonesia. *SPC Live reef fish information bulletin*, Vol. 4, pp. 28–36.

PITCHER, T.J.; AINSWORTH, C.; BUCHARY, E.; CHEUNG, W.L.; FORREST, R.; HAGGAN, N.; LOZANO, H.; MORATO, T.; MORISSETTE, L. In press. Strategic management of marine ecosystems using whole-ecosystem simulation modelling: The 'back to the future' policy approach. In: I Linkov (ed.), *The strategic management of marine ecosystems*. The Netherlands, NATO-ASI Kluwer.

PRINCE, J.D. 2003. The barefoot ecologist goes fishing. *Fish and Fisheries*, Vol. 4, pp. 359–71.

ROBERTS, C.M. 1997. Connectivity and management of Caribbean coral reefs. *Science*, Vol. 278, pp. 1454–6.

SADOVY, Y. 2001. Summary of regional survey of fry/fingerling supply for grouper mariculture in Southeast Asia. *SPC Live Reef Fish Information Bulletin*, Vol. 8, pp. 22–9. Full report available at http://www.spc.int/coastfish/News/LRF/8/LRF8-09-FrySadovy.htm (last accessed 29 June 2005).

STOBUTZKI, I.C.; BELLWOOD, D.R. 1998. Nocturnal orientation to reefs by late pelagic stage coral reef fishes. *Coral Reefs,* Vol. 17, pp. 103–10.

WROBLEWSKI, J. 2000. The colour of cod: Fishers' and scientists identify a local cod stock in Gilbert Bay, southern Labrador. In: B. Neis; L. Felt (eds.), *Finding our sea legs: Linking fishery people and their knowledge with science and management.* St John's, Newfoundland, ISER Books, pp. 72–81.

WROBLEWSK, J.; NEIS, B.; GOSSE, K. 2005. Inshore stocks of Atlantic cod are important for rebuilding the East Coast fishery. *Coastal Management,* Vol. 33, No. 4, pp. 411–32.

PART 1:
INDIGENOUS PRACTITIONERS AND RESEARCHERS

CHAPTER 2 Life supports life

Klah-Kist-Ki-Is; Chief Simon Lucas

ABSTRACT

Our people understand that our universe is full of the highest forms of life, and that our earth benefits from this. Our teachings and belief system celebrate the connection between all forms of life. We believe that life supports life, that we are one with the animals of the air, land and water. What you call 'biodiversity' is only a part of it. Long before sounders and geographical positioning systems were invented, our people knew the precise location of the offshore fishing banks. We hunted whales; we fished deepwater species. We fished for species that no longer exist in these waters. More recently we have become involved in high-tech fisheries that, in a few decades, depleted resources that had sustained our people for thousands of years. Our leadership is saying that we value technology but we want to combine it with our traditional values. So, when we harvest resources, we always keep in mind future generations of all living things.

INTRODUCTION

British Columbia (BC) is a unique place. It has 197 First Nations, each speaking its own distinct language. So, when you're talking about the six species of Pacific salmon (*Oncorhynchus* spp.), biologists reckon there are or were almost 10,000 stocks in 3,600 streams (Slaney et al., 1996). This count may well be low, as another leading scientist notes that intensive commercial fishing might have reduced diversity at the time of contact by 40 per cent (Carl Walters, UBC Fisheries Centre cited in Haggan et al., 2004). So you can just imagine the number of names there were for each species, each stage of their lifecycle, names for different ways of preparing and preserving them.

The map in Figure 2.1 shows the main tribal groupings to which the 197 First Nations belong. Each Nation has its own songs and dances relating to salmon and other species. We understand each other in terms of the philosophy that links us together. We use the saying 'everything is one' to describe things. We also say that we need to be 'respectful' of the skies and streams – what you call the 'environment'. If there is no respect for the things that make life possible, how can we respect ourselves? When I gave this talk in Vancouver, I started with a song in my language (Lucas, 2003). I can't do that in this book, but in English, it means that, 'We thank the Creator for the Day'. We ask for the power to be respectful of all living things, and the generations yet to come. The song summarizes a whole understanding of oneness between people and the natural and physical world that means a great deal more than 'biodiversity', 'ecosystem' and the 'environment'. This is what it means to us:

LIFE SUPPORTS LIFE

Life is a Treasure, Life is enormous.
Life is emotional, Life is mental.
Life is physical, Life is spiritual.
Hold on to life, Hold on to life.
Life enhances our feelings.
Life enhances our mind.
Life enhances our strength.
Life enhances our spirituality.

Life enhances how we live in our own environment.
Life enhances the first cry of a newborn.
Life teaches that we are here only temporarily.
Life teaches responsibility.
Life teaches respect and self-respect.
Life supports the circle of our journey here.

Today those chants still exist. Some of them were made while sitting on the shoreline, listening to ripples and waves. Some were made in the forest, listening to the movement of the trees, some while looking up at the stars.

Our history goes back a long time. In the territory of my tribe, the Hesquiaht, there were archaeological digs between 1971 and 1979 at fifteen sites where we laid our people to rest in a cave. After two years, the archaeologists were startled: 'This isn't changing and we have gone back five thousand years.' Among the remains were seventy-five different marine resources (McMillan, 1999).[1] Along with those remains were cedar bark, old masks and different rattles that our people used. I tell you that to make you think how far back the knowledge that we have to offer goes. I am not saying it is the best method, but it is an alternative that we have to offer. What we saw with our own eyes and what we learned from our grandfathers.

1. For more information on these marine species see Sumpter et al. (2002).

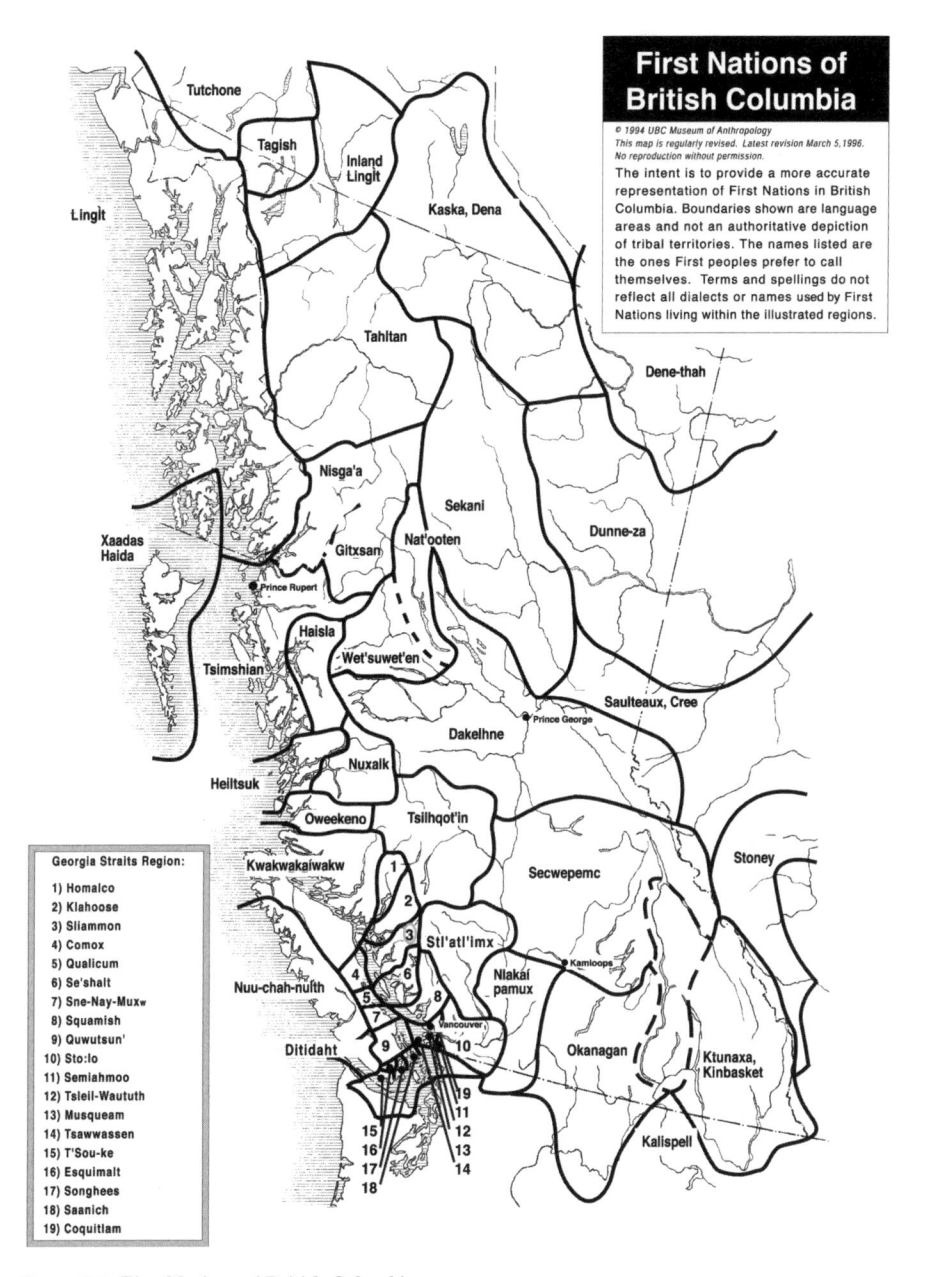

FIGURE 2.1 *First Nations of British Columbia.*
Source: Map courtesy of the Museum of Anthropology, University of British Columbia.

INDIGENOUS PRACTITIONERS AND RESEARCHERS

When I was 5 years old, my dad started taking me out on a 31 ft (10 m) salmon troller that he owned.[2] He said I was so active that he had to tie me to the mast for fear of my falling overboard. Before I was 10 years old, my father and grandfather had taught me all the *mit tuk* – what you would call 'landmarks' – to the fishing banks and the entire underwater landscape of our territory, where to get the cod and the halibut and the salmon. The first place I want to tell you about is called Oom piilts in our language.

OOM PIILTS

You had to know landmarks to find the banks. Oom piilts, which is about 3½ miles due west of Estevan Point (Figure 2.2A), is almost a perfect circle, 3 miles around with a gravel bottom. It is an incredible place for needlefish (*Ammodytes* spp.). My grandfather used to say that this place was important. When I got modern sounding gear, I found out he was right, because in the morning there is no sign of life, but at certain times of day the needlefish rise up. We knew that there would be lots of salmon hanging around because of the herring (*Clupea pallasii*), shrimp (*Pandalus* spp.) and needlefish; the whole food web is there in abundance, so it's a great place for halibut (*Hippoglossus stenolepis*), lingcod, (*Ophiodon elongatus*), snappers (*Sebastes ruberrimus*) and rockfish (*Sebastes* spp.).

The reason this area was so popular with fish was the tidal currents. My grandfather said that you have to understand the movement of the sea. The moon and tides are incredible indicators of when the fish start migrating (see also Hickey, Kalikoski and Vasconcellos, and Poepoe et al. this volume). My dad used to say: 'Don't ever go fishing during the flood tides because everything goes behind the reefs.' So the reef is important to us because it offers protection for migrating and local stocks.

TU QUIS

The first *mit tuk* for Tu Quis is an east–west line through two peaks on Flores Island; the second is a north–south line from a mountain called Tsawunaps to a sharp point on Conuma (Figure 2.2B). It was a great place for halibut; even the anthropologists knew about it (Drucker, 1951).

---

2. Trollers are one of three types of boat used in the Pacific salmon fishery. They use lines with multiple lures fished at varying depths.

LIFE SUPPORTS LIFE

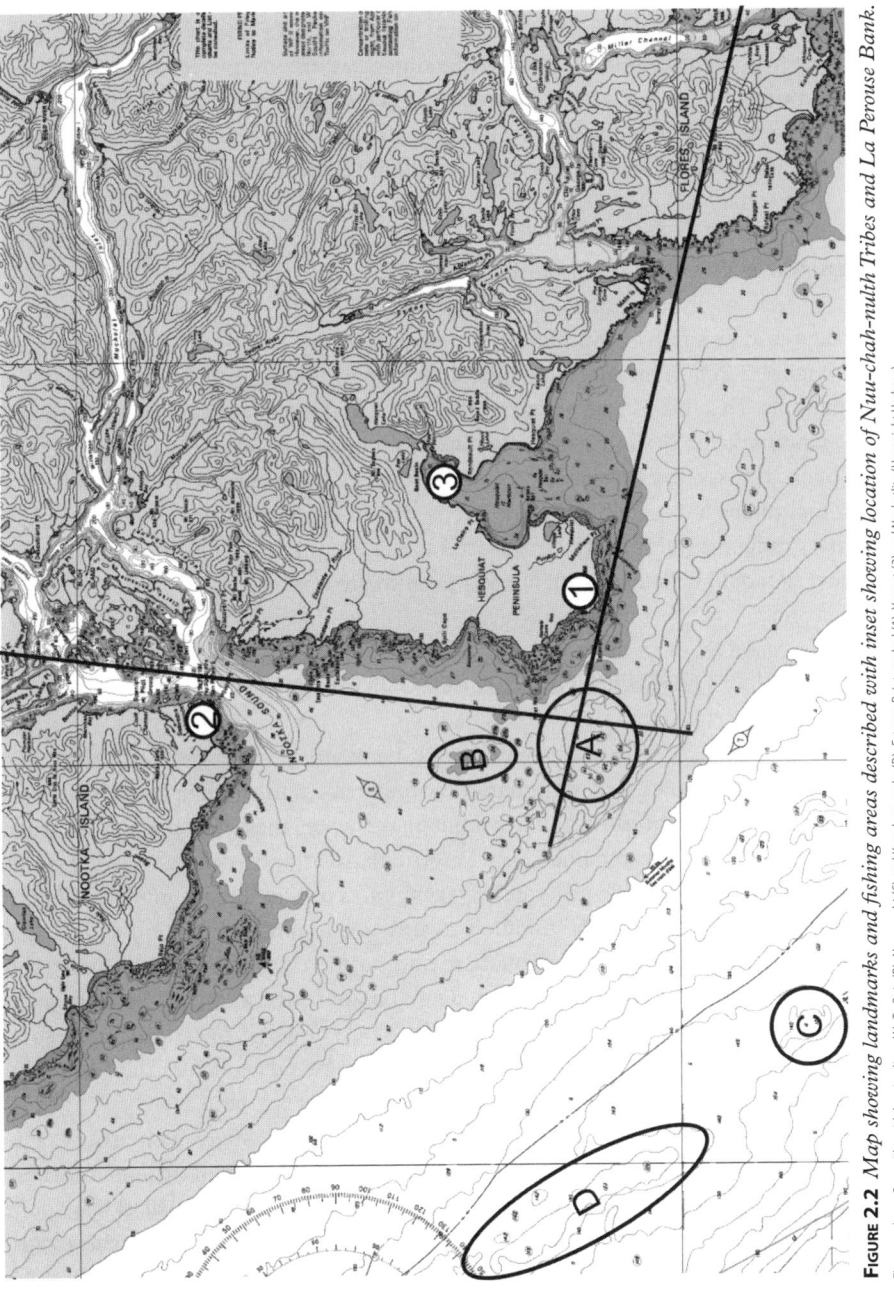

FIGURE 2.2 *Map showing landmarks and fishing areas described with inset showing location of Nuu-chah-nulth Tribes and La Perouse Bank.*
The map shows 00m pillts with bearing lines (A). Tu Quis (B), Na woo chi (C) and Kwa-kwa-wahs (D). Estevan Point marked (1), Yuquot (2) and Nuchatlitz (Hesquialht Harbour).

NAA WOO CHII

Naa woo chii is almost due south of Nootka sound and west-south-west of Estevan Point (Figure 2.2C). The landmark we use to find it is a mountain we call Conuma, which has a very sharp peak. Naa woo chii is important because this bank starts at 40 fathoms and drops down to 70 fathoms (73 to 146 m). Long before we had radar and sonar, our people knew that life is most abundant at that depth.

KWA-KWA-WAHS

South-west by south of Youquot is an incredible fishing place called Kwa-kwa-wahs (Figure 2.2D). This is not a bank, but a hole where the bottom falls from 50 to 80 fathoms. To get to Kwa-kwa-wahs, you line Conuma up with Nootka lighthouse. The landmark for Kwa-kwa-wahs is four mountains: as you approach the edge of the bank, you see the first mountain appear; when you are right in the middle, the four mountains look as if they had flat tops. This was an incredibly productive place in my younger days. It is where we used to see so much shrimp boiling up at the surface; it would be loaded with herring from top to bottom.

THLO-THLO-THOMLTH-NEE (LA PEROUSE BANK)

Thlo-thlo-thomlth-nee is an incredibly rich bank, 45 fathoms (~80 m) deep, that stretches for miles and was shared by four tribes. In my language, *Thlo-thlo-thomlth-nee* means the sound that halibut make when they slap their tails on the surface. The Europeans renamed it La Perouse Bank, after one of the early explorers. To find Thlo-thlo-thomlth-nee you keep one particular mountain in view. When it appears to be precisely in the middle between two mountains in the distance, that shows you are 22 miles (35 km) offshore and on the edge of the bank. Today, this landmark is called the 'gunsight'. It is there that many of the migrating salmon stocks from the Fraser River will be. You know that you are going to be catching lots of salmon, especially if the moon is right. Sometimes, two days before the full moon, the fish will be nuts, and two days after they will be a lot crazier.

We have gone from traditional fishing into a more high-tech fishery. The coastal tribes now own some huge fishing boats, with gillnetting and trawling. Some fish offshore for tuna (albacore, *Thunnus alalunga*). Our tribal elders spoke of harpooning big fish from canoes; nobody believed them until an archaeologist found bluefin tuna (*Thunnus thynnus*) bones in our territory (Crockford, 1994, 1997). Historical records tell of a feast given by Chief Maquinna on 5 September 1792 that consisted of 'large tunny and porpus'. Archibald Menzies, who was there, even describes how it was cooked in a bentwood box using heated stones (Menzies, 1792).[3] Some non-native people have argued that we did not go far out to sea, but now the archaeologists confirm that our people were there. Some commercial fishers have argued that we never ate black cod (*Anoplopoma fimbria*) because they were too deep down, but our elders told us that that black cod was the daily diet of our tribe. In particular, women who were pregnant drank broth made from the black cod to make their milk ready and rich for when they started breast-feeding (Drucker, 1951). Also, there are several references to 'coalfish', later identified as black cod, in accounts of the voyage of James Cook.

The scientists do not understand that our connection goes back long, long before they came here. They do not want to accept the simple definition of what we talked about; they like to speak of 'biodiversity' but our grandfathers linked it all together in one sentence: 'It is because of life that we live'.

There was a spiritual component in the way we harvested those resources, as well as physical and mental reasons. Our grandfathers say: 'Always look at the day'. You just do not do things without thinking about what the consequences might be for generations to come. Our grandfathers say: listen to the day – sometimes it talks to us. Do we really take our time to listen and to look and see what is happening in our area, and what the consequences of our actions will be?

When the herring-roe industry started,[4] we went to see an old chief, called Felix Michael, and took him to the beach on the first day to see how he was going to react. There were hundreds of nets in his territory, which has fed his people for thousands of years. He said, 'What are you people doing? What are you involved in? You are fishing these fish when they are near spawning!'. He told us it was the ultimate crime. He was right (Lucas, 2004). The herring spawn on giant kelp (*Macrocystis spp.*) and eelgrass (*Zostera*), and the spawn used to be really thick.[5] This is one of our important traditional foods; we still put out kelp fronds to collect eggs, but now we're lucky if it is half an inch (~1cm) thick.[6]

3. Bentwood boxes were made of red or yellow cedar. The four sides were made of a single plank, deeply grooved at three corners then steamed and bent into a square. The bottom was bevelled or grooved and the box was put together by sewing the fourth corner and four sides of the bottom. Bentwood boxes were waterproof and used for cooking, keeping food and goods dry on canoe journeys, storing regalia and other uses. The top and sides were frequently painted with traditional designs. See Stewart (1984) for more on bentwood boxes.
4. A fishery targeted on ripe female herring. The roe is removed whole and sold primarily on the Japanese market as kazunoko. The male and female carcasses go for fishmeal or pet food. In 1999, the total BC catch for roe herring was 26,525 tonnes valued at C$100 million (http://www-comm.pac.dfo-mpo.gc.ca/publications/factsheets/species/herring_e.htm, last accessed August 30, 2004).
5. Spawn might accumulate to as thick as four inches. Indeed, as the quality is best when it is between one and two inches, we had to be careful to collect it before it became too thick.
6. Fresh and dried herring spawn continues to be an important food and trade item for coastal First Nations. The eggs are collected from kelp and eelgrass or by hanging large branches of the hemlock (*Tsuga heterophylla*) or cedar tree (*Thuja plicata*). The modern herring spawn on kelp fishery mirrors traditional harvest methods of hanging kelp fronds in the water. In commercial operations, many owned by First Nations, kelp fronds are suspended from lines stretched across rectangular log frames. In the 'closed pond' system, herring are caught by seine and transferred to netpens. In the 'open pond' system, the frames are moved to areas of high spawning intensity, with lower mortality of the herring. Commercial licences have an 8 t limit. The product is mostly exported to Japan for sale as Komoichi kombu, but the price declined from a high of $US75/kg in the mid-1970s to $US17/kg in 1999.

NUCHATLITZ (HESQUIAHT HARBOR)

To the south of Chief Felix Michael's territory, there is a place called Nuchatlitz where we always harvest herring for our own use (Figure 2.2). Nuchatlitz is a huge pool, a mile (1.6 km) long and three feet (1 m) deep. There was so much herring there when I was young that we would just take a bucket and scoop them up. Then a herring gillnet commercial fishery took so many that it almost wiped the herring out.

Our tribe negotiated for two years with the Canadian Department of Fisheries and Oceans (DFO) to get that harbour closed to fishing. They asked us if we had a plan. No, we said, we just wanted it closed. Our elders believed that the herring was one of the ocean's most important stocks. It fed all of the different species that went through our part of the world, like rockfish (*Sebastes* spp.). So if I had herring, I also had the fish that ate herring. Fortunately, the DFO supported our argument that we wanted 'no take' on herring in Hesquiaht Harbor. We were of one mind that we did not need a Marine Protected Area (MPA). We are the ones who understand what is going on. Creating an MPA is good, but who is going to enforce it? Who is going to watch it? They closed it for us. And I saw what my grandfather saw. When the herring came in, there were hundreds of seagulls and ducks of every kind. Everything was there, without a plan, without an MPA. There is a time when things have to be totally natural, and how we fit into that scheme is important.

What we found out is that, as we try to implement the ways our grandfathers understood, we face a tremendous struggle. Some tribes are affected by the development of dams and can never fish again. So I think it is important for you to listen. We have talent. We have educated First Nations people in BC. We have people who understand about our grandfathers. We have biologists working for our tribes who understand and listen to the teachings of our forefathers. Our people are not talking about total isolation, because we recognize the fact that the people that are here now are here to stay. We do not want to create an imbalance. The sea otters (*Enhydra lutris*) that have been re-introduced to Nuu-chah-nulth territory are an example of what happens when we create imbalance.

We used to be the dominant species over the things that moved in Nuu-chah-nulth territory. Now there is another dominant species: sea otters. Because of them, there are no more clams and no more sea urchins. The DFO said that the sea otters were extinct on the west coast of Vancouver Island, so they brought some from Alaska. There are more sea otters now in Kyuquot and they are eating all the sea urchin (Watson, 2000). When the sea otters were re-introduced, nobody asked us how to control them. Sea otters and sea lions (*Eumetopias jubatus*) are in decline in the Aleutian Islands, but in our territory they are too plentiful. The animal rights people say sea otters have a right to live, but the animal rights people do not live in our territory. We have a right to live too. We do not want people to forget that there is a human aspect to whatever decision is made. We want to be part and parcel of the decisions about our home. The Kyuquot people are almost extinct! Compared to the sea otters, we are now the endangered species. We as humans are seen as less important than sea otters, and the sea otters now in my territory were brought from somewhere else.

Our leadership is saying that we value technology but we want to combine it with our traditional values. Some of our people have done very well. They have been very competitive; that has become part of us over the past ninety years. Over the years, our people have been badly affected by regulations, but they are still out there. We have one person who has a halibut licence, one involved in the black cod fishery, another in the crab fishery, and the list goes on. Our people lived off the sea and we sustained ourselves.

So I leave you with this: think for a moment. You are in a forest. Listen to what it might be saying to you. As you are in the forest, you are beside a little brook making these little sounds. We have been of the same people as long as the rocks have been here.

REFERENCES

CROCKFORD, S. 1994. New archaeological and ethnographic evidence of an extinct fishery for giant bluefin tuna on the Pacific Northwest coast of North America. In: W. Van Neer (ed.), *Fish exploitation in the past. Proceedings of the 7th meeting of the ICAZ Fish Remains Working Group.* Tervuren, Annales du Musee Royal de l'Afrique Centrale, Sciences Zoologiques No. 274, pp. 163–8.

——. 1997. Archeological evidence of large northern bluefin tuna in coastal waters of British Columbia and northern Washington. *Fishery Bulletin,* Vol. 95, pp. 11–24.

DRUCKER, P. 1951. *The northern and central Nootkan tribes.* Washington, D.C., Bureau of American Ethnology Bulletin. Smithsonian Institution, Vol. 144, 480 pp.

HAGGAN, N.; TURNER, N.J.; CARPENTER, J.; JONES, J.T.; MENZIES, C.; MACKIE, Q. 2004. 12,000+ years of change: Linking traditional and modern ecosystem science in the Pacific Northwest. *Proceedings of the 16th Society for Ecological Restoration Conference,* Victoria, BC.

LUCAS, CHIEF SIMON. 2003. A native chant. In: N. Haggan, C. Brignall and L. Wood (eds.), *Putting fishers' knowledge to work.* Fisheries Centre Research Reports Vol. 11, No. 1. Vancouver, University of British Columbia, pp 13–14.

——. 2004. Aboriginal values. In: T.J. Pitcher (ed.), *Back to the future: Advances in methodology for modelling and evaluating past ecosystems.* Fisheries Centre Research Reports Vol. 12, No. 1. Vancouver, University of British Columbia, pp. 114–15.

McMillan, A.D. 1999. *Since the time of the transformers: The ancient heritage of the Nuu-chah-nulth, Ditidaht, and Makah*. Pacific Rim Archaeology. Vancouver, UBC Press, 264 pp.

Menzies, A. 1792. *Menzies' journal of Vancouver's voyage, April to October, 1792*. Edited, with botanical and ethnological notes, by C.F. Newcombe, M.D., and a biographical note by J. Forsyth. William H. Cullin, printer to the King's most excellent majesty, Victoria, BC, 155 pp.

Poepoe, K.K.; Bartram, P.K.; Friedlander, AM. 2006. The use of traditional knowledge in the contemporary management of a Hawaiian community's marine resources. In: N. Haggan, B. Neis, and I.G. Baird (eds.), *Fishers' knowledge in fisheries science and management*. Oxford, Blackwell Science, and Paris, UNESCO.

Slaney, T.L; Hyatt, K.D.; Northcote, T.G.; Fielden, R.J. 1996. Status of anadromous salmon and trout in British Columbia and Yukon. *Fisheries*. Special issue on Southeastern Alaska and British Columbia salmonid stocks at risk, Vol. 21, No. 10, pp. 20–35.

Stewart, H. 1984. *Cedar: Tree of Life to the northwest coast Indians*. Vancouver, BC, Douglas and McIntyre.

Watson, J. 2000. The effects of sea otters (*Enhydra lutris*) on abalone (*Haliotis* spp.) populations. In: A. Campbell (ed.), Workshop on Rebuilding Abalone Stocks in British Columbia. *Canadian Special Publication Fisheries and Aquatic Science*, Vol. 130, pp. 123-32.

CHAPTER 3   My grandfather's knowledge
First Nations' fishing methodologies on the Fraser River

Arnie Narcisse

ABSTRACT

THE author is a member of the Stl'atl'imx First Nation, whose traditional territories are on the Fraser River in British Columbia, Canada. This chapter describes traditional fishing methods, and the importance of fishing for the cultural and physical existence of aboriginal people. Small-scale fisheries and the mechanisms of intergenerational transfer of traditional knowledge are vital not only to the cultural and physical existence of the Stl'atl'imx people, but also in the global context: indigenous and artisanal fishers catch 50 per cent of the world's food fish, and a billion people depend on fish and seafood for a major part of their diet.

INTRODUCTION

THIS chapter deals with the knowledge of the Stl'atl'imx First Nation in British Columbia, as handed down to me by my grandfather. His world was 10 miles (16 km) of the Fraser River, in the heart of Stl'atl'imx territory (Figure 3.1), the three ranches that he ran, and the livestock that he owned. For the purposes of this chapter, I will try to describe my grandfather's role as a fisher (Figure 3.2), although my grandmother was also a big influence in my life (Figure 3.3). My earliest recollections of going to the river include riding on the old two-wheeled horse-drawn cart. On the way, my grandfather would point out various plants and animals in the ecosystem around us.

FISHING FOR *ZUMAK*, SPRING SALMON

THE earliest fishery of the year was the *Zumak* or Chinook salmon (*Oncorhynchus tshawytscha*), which were the first to swim upriver. I can recall my grandfather making his nets and getting ready for this first fishery. We lived in a house with one room, which would be full of gillnets in various stages of completion, dipnets,

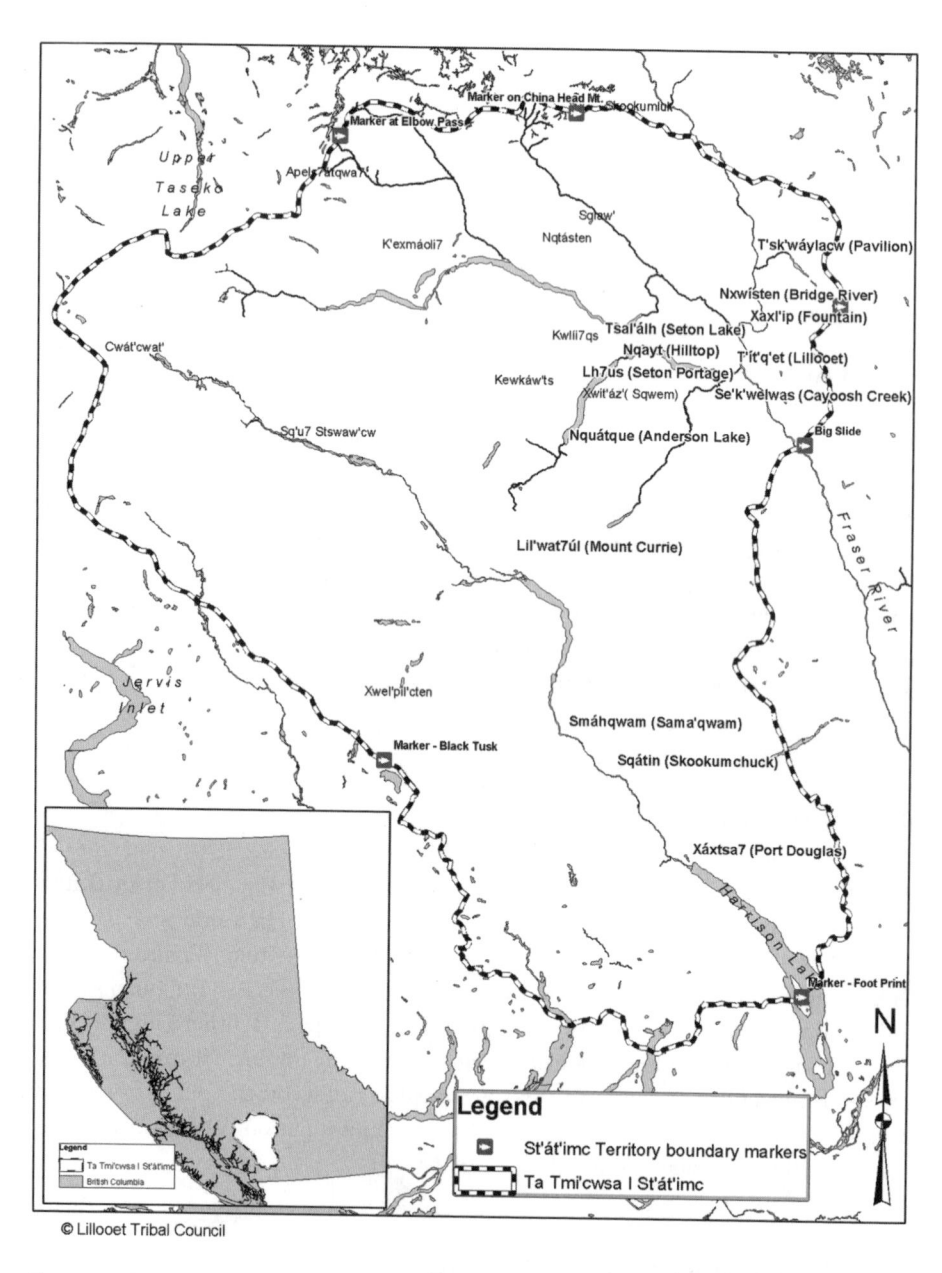

FIGURE 3.1 *Map of Stl'atl'imx territory, courtesy of Lillooet Tribal Council. The Stl'atl'imx people fished throughout the entire area delineated. The fishing sites described here are located between Bridge River to the north to just south of the Cayoosh Creek/Fraser confluence.*

hoops and poles. Everywhere you looked there were needles and wooden spacers of different sizes for different nets.

My grandfather had a sense of excitement about him at that time of year and he would speak in hushed tones 'The *Zumak* are coming, they are coming!'. You could sense his excitement. He was my whole world. When he had the spring *Zumak* gear ready – nets with 6–8 inch (15–20 cm) mesh – we would head down to the river. That is when he pointed out the various bushes along the way. We would catch enough for supper and for a couple of days. We had no refrigeration back then and we were not big on canning and spring salmon is hard to dry. The spring was a break from the salt and dry salmon that sustained us through the winter months.

Figure 3.2 *Arnie and grandfather. West bank of the Fraser River, immediately south of the Bridge River confluence, ca. 1959.*
Photo: Malcolm Parry.

INDIGENOUS PRACTITIONERS AND RESEARCHERS

FIGURE 3.3 *Arnie Narcisse with grandmother. West bank of the Fraser River, immediately south of the Bridge River confluence, ca. 1959.*
Photo: Malcolm Parry.

FISHING FOR SOCKEYE

THE second fishery my grandfather taught me was for sockeye salmon (*Oncorhynchus nerka*). During the interval from spring to the sockeye fishing season he would work on his ranch. I recall the water system he built, a ditch which was probably about 5 miles long to catch the water from the mountain to irrigate his fields. The man was a magician: the water ran uphill, following him. There was constant activity of fishing and farming in the summer months. And so in the early summer when the rose petals began to bloom, he would go fishing for sockeye. I remember him pointing out the rose petals to me.

WINTER FOOD PREPARATION

SUMMER was a really busy time because we had to dry and put away enough salmon for the winter months. I don't know how many racks of dry salmon we had when I was little. But each rack would hold about 200, and we would replenish them three times over, so there would be about 600 dried salmon per rack. Salting and drying were the main preserving methods. Catching the fish and cutting it up was a lot of work, from the crack of dawn right through to dusk. In this way you went through an apprenticeship as a young person. My first recollections of this come from when I was 4 or 5 years old, when my main jobs were carrying fish guts and hanging up the smaller strips of salmon to dry, as my family is doing in Figure 3.4. I was a productive little guy back then. It is amazing what a 4-year-old can do!

As you get a little older, you begin to carry the salmon to the drying rack. When you are older still, you can handle the ropes and gillnet (Figure 3.5), which are not as dangerous as the dipnet style (Figure 3.6). The crowning moment of glory would

FIGURE 3.4 *Drying racks.*

be when you are 12 or 13 when you handle the dipnet. This was dangerous work because of the fast flowing river (Figure 3.7). You were then considered a man.

I regretted having to learn to cut and dry salmon because I got stuck up there with the old ladies. My young buck buddies were down there fishing and I was stuck up on the bank with the old ladies teaching me the proper way of cutting – but it came to be a useful skill and I hope I will teach my grandchildren to do that.

In the later part of the year the Hane, pink salmon (*Oncorhynchus gorbuscha*), would start to arrive. By the time they got to our territory they were basically useless for human consumption. I hated them because they died right outside my doorstep. I lived just halfway between the confluence and the spawning ground and they were dying everywhere. It took me a while to learn they had a role in the ecosystem, but that is what our people recognize in the First Salmon Ceremony (Swezey and Heizer, 1993; Haggan and Neis, this volume). Western fisheries science is only now rediscovering the key role that marine nitrogen and phosphorous from salmon carcasses play in the health of forests and watersheds (Reimchen, 2001; Watkinson, 2001; Stockner, 2003).

In the grander scheme of things my grandfather's knowledge might seem a very small matter, but it has given much to my family. It allowed us to survive and thrive and continue to exist. For that, it is very useful. It has given me knowledge of the benefits of hard work and perseverance, the simple pleasure that you get from feeding yourself and your family. I am sure my grandfather was very proud of that.

FIGURE 3.5 *Arnie with gillnet rigged to gin pole.*

FIGURE 3.6 *Arnie wants to use a dipnet. West bank of the Fraser River, immediately south of the Bridge River confluence, ca. 1959.*
Photo: Malcolm Parry.

A SHORT HISTORY OF DISPOSSESSION

I would like to say a little about my great-grandfather and his role in Stl'atl'imx history. His name was Uthla and he was one of the chiefs that signed the 1911 Lillooet declaration of sovereignty (Terry-Drake, 1989), under his English name of John Baptiste. The declaration stated that my great-grandfather's people had owned and defended their lands against all comers since ancient times and had never been conquered or ceded them to the settlers. It required:

INDIGENOUS PRACTITIONERS AND RESEARCHERS

FIGURE 3.7 *Dipnetting. West bank of the Fraser River, immediately south of the Bridge River confluence, ca. 1959.*
Photo: Malcolm Parry.

'that all matters of present importance to the people of each of our tribes be subject to [treaties], so that we shall have a definite understanding regarding lands, water, timber, game, fish, etc.'

The start of industrial fishing on the Fraser River had been signalled in 1888 by legislation that forbade aboriginal people to sell salmon. A mere twenty-three years after the Lillooet Declaration, the Fraser River fishery had been depleted to such a degree that interventionist measures such as fish hatcheries were necessary. In the 1911 declaration, my great-grandfather questioned why we had arrived at such a state in so short a time. Today, the great salmon runs of the past are dwindling to the point where the future of Pacific salmon is in question. Desperate measures are being put forward, such as the privatization of the salmon fisheries that sustained our people for over ten thousand years (McRae and Pearse, 2004). As I look to the future, I foresee continuing problems due to the impact of the salmon aquaculture industry on wild salmon.

KNOWLEDGE IN PLACE AND TIME

THE tools of the trade that my grandfather made were specific to those 10 miles of the Fraser River. He knew every back eddy, riffle and run in that 10-mile stretch. He knew which net should be used in which specific spot. He moved upriver as the level of the river receded, and when the fish were very plentiful he would just use his dipnet, and then he could catch as many as with his gillnet. I guess this could be viewed as adaptation to specific requirements. And again that is very much the nature of most artisanal fisheries.

The amazing thing about these simple technologies of small gillnets and dipnets is that they are still as useful today as they were in my grandfather's time. I still make my dipnets the way he taught me. My sons and I now use dipnets and gillnets in the same spots he showed me while I was growing up.

I see my great-grandfather's political stand, my grandfather's actions and my own work as efforts to preserve wild stocks of salmon that our family has always depended upon for sustenance. My family still dry and process salmon for our winter food (Figure 3.4). The baby you see in Figure 3.8 is my grandson, who my family and I will teach as I was taught by my grandfather. In the modern-day vernacular this is termed intergenerational equity – simply, the right of future generations to enjoy at least the same resources and benefits as we do, and the passing down of knowledge from one generation to the next (Figure 3.8).

THE SMALL-SCALE WORLD

I said earlier that my grandfather's knowledge might seem a small thing in the great scheme of things, but I have come to learn that over 50 per cent of the world's fish and seafood are caught in small-scale indigenous and artisanal fisheries. Like my grandfather's, these fisheries only involve small pockets of the ecosystem, but around the world a billion people depend on them for food, livelihood (Berkes et al., 2001 and references therein) and, like my Stl'atl'imx people, for cultural survival. The trick for us is to figure out how all of us who depend on these small-scale fisheries can reach out to each other and convince the world just how important these fisheries and the ecosystems that sustain them, are to the future of the planet.

FIGURE 3.8 *Intergenerational equity. My grandparents, my immediate family and my grandson.*

In retrospect, I have been very fortunate. I had a very good teacher, and all I hope to do is to pass on my grandfather's knowledge to my grandchildren. All I want is the same as yesterday, just like my grandfather. I always think that if we were to look after our respective backyards and work together in concert we could indeed make this world a better place.

REFERENCES

BERKES, F.; MAHON, R.; MCCONNEY, P.; POLLNAC, R.; POMEROY, R. 2001. *Managing small-scale fisheries:Alternative directions and methods*. Ottawa, International Development Research Center, 320p. Available online at: http://www.idrc.ca/fr/ev-9328-201-1-DO_TOPIC.html (last accessed on 24 June 2005).

McRae, D.M.; Pearse, P.H. 2004. *Treaties and transition: Towards a sustainable fishery on Canada's Pacific coast.* Vancouver, Hemlock Printers. 58 pp.

Reimchen, T. E. 2001. Salmon nutrients, nitrogen isotopes and coastal forests. *Ecoforestry,* Vol. 16, pp. 13–17.

Stockner, J.G. (ed.). 2003. *Nutrients in salmon ecosystems: Sustaining production and biodiversity.* Bethesda, Md., American Fisheries Society, Symposium 34, 285 pp.

Swezey, S.L.; Heizer, R.F. (1993) Ritual management of salmonid fish resources. In: T.C. Blackburn and K. Anderson (eds.), *Before the wilderness: Environmental management by native Californians.* Menlo Park, Calif., Ballena Press, pp. 299–327.

Terry-Drake, J. 1989. *The same as yesterday: The Lillooet chronicle the theft of their lands and resources.* Lillooet Tribal Council. Published 1991 in *Canadian Historical Review,* Vol. 72, No. 2.

Watkinson, S. 2001. *The importance of salmon carcasses to watershed function.* Vancouver, University of British Columbia Fisheries Centre, MSc. thesis, 111 pp.

CHAPTER 4 Indigenous technical knowledge of Malawian artisanal fishers

Edward Nsiku

ABSTRACT

COMMERCIAL, semi-commercial, artisanal and ornamental fisheries provide employment, income and food security for many people in Malawi. The fish and ecosystems that support them are of scientific and educational value, and represent a natural heritage of aesthetic beauty. Most fisheries are artisanal and rely extensively on indigenous technical knowledge (ITK). ITK informs fishing methods and devices, fishing craft, and artisanal fishers' understanding of events and issues in ecology and climate. ITK could be applied to fish conservation through mechanisms such as co-management and fisheries monitoring schemes.

INTRODUCTION

THE fishery system in Malawi covers about 29,000 km^2 (approximately 25 per cent of the total area of the country) and consists of four lakes, various rivers and other water bodies (Figure 4.1). The annual fish catch averages over 60,000 tonnes (FD, 1996; Nsiku, 1999). Fish provide about 70 per cent of protein from animal sources and 40 per cent of overall protein intake, and the fisheries are a scientific and educational asset as well as a natural heritage of aesthetic beauty (GOM, 1989; ICLARM/GTZ, 1991). Fisheries contribute 4 per cent of the gross domestic product of Malawi and employ over 230,000 people directly and indirectly. In addition, they support the livelihoods of about 250,000 to 300,000 people who live in the households of fish workers (Munthali, 1997; Scholz et al., 1998).

There are four types of fishery in Malawi: commercial, semi-commercial, artisanal (traditional) and a small but economically important aquarium or ornamental fishery for export. The artisanal fisheries described in this chapter account for 90 per cent of annual fish landings, or more than 50,000 tonnes (GOM, 1999). Artisanal fishing is carried out primarily from dugout canoes, but also from plank boats and bark platforms or canoes. The Fisheries Department (FD) oversees the fisheries sector, implementing activities related to research, extension and development, training, fish farming, and management and administration.

Artisanal fishers rely extensively on indigenous knowledge in their harvesting, processing, marketing/distribution and consumption of fish. There was limited government influence in the fisheries during the colonial era (1891–1963), and until the establishment of the FD in 1971. Since that time, fisheries have been managed through the centralized enforcement of regulations relating to licensing, closed seasons, prohibited methods of fishing, prohibited fishing devices, and the dimensions, and minimum sizes or lengths of fish that may be landed (Scholz et al., 1998; Nsiku, 1999). However, funds for enforcement programmes have been inadequate and extension services to teach fishers about regulations have also been weak. As a result, fisheries laws have not been followed and national catch figures have declined. There are some known cases of localized overfishing (Chirwa, 1996; Munthali, 1997; Donda, 1998; Hara, 1998; Scholz et al., 1998).

This chapter explores Malawian artisanal fishers' knowledge of fishing methods and equipment, fishing craft, ecological issues and climatic parameters; it discusses the potential application of this ITK to the implementation of co-management and alternative fisheries monitoring schemes intended to improve conservation. The data summarized in the chapter are derived from the author's experiences and discussions with fishers and colleagues while working with the FD between 1986 and 1997.

ARTISANAL FISHERS' ITK

General information on indigenous knowledge

Sources of knowledge for most rural communities include traditional and modern or scientific knowledge, or a blend of the two. Interest in ITK gained prominence during the 1970s and 1980s when professionals recognized that rural people were well versed in many subjects that touched their lives, and that ITK could provide rich and valuable insights during rapid rural appraisals (Chambers, 1993). Further, Sambo and Woytek (2001) noted that:

> 'There is a global recognition (e.g. United Nations Conference on Environment and Development Chapter 26 of Agenda 21, International Convention on Biodiversity, World Bank and International Union for Conservation of Nature and Natural Resources based on Keating (1993), Matowanyika and Sibanda (1998) and World Bank (1998)) that indigenous knowledge has an important role to play in consonance with modern scientific and technological intervention in social and economic development and cultural and political transformation (p. 83).'

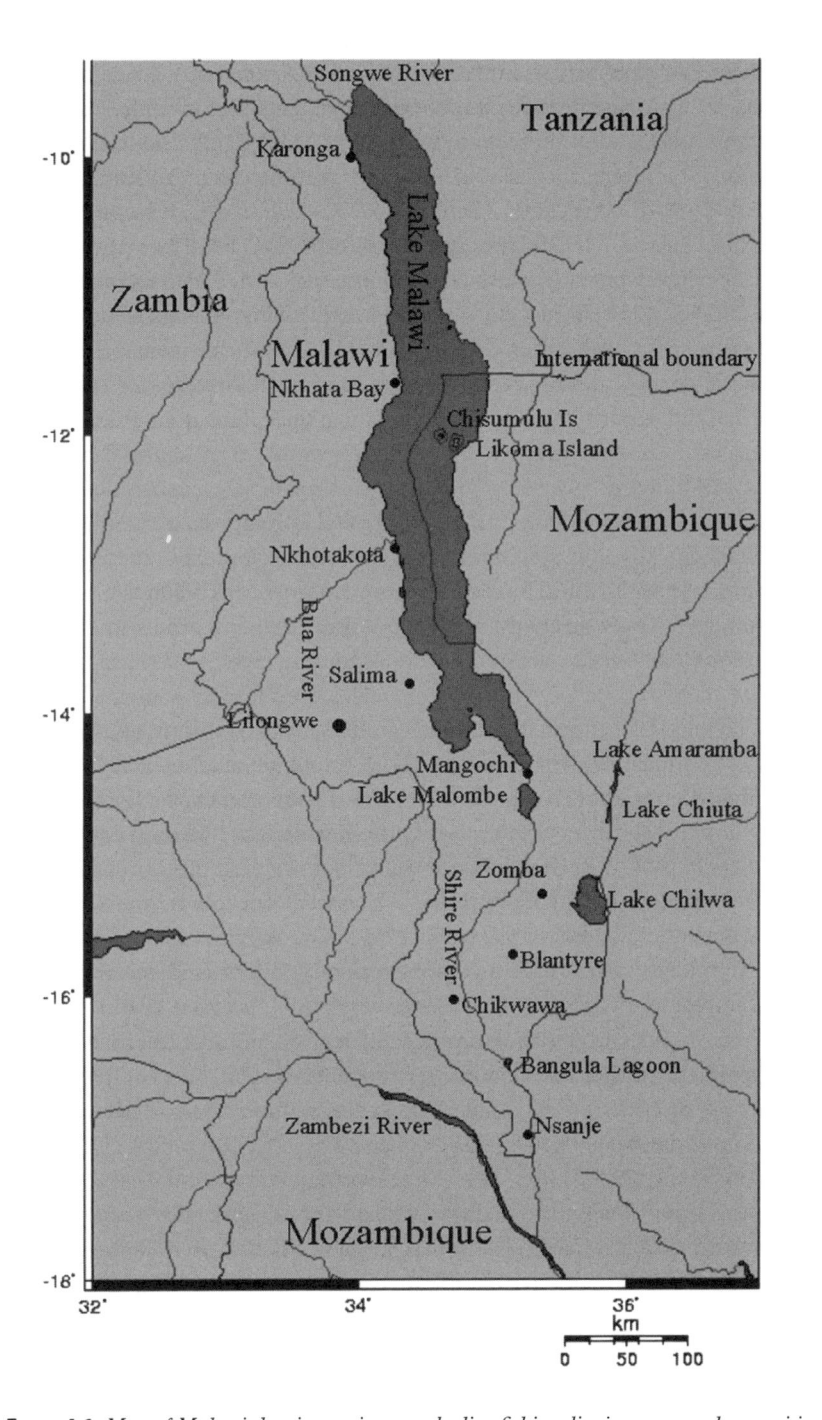

INDIGENOUS TECHNICAL KNOWLEDGE OF MALAWIAN ARTISANAL FISHERS

FIGURE 4.1 *Map of Malawi showing main water bodies, fishing district towns and some cities.*
Source: www.aquarius.geomar.de/omc (last accessed 26 June 2005).

In ITK the information or understanding related to environment, science or technology is local in origin and has accumulated over long periods. The resulting experiences and adaptations or techniques and practices differ between social groups or cultures, and are generally in harmony with the environmental conditions and responsive to constraints (Sambo and Woytek, 2001). The knowledge – in the form of concepts, beliefs, perceptions, information, facts or evidence – is one facet of sociocultural arrangements that include political, economic and social organization. These enable local people to cope with many ecological processes, and with environmental and other events, and thus to survive as societies. The 'technical' aspect of ITK relates to its use in acquiring desired goods and conditions such as food security through manipulating the physical environment to acquire resources. ITK also integrates information drawn from many aspects of people's lives, including: fish harvesting and processing; ecology, especially animal behaviour; botany or ethnobotany; hydrology (parts of the rain cycle); geomorphology (knowledge of local soils and terrains); climatology and meteorology; knowledge about seasons (lunar calendars, rain or storm predictions); oceanographic knowledge (waves and water currents); and religious beliefs and explanations for things not fully understood (Matowanyika, 1994; Poepoe et al., this volume).

Artisanal fishers, like others in rural communities learn informally. Folklore serves as one of the channels for passing ITK to future generations and is rich in beliefs, customs and practices (Kalipeni, 1996). Information on the traditional value systems, which influence and guide the activities of a majority of rural Malawians (GOM/UN, 1992), is usually transmitted orally in stories, riddles, proverbs and songs related to experiences of daily life as well as taboos, totem animals, place names and nicknames (Sambo and Woytek, 2001). The activities by which people fashion their livelihoods are learnt and passed on to subsequent generations through observation and practice (Berlin, 1992; Matowanyika, 1994; Dawson, 1997). Fishing in shoreline communities exemplifies this type of instruction. For example, Hoole (1955) describes the strong link fishing has to the transition into adulthood for the Tonga people of Nkhata Bay District along the north-western shore of Lake Malawi:

'The male Tonga is wedded to the lake almost from the day he is born. ... [He] learns to tumble in it, to swim like a fish, to exult his skill on it, and love it in all its moods. His main ambition in life then becomes to own his own net, and paddle his own canoe. In the hot season the boys of the village build themselves *mphara*, roofless shelters of reeds on the shore and at all times they are assisting their elders, and learning from them the many details of the fisher's craft. In the kindergarten stage they become adept at catching small fish with a *matete* reed (*Phragmites mauritianus*) for a rod (pp. 26–7).'

FISHING METHODS AND DEVICES

TRADITIONAL fishing methods fall into five categories: netting, trapping, line or hook fishing, simple manual techniques, and using fish stupefacients or piscicidal plants. Fishing methods vary widely and are closely adapted to the local details of fishing grounds and behavioural patterns of the species present (ICLARM/GTZ, 1991).

Net fishing

Net fishing equipment includes *machera, ndangala or chilepa* (gillnets); *chilimila* and *nkacha* (open water seines); *mkwau, (n)khoka* or *ukonde* (shore seines); *pyasa* (scoop or dipnets); and *chabvi* (cast-nets). Artisanal fishers made nets from fibres of shrubs or tree bark before synthetic netting materials appeared in Lake Malawi in the early 1900s, and spread to other water bodies during the late 1950s and early 1960s (Mzumara, 1967; Chirwa, 1996). The preferred fibre plant materials used were the evergreen sisal shrub *(Pouzolzia hypoleuca)*, variously called *mulusa, t(h)ingo, lu(i)chopwa, lukayo, (b)wazi, gavi* or *khonje*; tree violet *(Seciridaca longendunculata)*, called *nakabwazi, chosi, chiguluka, njefu, muluka* or *mu-uruak*; and baobab *(Adansonia digitata)*, locally known as *mlambe*. The cultivated sisal (*P. hypoleuca*) was the most popular because its fibres are particularly durable. The outer layers of its leaves are scraped off and then sun-dried before soaking in water. The wet leaves are partially dried again so that the white inner fibres can be easily removed. Nets are woven from strings that are produced when fibre strips are skilfully spliced and rolled together by hand on the thigh (Hoole, 1955). Although the nets made from local materials did not keep as well as those of synthetic fibres, their working life could be improved by proper maintenance. Fishers often repaired their nets, kept them away from direct sun, and stored them in *mphara* or *khumbi* huts – small, wall-less round shelters with good air circulation. Nets were also protected from rotting by oils from plants or trees such as cashew nut *(Anacardium occidentale)*, known in different areas as *mbibu, msololikoko* or *nkoloso* (ICLARM/GTZ, 1991).

Gillnets are the most common contemporary gear. They can be used actively but are mainly used passively. To improve the effectiveness of gillnets, fishers sometimes dye them brown or reddish brown with a bark or root preparation (usually boiled in water) from local trees or herbs such as *Newtonia* spp., *Elephantorrhiza goetzei, Acacia macrothyrsa* and *Lannea stuhlmanni*, which respectively are known as *chanilama, chanima, chirima* and *chilusa*. When gillnets are used actively, it is in the fishing method known as *chiombela*, whereby fish are driven into nets or other devices by beating the water surface with poles or paddles to make noise (Mills, 1980; Ojda, 1990; ICLARM/GTZ, 1991;

M. Hara,[1] personal communication). In the northern part of Lake Malawi *chilimila*[2] seines are used effectively over *virundu* (sing. *chirundu*) – rocky prominences, pinnacles or 'reefs' that occur in some fishing grounds, where they protrude from the lakebed. These are usually rich with species such as *utaka* (cichlid, *Copadichromis* spp.). The fish orient themselves toward the current, which concentrates their planktonic food around the *virundu* where the regime of the currents fluctuates, both annually and diurnally. The *chilimila* thus functions as a diver-operated lift net that requires thorough knowledge of the current pattern and bottom topography to be successful (ICLARM/GTZ, 1991).

A small number of fishers make use of *vuu*, precarious stands on rocky ledges on falls or rapids, where they use *khombe* – specialized scoop nets – to target anadromous fish species, *sanjika* (lake trout *Opsaridium microcephalus*) and *mpasa* (lake salmon *Opsaridium microlepis*), particularly during upstream spawning migrations from Lake Malawi. *Khombe* were most commonly used at Chiwandama falls on Luweya River in Nkhata Bay District, but are not often seen nowadays (Hoole, 1955). The nets can also be used at night; an example is the fishing method called *kauni*, in which light from *mwenji* torches of wood, grass bundles or oil lamps is used to attract the fish. The method is also known as *chi(w)u*, if a scoop net is used to target *usipa* (lake sardines, *Engraulicypris sardella*).

The net accessories such as *zingwe/luzi* (ropes), *mabungwa/masila* (floats) or *kalanje* (a large centre-float), sinkers, *mabingo* (buoys or large floats serving as markers), *nchonjolo* (sturdy poles) and *nthepa* (stakes) were also made from locally available materials. The last two items were very useful for setting nets in shallow waters (Mzumara, 1967). Ropes were woven from *milaza*-leaves of *ngwalangwa* (the *Hyphaene crinita* palm), stems of common creepers or climbers – *chilambe* (*Helichrysum chrysophorum*), *msaula/malandalala* (*Ipoma pes caprae*) and *lulisi* (*Tinospora caffrara*) – or fibres of other plants. Floats were made from softwood like *chiumbu/sidyatungu* (livelong *Lannea discolor*), and *chiwale* (*Raphia vinifere*) and *mvumo* (*Borassus aethiopum*) palms. Medium-sized stones and fire-baked clay balls or rings were usually used as sinkers.

Fish trapping

Fish trapping uses two main types of equipment, *mono/chisako* (basket traps) and *psyailo/beyu* (fence traps). *Mono* basket traps are the most common of all fishing devices (FD, 1996). They are made from split bamboo canes, reed stems or thin

---

1. Dr Mafaniso Hara, Centre for Southern African Studies, University of Western Cape, South Africa. Former Principal Fisheries Officer in the Malawi Fisheries Department.
2. Variations of the word are chilimira and chirimila.

branches (twigs, wicker) held together by twisted *milaza* (leaves of *H. crinita*), fibres of *P. hypoleuca* or creepers, and then tied to hoops of *nthepa* staves or lengths of supple branches as frames to give the *mono* their shape and full size. They may be tapered, with valves placed on their larger front ends to allow fish in but not out. The back ends are closed when the traps are set and opened to remove catches. *Mono* may be used either on their own or in association with weirs constructed of poles or other vegetative materials to close sections of rivers or other water bodies. The traps are set in gaps left within the weirs. When used separately, traps are usually baited with other smaller fish or meal remains. *Mono* may be weighted so that they can be set on the lake or river bed, attached by a creeper or rope of other material to a *bingo* marker (Hoole, 1955).

The *psyailo/beyu* (fence trap) is a fish fence that is used to encircle and trap the fish. It is made of *bango* (*P. mauritianus*), *nsenjere* (*Pennisetum purpureum*) or other reeds and grasses such as *manjedza* (*Typha capensis*), and stakes or poles. The materials are bound together by *milaza* palm leaves, or stems of *chilambe*, a common creeper (*H. chrysophorum*). Six or more people may operate a *psyailo* in shallow water (Mills, 1980; ICLARM/GTZ, 1991; Brummett and Noble, 1995). Another form of fish encircling is used at Bangula Lagoon and Ndinde Marsh in the southern part of Malawi. This involves making banks of submerged macrophyte *Ceratophyllum demersum* or related species and mud, scooped out to form enclosures. Traps are set in the gaps left in the vegetative walls to catch fish as they try to escape. Grasses such as hippo grass (*Vossia cuspidata*), locally known as *dunvi*, *nsanje* or *msali*, are also used for fish-trapping devices (ICLARM/GTZ, 1991).

Hook and line fishing

There are three types of hook fishing (*kuwedza* or *kuweja*): long lines (*khuleya*), single lines (*chomanga*), and pole-and-lines (*mbedza*). *Khuleya* are long main ropes with 50 to 800 or more short sidelines, each with its own hook (Mzumara, 1967). They are held in place by poles fixed in the water at each of the ends. A few *zichili* (staves) are sometimes included at intervals between the two ends. Anchored *masila* (floats or weights) are set on the bed in deep waters. *Chomanga* are single hooks on short lengths of line attached to anchored floats or fixed stakes. A *chomanga* fisher may set several of these on the fishing ground. Pole-and-lines are also single-hook sets on hand-held rods about a metre long, with lines twice that length. Small floats are usually attached so that the baits are set around 30 cm below the surface. Pole-and-lines are also used as a partial harvesting technique for fishponds (Brummett and Noble, 1995). In Lake Chilwa hook fishing is also used in association with a specialized fishing technique called *magalaji* (translated as 'garages'), where fish are kept in floating baskets suspended in the water near the fishing site for up to a week or even longer, fed on maize meal or *matemba* (small barbs, *Barbus* spp.). Fishers target *mlamba* (catfish, *Clarias* spp.) with this technique (Mzumara, 1967).

In Lake Chilwa, some fishers use *zimbowela* (temporary huts on floating islands) to operate offshore. The lake is fringed by an extensive and dense growth of macrophytes, particularly *manjedza* bullrush (*T. capensis*), which in some parts extend up to 15 km from fishing villages to the open water area. *Manjedza* and other floating weeds are often dislodged from marsh areas and cover up beaches, landing points, jetties and fishing grounds during windstorms, sometimes blocking them for long periods (Landes and Otte, 1983). Fishers use floating *manjedza* to make platforms on which they build *zimbowela* where they can use *machera* gillnets, *khuleya* long-lines, and *mono* basket traps. *Zimbowela* can be built as deliberately planned fishing camps or temporary shelters when stranded in rainstorms.

Baits are used for all types of hook fishing, and vary with the species targeted and water conditions; baits include small fish such as *usipa* (*E. sardella*) and *matemba* (*Barbus* spp.), worms, insects, frogs or other amphibia, pieces of meat and remains of *nsima* meal. Pieces of tablet soap have been reported among the baits for *mlamba* in Lake Chilwa (Mzumara, 1967).

Manual fishing devices

Simple manual fishing techniques include plunge baskets, *mkondo* spears, and *uta* bows and *mubvi* arrows. Plunge baskets are constructed like *mono* but they do not have valves and are conical in shape with openings on the sides of their apexes. They are mainly found in the Lower Shire Valley. They are operated in shallow water by driving or plunging them downward over an observed fish or disturbance in the water. Both spears and bows and arrows are common in the marshes and flood plains of the Shire River, and spears in Lakes Chilwa and Chiuta. Spears are hard metal blades, usually sharpened, that are fixed to thin but strong shafts about a metre and a half in length. The blades may, in rare instances, be winged or have barbs at their tips. Manual fishing devices are used in very shallow waters, mainly for subsistence fishing during both the dry and the wet season. During the floods of the wet season, the main target species, the *mlamba* (catfish), is on spawning migration. Spear fishing may be conducted by day or by night (Mzumara 1967; Mills, 1980; Ojda, 1990; ICLARM/GTZ, 1991).

Stupefacients and poisons

The use of stupefacients and piscicidal plants or fish poisons has been banned since 1934 (ICLARM/GTZ, 1991), although some cases still occur and the FD maintains surveillance against the technique, most intensively in the dry season. Such methods were used mainly along seasonal rivers with pools that dry late, and at small, isolated swamps and marshes. The stupefacients and poisons were associated with certain ceremonial occasions, particularly during the dry season, and large fish kills provided fish for whole communities. Many plants in the country were used for this purpose (ICLARM/GTZ, 1991), and some of them are indeed very potent. Brummett and Noble (1995) report on recent research on piscicidal plants at the University of Malawi:

> 'Fifty potential candidates were investigated by Chiotha *et al.* (1991). Of these, 14 (*Agave sisalana, Aloe swynnertonii, Bridelia micrantha, Breonadia microcephala, Ensete livingstonianum, Erythrophleum suaveolens, Euphorbia* (unidentified species), *Neorautenenia mitis, Opuntia vulgaris, Phytolacca dodecandra, Sesbania macrantha, Swartizia madagascariensis, Tephrosia vogelii, Xeromphis obovata*) were found to kill 95–100% of *Tilapia rendalli* and *Oreochromis shiranus* within 24 hours at a concentration of 100 mg·l^{-1}. The potential risks to humans of eating fish killed in this manner have yet to be determined (p. 32).'

The local communities where these plants grow naturally have long known of their potency, and were probably aware of some effects if people ate the dead fish. Fish poisons were only used in still or slow moving waters where their effectiveness was limited to short periods of time, and the amounts, depths and flows of water easily reduced their strength (Chirwa, 1996).

FISHING CRAFT

ARTISANAL fishing craft, mostly dugout canoes called *wato*, *bwato* or *ngalawa*, are moved by poling in shallows, paddling in deeper waters, and occasionally sailing. Strengthened bark platforms or canoes are also used to a limited extent for fishing in Lake Chilwa and the Lower Shire (Mills, 1980; ICLARM/GTZ, 1991). At one time it was expected that plank, fibreglass or other types of boats with outboard motors, as well as relatively large inboard vessels, would replace the canoe (Emtage, 1967). However, it has prevailed as the main fishing vessel, representing 85 per cent of all traditional fisheries craft in the country between 1985 and 1995 (Nsiku, 1999). In Lake Malawi in 1986, canoes made up 78 per cent of fishing craft (ICLARM/GTZ, 1991), a proportion that rose by 1994 to 81 per cent (Banda and Tomasson, 1997). The simplicity of design and limited cost seem to have bolstered the canoe's resilience, despite declines in lifespan and size due to the lack of suitable tree species for making canoes (ICLARM/GTZ, 1991). The in-curved lips of the hull make well-made dugouts virtually impossible to turn over; they can roll through 90° to lie on their sides and still recover, without shipping water. Slight lifts at the bow and stern reduce rolling and enhance recovery (Emtage, 1967). Besides the two square projecting knobs at the prow (*mushyio*) and stern (*chisiuka* or *matambi*), the only other essentials and appropriate accessories are *nkhafi* or *mphondo* (paddles) or *mchonjolo* (poles), depending on the water body, and *lupu* (balers). *Lupu* used to be wooden, but tin cans or pails are now common. *Ziwo*, which are specially constructed compartments of compressed and tied bundles of grass or creepers, keep fish or other items confined in compartments, increasing stability and ensuring they are not stepped on (Hoole, 1955).

The best hardwoods for canoe making include *chonya* or *mung'ona* (*Adina microcephala*), *mlombwa* (*Pterocarpus angolensis*), *mbawa* (*Khaya nyasica*), *mvunguti* (sausage tree, *Kigelia* sp.), *mkuru* (*Pterocarpus stolzii*), *muawanga* (*Afrormosia* sp.), *nsangu* (*Acacia albida*), *mtondo* (*Cordyla africana*) and *ntondo-oko* (*Sclerocarya caffra*). These are favoured because they have long lifespans and relatively high oil content, and require little protective maintenance, similar to Mulanje cedar (*Widdringtonia whytei*), currently the most popular wood for boat construction (Mills, 1980). Canoe makers now resort to using inferior trees such as softwoods, acacia and blue gum, palm trees, and even mango trees. The use of canoes is now limited to inshore areas or periods of good weather. Sometimes fishers or other users drown in bad weather and in open water offshore areas where waves are high.

ECOLOGICAL ISSUES AND CLIMATIC PARAMETERS

ARTISANAL fishers have extensive ecological knowledge of fish resources (O.V. Msiska,[3] personal communication) and climate in their local areas. However, their knowledge system is very different from that associated with international scientific work. Data and interpretations of those data tend to be place-based; transmission of knowledge does not depend on mechanized recording; and information systems are integrally linked to local sociocultural and human ecological arrangements, i.e., not removed from these cultural phenomena (Matowanyika 1994; also Hickey and Satria, this volume). Effective sharing of information between fishers and outsiders generally requires understanding and sharing of appropriate behaviours and attitudes that are not always understood by outsiders and scientific researchers (Chambers, 1993). That said, ITK incorporates substantial ecological information related to fish species, migration, feeding, breeding, conservation and interpretation of signs pertaining to climate or weather changes that are of interest to fisheries managers and scientists.

---

3. Dr Orton V. Msiska, affiliated to the Universities of Namibia and Malawi; former Senior Fish Farming Officer in the Malawi Fisheries Department; and former Principal Scientific Officer in the Ministry of Research and Environmental Affairs.

Fish classification system

Artisanal fishers know the many different fish species they see or catch, and assign names to them. More than 700 fish species have been described or identified in Malawi waters (Nsiku, 1999). Snoeks (2000) records 845 species for Lake Malawi alone. Some researchers believe there may be up to 2,000 such species due to the high diversity of cichlids in the lake (Turner, 1995). Fishers have local names for a majority of the species, or at least a label for the group to which species are perceived to belong (Berlin, 1992; F.M. Nyirenda, personal communication). These names refer to physical characteristics such as shape of mouth (*samwamowa*, does not drink beer, for *Mormyrus deliciosus*) or head (*mgong'u*, heavy/big head, for *Pseudotropheus* 'acei' or *Protomelus annectens*); habitat (*mbuna*, holes or rock crevices, for rock dwelling cichlids *Pseudotropheus* spp. complex and others); and the feel of the fish's skin (*nyesi*, static or chemical electric shock, for the catfish *Malapterurus electricus*). The offshore occurrence of a species seems to limit the number of local names assigned to it (Ambali et al., 2001).

Judging from the names assigned, some fish groups such as *chambo* (*Oreochromis* spp.) and *utaka* (*Copadichromis* spp.) are accurately distinguished to the species level. Fishers also know the bycatch species associated with their fishing devices and have information relevant to general stock statuses: that is, they know when they are experiencing declines or increases in their catch rates in particular areas (Smith, 1998). Local fishers are far better at identifying fish species than government fisheries personnel, as there are so far only a few people with any training in formal ichthyology in Malawi.

Fish migrations are associated with corresponding movements of fishers, who know whether such migrations are for feeding or spawning, or due to a change in water conditions and water body productivity (ICLARM/GTZ, 1991; Munthali, 1997; Baird, this volume). They are aware of the feeding relationships associated with their target species' predators (Munthali, 1997; Smith, 1998; F.M. Nyirenda, personal communication). Local fishers keenly observe specialized feeding relationships among some species such as *Corematodus shiranus* (whose local name, *yinga*, means to chase or herd) and *chambo* (*Oreochromis squamipinnis*), in which the former feeds on the tail fins of the latter (biting small pieces). Since *C. shiranus* follows the other species, it is aptly named *kapitawo*, supervisor or 'captain', of *chambo* (Lowe, 1948). Fishers are also aware of how some species breed. They know the seasons, colours and sites associated with breeding (F.M. Nyirenda, personal communication). Lake Malawi fishers have long observed *chambo* and other mouth brooders, but they tend to think that the young, which are from eggs laid and fertilized in sand-scrape 'nests' and then picked up by the females, are born through the mouth (Lowe, 1948). See Johannes and Neis (this volume) for other examples of how ITK may contain valuable information, even when not all conclusions drawn from it are correct. Some fishers know that anadromous species, *sanjika* (*O. microcephalus*), *mpasa* (*O. microlepis*), *chimwe/ngumbo* (*Barbus johnstonii*) and *kadyakolo/kuyu* (*B. eurystomus*) spawn with first rains and thus migrate to upstream tributaries (Hoole 1955; F.M. Nyirenda, personal communication).

Traditional conservation

Before state fishing rules were introduced in 1931, conservation of fish resources was promoted by traditional controls. Chiefs and village headmen controlled fish exploitation in their areas (Chirwa, 1996; Sambo and Woytek, 2001). There were also ritual prohibitions at some sites (Hickey and Satria, this volume). The number of fishers was also limited by magic and taboos for certain fish species. Low population densities and the types of fishing equipment they used further contributed to conservation by permitting the fish resources to resiliently absorb fishing pressure from traditional fishing operations (Munthali, 1997). A typical example of the technological constraints associated with traditional fishing equipment is the way canoe technology limited access to offshore parts of large lakes, particularly Lake Malawi (ICLARM/GTZ, 1991). Thus, while inshore areas may have been heavily fished, the overall impact of fishing on the lake may have been minimized.

Reading the environment

Fishers in Malawi have acquired the ability to interpret natural signs to predict weather or other climatic conditions, and thereby enhance the effectiveness of their fishing operations. They know the influence of lunar cycles (Smith, 1998; Poepoe *et al.* and others, this volume), as well as currents or waves on their catch. In the northern part of Lake Malawi, the combination of rising clouds and winds from the western mountains that follows a period of *chimphungu* (absolute calmness) is a sure sign of an impending heavy downpour or windstorm. *Mupungu* refers to rain brought by easterly winds, starting from the lake's Mozambican shore, in the evening (between 5 and 8 p.m.) and early morning (from around sunrise until 10 a.m.). The wind may be short-lived, blowing for an hour or less. Winds that blow in the direction of home are observed carefully. If they are not too fierce, the fishers continue their work and 'ride on the winds' as they paddle home, exerting little or no effort. On the south-eastern side of Lake Malawi, some fishers sail their fishing craft (ICLARM/GTZ, 1991). If the winds are strong or blow away from home, the fishers rush for safety. Rising water levels in swamps or wells, particularly in areas of clay soils, which usually crack during the dry season, indicate the onset of *m(u)wera* (southerly trade winds) that can bring rains for several weeks (F.M. Nyirenda, personal communication). People, including fishers in the Lower Shire and probably other areas as well, have many ways to anticipate the onset of rains, and therefore prepare accordingly. Signs include phases of the moon, croaking of frogs, sounds of birds or crickets, presence and strength of whirlwinds, mosquito activity and frenzied food collection by certain insects and ants. High temperature, wind direction and the type or colour of clouds are also used to predict amounts of rain.

ITK AND FISHERIES CO-MANAGEMENT IN MALAWI

THE FD's major monitoring tool for traditional fisheries is through frame surveys (FS) every August, when the number and types of all craft and equipment are recorded, and monthly data collection in ten management zones. Four of these are associated with the fisheries of the Lower Shire Valley, Lake Chilwa, Lake Chiuta and Lake Malombe/Upper Shire, and the remaining six relate to artisanal fisheries of Lake Malawi where fishing areas are also used. There are nine areas where entry is regulated, at least in principle (ICLARM/GTZ, 1991). The tenth area (split into six zones) is mainly inshore, available to traditional fishing operations, and is open access.

Two systems of monthly data collection, the Catch Assessment Survey (CAS) and the Malawi Traditional Fisheries (MTF), are currently in place for sampling of catch. Aspects covered are: weight by type of fish; type, size and number of devices; number of fishers involved; and type of craft. For MTF, fishing effort is also sampled and recording is done during both day and night (FAO, 1993). The MTF is applied in the Mangochi area, the most heavily fished part of Lake Malawi, while CAS is used in the rest of the country. These systems entail elaborate record keeping and relatively high costs, which unfortunately are sometimes not met due to the underfunding of the FD.

The *chambo* fishery, which is the most lucrative in the country, collapsed in Lake Malombe in the 1980s, with catches plummeting from an estimated 8,484 tonnes in 1982 to only 545 tonnes ten years later. This collapse was linked to increased use of small-mesh-size fishing nets that catch immature fish, nonobservance of fishing regulations, and the destruction of fish nesting and feeding habitats by bottom-dragging nets (Hara, 1998). A number of other fisheries, such as those for anadromous fishes like *mpasa* (*O. microlepis*), *sanjika* (*O. microcephalus*) and *ntchila* (*Labeo mesops*) were also in decline (Chirwa, 1996; Munthali, 1997; Donda, 1998; Scholz et al., 1998).

In 1993, the FD, with financial assistance from German Technical Cooperation (GTZ), recommended a management programme for Lake Malombe with a number of components that included community participation (Scholz et al., 1998). Dialogues between fishers, government and other interested parties have taken place for the fisheries of Lakes Malombe, Chilwa and Chiuta (Dawson, 1997; Donda, 1998; Hara, 1998; Scholz et al., 1998). As part of its commitment to the co-management initiatives, the government enacted legislation encouraging other stakeholders, particularly fishers, to take part in decision-making (Scholz et al., 1998). External agencies have played a role in sensitizing the local fishers (Hara, 1998) who, like other stakeholders, have to participate in transformative learning (Pinkerton, 1999).

Empowering fishers to participate in decision making through the new legislation has been an important step towards the development of fisheries science and management that combine ITK and Western science in an effective way. A critical part of this success was legal recognition of ITK by the government, which enacted some fisheries regulations proposed by fishers in Lakes Malombe, Chilwa and Chiuta in 1996 (Scholz et al., 1998). Care has also been taken to avoid aligning the local committees with political parties or people without a direct interest in the community's welfare or value system. There is some concern, however, that the process is being rushed and may result in the creation of institutional structures that are as bureaucratic as traditional government institutions. This suggests a need to critically examine current non-political leadership arrangements, and to come up with flexible ways to incorporate their input into discussions about proposed changes to fisheries management. This will help to ensure that local leaders are not alienated from the co-management process.

Elsewhere in Malawi, the process of developing co-management is in its early stages. It is hoped that the experiences in Lakes Malombe and Chiuta will serve to catalyze crucial aspects of the process elsewhere. Regional community-based natural resource management (CBNRM) experiences like those of CAMPFIRE in Zimbabwe and ADMADE in Zambia (SADC, 1996) have provided important lessons. When the mechanisms supporting sustained dialogue are in place and functional and co-management institutions have been developed, this should provide an appropriate base for incorporation and effective use of ITK in management.

The involvement of fishers in providing information for the monitoring schemes (FS, CAS, MTF or specialized research programmes) has strengthened their role in appropriate management of fish resources in the period of co-management in Malawi that began in 1994. It is widely recognized that effective management of natural resources, especially fisheries, must include all user groups. Effective fisheries management thus requires mutually agreed systems of controls, with appropriate forms of enforcement to ensure responsible use of resources (FAO, 1986; Taylor and Alden, 1998).

For ITK to be effectively incorporated into fisheries management in Malawi, the co-management process will need sustained and long-term institutional support. Such support will help to ensure that the fishers' ITK, based on their interests and values as a group, will be 'crystallized' and then appropriately harnessed for fish resource conservation. However, sustainable resource management requires more than the crystallization of ITK. It also requires that fishers participate in decisions related to resource-use rights, assist in determining resource usage modes, receive the full benefits from resource use, and assist in setting rules of access and determining how benefits should be distributed (Murphree, 1993).

PUTTING ITK TO WORK

Lakes, rivers and other wetlands cover about one-quarter of Malawi. Artisanal fishers account for 90 per cent of annual fish landings and make extensive use of ITK. Their knowledge resides in diverse fishing methods, devices and craft. Artisanal ecological ITK relates to species identification, migrations, feeding and breeding behaviours among many species, and to conservation. The ITK of climatic parameters relies on many natural signs such as phases of the moon, sounds of animals and insects, and other behaviours of insects and ants, wind and clouds that are used to predict weather or other climatic conditions (see also Baird, this volume).

Two areas where ITK should be effectively applied relate to using it to strengthen user involvement in the decision-making process for the management and development of fish resources, and to develop an inexpensive but robust monitoring system. Fishers' knowledge of species, bycatches, processing practices and other areas could play an important future role in fisheries management in Malawi. Incorporating the rural poor, including artisanal fishers who are very knowledgeable about ITK, into fisheries will pay dividends for Malawi's fisheries. However, it may take some time before these dividends become apparent.

Drawing on ITK to design alternative fisheries monitoring schemes for Malawi's artisanal fisheries that are both less costly and more robust would be worthwhile, as it would bridge gaps that are sometimes inevitable in the other methods (Johannes and Neis, this volume). This may be in the form proposed by Smith (1998): determining species composition from surveying drying racks and identifying fish found by using their local names (known to fishers). Fishers already help identify species in their catches in the CAS and MTF data-recording systems. Starting with a small study in at least one of the fishing areas in Malawi in order to establish some basic statistical parameters – such as estimating weights from local measuring containers, identifying the main species caught and processing methods used with input from fishers – could facilitate the development of a model for an effective science and co-management regime informed by ITK.

Artisanal fishers have long been sceptical of the government's ability to conserve and manage resources. However, in the face of government policies, there was no motivation or real authority for fishers to manage the resources themselves. The current legislation is an important milestone in support of community participation in fisheries management. A lot of effort will be required to ensure fishers make appropriate management decisions and to ensure that they feel fully part and parcel of the fish resources 'ownership'.

While it is now fairly well accepted that the ITK of artisanal fishers can be very useful to help achieve sustainable fisheries management and conservation, there are some important challenges associated with effective application of ITK. Like science and other knowledge systems, ITK is influenced by social relationships and other influences that exist within communities of fishers (Murphree, 1993). Knowledge is generally differentially distributed within groups. That said, differential distribution and social influences are issues for all knowledge systems, including international science (Matowanyika, 1994). In addition, outside change agents often have inappropriate attitudes and behaviours that hamper their ability to work effectively with local people. Change agents, including extension and development personnel, scientists and other professionals, have to take time to learn the required local etiquette to fully benefit from ITK.

ACKNOWLEDGEMENTS

The author wishes to thank two anonymous reviewers who made extensive comments and suggestions to improve the earlier version of the chapter titled 'The use of fishers' knowledge in the management of fish resources in Malawi'. I also thank Ian G. Baird of the University of Victoria, Geography Department, Barbara Neis of Memorial University of Newfoundland, Department of Sociology, and Nigel Haggan of the University of British Columbia Fisheries Centre for their encouragement, comments and assistance with the general editorial work; and this volume's copy editors for helping with clarity of the chapter. However, any errors in this chapter are mine.

REFERENCES

Ambali, A.; Kabwazi, H.; Malekano, L.; Mwale, G.; Chimwaza, D.; Ingainga, J.; Makimoto, N.; Nakayama, S.; Yuma, M.; Kaka, Y. 2001, Relationship between local and scientific names of fishes in Lake Malawi/Nyasa. *African Study Monographs*, Vol. 22, No. 3, pp. 123–54.

Baird, I.G. 2006. Local ecological knowledge and small-scale freshwater fisheries management in the Mekong River in Southern Laos. In: N. Haggan, B. Neis; I.G. Baird (eds.), *Fishers' knowledge in fisheries science and management*. Chapter 12, this volume. UNESCO-LINKS, Paris.

Banda, M.C.; Tomasson, T. 1997. Demersal fish stocks in southern Lake Malawi: Stock assessment and exploitation. *Government of Malawi, Fisheries Department, Fisheries Bulletin* No. 35, 39 pp.

BERLIN, B. 1992. *Ethnobiological classification: Principles of categorization of plants and animals in traditional societies*. Princeton, Princeton University Press, 335 pp.

BRUMMETT, R.E.; NOBLE, R. 1995. *Aquaculture for African smallholders. ICLARM Technical Report*, No. 46, Manila, ICLARM, 69 pp.

CHAMBERS, R. 1993. Participatory rural appraisals: Past, present and future. UPPSALA: *Forests, trees and people newsletter*, No. 15/16, pp. 4–8.

CHIRWA, W.C. 1996. Fishing rights, ecology and conservation along southern Lake Malawi, 1920–1964. *African Affairs*, Vol. 95, 351–77.

CHIOTHA, S.S.; SEYANI, J.H.; FABIANO, E.C. 1991. Mulluscidal and piscicidal properties of indigenous plants (abstract), p. 22. In: B.A.Costa-Pierce, C. Lightfoot, K. Ruddle and R.S.V. Pullin (eds.), *Aquaculture research and development in rural Africa*. International Center for Living Aquatic Resources Management Conf. Proc. 27. Manila, Philippines, ICLARM, 52 pp.

DAWSON, K.A. 1997. *Applying cooperative management in small-scale fisheries: The cases of Lakes Malombe and Chiuta, Malawi*. Michigan, Michigan State University, MSc thesis, 53 pp.

DONDA, S. 1998. Fisheries co-management in Malawi: Case study of Lake Chiuta fisheries. In: A.K. Norman, J.R. Neilsen; S. Sverdrup-Jensen (eds.), *Fisheries co-management in Africa*. Mangochi, Malawi, Fish. Co-mgmt. Res. Project No. 12, pp. 21–39.

EMTAGE, J.E.R. 1967. The making of a dugout canoe. *The Society of Malawi Journal*, Vol. 20, No. 2, pp. 23–5.

FAO (FOOD AND AGRICULTURE ORGANIZATION). 1986. *Strategy for fisheries management*. Rome, Food and Agriculture Organization of the United Nations, 26 pp.

——. 1993. Fisheries management in the southeast arm of Lake Malawi, the Upper Shire River and Lake Malombe, with particular reference to the fisheries on chambo (*Oreochromis* spp.). *CIFA Technical Paper*. No. 21. Rome, FAO, 113 pp.

FD (FISHERIES DEPARTMENT). 1996. *Fisheries statistics* (unpublished), 12 pp.

GOM (GOVERNMENT OF MALAWI). 1989. *Statement of development policies, 1987–1996*. Zomba, Malawi, Government Printer, 198 pp.

——. 1999. *Fish stocks and fisheries of Malawian waters: Resource report 1999*. Fisheries Bulletin No. 39. Lilongwe, Fisheries Department, 53 pp.

GOM/UN (GOVERNMENT OF MALAWI AND UNITED NATIONS IN MALAWI). 1992. The situation analysis of poverty in Malawi (Draft). UNICEF. Lilongwe, 202 pp.

HARA, M. 1998. Problems of introducing community participation in fisheries management: Lessons from Lake Malombe and Upper Shire River (Malawi) Participatory Fisheries Management Programme. In: A.K. Norman, J.R. Neilsen; S. Sverdrup-Jensen (eds.), *Fisheries co-management in Africa*. Mangochi, Malawi, Fish. Co-mgmt. Res. Project No. 12, pp. 41–60.

HICKEY, F. 2006. Traditional marine resource management in Vanuatu: Worldviews in transformation. In: N. Haggan, B. Neis; I.G. Baird (eds.), *Fishers' knowledge in fisheries science and management.* Chapter 7, this volume. UNESCO-LINKS, Paris.

HOOLE, M.C. 1955. Notes on fishing and allied industries as practiced among the Tonga of the West Nyasa District. *The Nyasaland Journal,* Vol. 8, No. 1, pp. 25–38.

ICLARM/GTZ. 1991. The context of small-scale integrated agriculture–aquaculture in Africa: A case study of Malawi. *ICLARM Studies Review,* No. 18, 302 pp.

JOHANNES, R.E.; NEIS, B. 2006. The value of anecdote. In: N. Haggan, B. Neis; I.G. Baird (eds.), *Fishers' knowledge in fisheries science and management.* Chapter 1, this volume. UNESCO-LINKS, Paris.

KALIPENI, E. 1998. Chewa. In: T.L. Gall (ed.), *Worldmark encyclopedia of cultures and daily life. Volume 1 (Africa).* Detroit, Michigan, Gale Research, pp. 98–102.

KEATING, M. 1993. *Agenda for Change: A plain language version of Agenda 21 and the other Rio Agreements.* Geneva, Centre for Our Common Future, 70 pp.

LANDES, A.; OTTE, G.V. 1983. *Fisheries development in the Lake Chilwa and Lake Chiuta region in Malawi:* Report prepared for the German Agency for Technical Cooperation (GTZ). Zomba, MAGFAD/GTZ, 35 pp.

LOWE, R.H. 1948. Notes on the ecology of Lake Nyasa fish. *The Nyasaland Journal,* Vol. 1, No. 1, pp. 39–50.

MATOWANYIKA, J.J.Z. 1994. Lecture on indigenous resource management: Six-week course in human and social perspectives in natural resources management: 6 February to 18 March 1994, Harare, Zimbabwe, 17 pp.

MATOWANYIKA, J.Z.Z.; SIBANDA, H. (eds.). 1998. *The missing links: Reviving indigenous knowledge systems in promoting sustainable natural resource management in southern Africa.* Proceedings of a Regional Workshop, Midmar, South Africa, 23–28 April, 1995. Harare, IUCN-ROSA, 75 pp.

MILLS, M.L. 1980. CIFA visit to the Lower Shire Valley, 12th December, 1980. *Malawi Fisheries Department Reports* (unpublished), 9 pp.

MUNTHALI, S.M. 1997. Dwindling food-fish species and fishers preference: Problems of conserving Lake Malawi's biodiversity. *Biodiversity and Conservation,* Vol. 6, 253–61.

MURPHREE, M.W. 1993. Communities as resource management institutions. London, International Institute for Environment and Development, Gatekeeper Series No. 36, pp. 1–15.

MZUMARA, A.J.P. 1967. The Lake Chilwa fisheries. *The Society of Malawi Journal,* Vol. 20, No. 1, pp. 58–68.

NSIKU, E. 1999. *Changes in the fisheries of Lake Malawi, 1976–1996: Ecosystem-based analysis.* Vancouver, Canada, University of British Columbia, MSc thesis, 217 pp.

OJDA, L. 1990. *Lake Chiuta study*. Report for Malawi-German Fisheries and Aquaculture Development Project (MAGFAD) in Zomba, Malawi, Hamburg, 50 pp.

PINKERTON, E. 1999. Factors in overcoming barriers to implementing co-management in British Columbia salmon fisheries. *Conservation Ecology*, Vol. 3, No. 2, Article 2. Available online at: http://www.consecol.org (last accessed on 16 May 2006).

POEPOE, K.K., BARTRAM, P.K.; FRIEDLANDER, A.M. 2006. The Use of Traditional Knowledge in the Contemporary Management of a Hawaiian Community's Marine Resources. In: N. Haggan, B. Neis; I.G. Baird (eds.), *Fishers' knowledge in fisheries science and management*. Chapter 6, this volume. UNESCO-LINKS, Paris.

SADC. 1996. Resource Africa. *SADC NRMP Newsletter*, Vol. 1, No. 1, pp. 1–8.

SAMBO, E.Y.; WOYTEK, R. 2001. An overview of indigenous knowledge as applied to natural resources management. In: O.L.F. Weyl and M.V. Weyl (eds.), *Proceedings of the Lake Malawi Fisheries Management Symposium, 4th–9th June 2001*, Capital Hotel, Lilongwe. National Aquatic Resource Management Programme (NARMAP), Technical Cooperation Republic of Malawi (Department of Fisheries) and Federal Republic of Germany (Deutsche Gesellschaft für Technische Zusammenarbeit), pp. 80–4.

SATRIA, A. 2006. Sawen: Institution, local knowledge and myth in fisheries management in North Lombok, Indonesia. In: N. Haggan, B. Neis; I.G. Baird (eds.), *Fishers' knowledge in fisheries science and management*. Chapter 10, this volume. UNESCO-LINKS, Paris.

SCHOLZ, U.F.; NJAYA, F.J.; CHIMATIRO, S.; HUMMEL, M.; DONDA, S.; MKOKO, B.J. 1998. Status and prospects of participatory fisheries management programmes in Malawi. In: A.K. Norman, J.R. Neilsen; S. Sverdrup-Jensen (eds.), *Fisheries co-management in Africa*. Mangochi, Malawi, Fish. Co-mgmt. Res. Project No. 12, pp. 5–19.

SMITH, L.W. 1998. Use of traditional practices and knowledge in monitoring Lake Malawi artisanal fishery. *North American Journal of Fisheries Management*, Vol. 18, pp. 982–8.

SNOEKS, J. (ED.). 2000. *Report on systematics and taxonomy*. Lake Malawi/Nyasa Biodiversity Conservation Project, SADC/GEF, 363 pp.

TAYLOR, L.; ALDEN, R. 1998. *Co-management of fisheries in Maine: What does it mean?* Summary Report based on J.A. Wilson, J. Acheson; W. Brennan (1998). Draft Report prepared for the Department of Marine Resources (DRM). Maine, DRM, 8 pp.

TURNER, G.F. 1995. Management, conservation and species changes of exploited fish stocks in Lake Malawi. In: T.J. Pitcher; P.J.B. Hart (eds.), *The impact of species changes in African lakes*. London, Chapman and Hall. Fish and Fisheries Series, 18, pp. 335–95.

WORLD BANK. 1998. *Indigenous knowledge for development: A framework for action.* New York, The World Bank, Knowledge and Learning Center, Africa Region, 38 pp.

CHAPTER 5 Application of Haida oral history to Pacific herring management

Russ Jones

ABSTRACT

THE Haida are the aboriginal inhabitants of Haida Gwaii (Queen Charlotte Islands) on the west coast of Canada. Pacific herring have been integral to their culture and economy for countless generations. In the past decade, stocks have been depleted because of low recruitment and excessive exploitation by commercial fisheries. The present harvest policy, established in the mid-1980s, does not take account of Haida traditional knowledge. Such knowledge provides information about stocks prior to the development of commercial fisheries and may contribute to reassessing reference points for the management of herring in Haida territory.

INTRODUCTION

RECENT herring (*Clupea pallasi pallasi*) fisheries in Haida Gwaii (Queen Charlotte Islands, British Columbia (BC) are managed with the aid of population assessment models and a harvest policy established in the mid-1980s. Over the past decade, however, stocks have been depleted and the validity of assessments and accuracy of forecasts are frequently questioned by Haida fishers. The Haida, the indigenous peoples of Haida Gwaii, blockaded the reopening of roe herring fisheries in 1998 and 2002, causing delays in those fisheries.[1] Fisheries and Oceans Canada (DFO) is currently undertaking a review of herring conservation policies that will address problems such as why some stocks, including those around Haida Gwaii, are not as abundant as expected (Anon. 2003). This chapter reviews Haida traditional knowledge of herring and its potential applications to herring modelling and harvest policy.

The Haida Nation, whose traditional territory includes the Haida Gwaii archipelago (Figure 5.1), is a distinct linguistic and cultural group that is well known for its maritime traditions. Historically, family groups held ownership rights to

---

1. Pers. comm. Gary Russ, Chair, Haida Fisheries Committee, Council of the Haida Nation, Skidegate, BC.

marine resource-harvesting sites such as salmon streams, halibut and groundfish fishing banks, and shellfish beds. The Haida Nation's current involvement in fisheries management ranges from formal to informal and includes management of Haida sockeye salmon (*Oncorhynchus nerka*) and traditional herring spawn-on-kelp fisheries, co-management of a razor clam (*Siliqua patula*) fishery, and monitoring of other herring fisheries. The Haida's cultural, social and economic interests in ocean resources are expressed through their continuing stewardship and use of fisheries resources in their territory, including herring.

Traditional ecological knowledge (TEK) has been increasingly applied and proving its validity in fields such as management of fisheries and wildlife and environmental impact assessment (see, for example, Freeman, 1992; Johannes, 1993; Berkes, 1999; Berkes et al., 2001). Writers such as Nadasdy have pointed out the difficulties of integrating traditional and scientific knowledge, including the politics of TEK's use by different actors in the management process (Nadasdy, 1999). This study does not attempt to summarize the full scope of Haida traditional knowledge about herring and ecosystems; rather, it focuses on information that can be used to test assumptions in modern herring management. It is recognized that this approach has limitations but it is hoped that it will provide some insight into herring and ecosystems in Haida Gwaii prior to the introduction of industrial scale fisheries.

Herring fisheries in British Columbia, Canada are relatively well documented, and stock estimates going back to 1950 have been generated for major herring producing areas, including Haida Gwaii (Schweigert, 2002). Data are regularly collected on herring spawns and age structure. These data are used in analyses to establish herring harvest policies that provide reference points for fisheries management. Stock estimates prior to the advent of large-scale industrial fisheries are not available and there are significant data gaps, particularly before the 1950s. Traditional knowledge is a possible source of information about herring abundance in this earlier period. The objectives of this chapter are, first, to summarize local observations about past Haida Gwaii herring abundance and past fisheries, and second, to discuss the application of traditional knowledge to modern herring fishery management.

Herring in Haida Gwaii have been exploited by traditional Haida fisheries and various industrial fisheries. *K'aaw*, or herring spawn-on-kelp, was an important Haida staple and trade commodity. Herring were also used as bait in halibut (*Hippoglossus stenolepis*) and blackcod (*Anoplopoma fimbria*) fishing, and less commonly as food and a source of oil. Industrial fishing for herring began about the beginning of the twentieth century, and can be divided into three main periods: the dry salt, reduction (with oil and meal products) and herring-roe periods. Exploitation rates during the reduction period in the 1950s and 1960s were 50–90 per cent (Hourston, 1980) compared with current target rates of less than 20 per cent (Stocker, 2001).

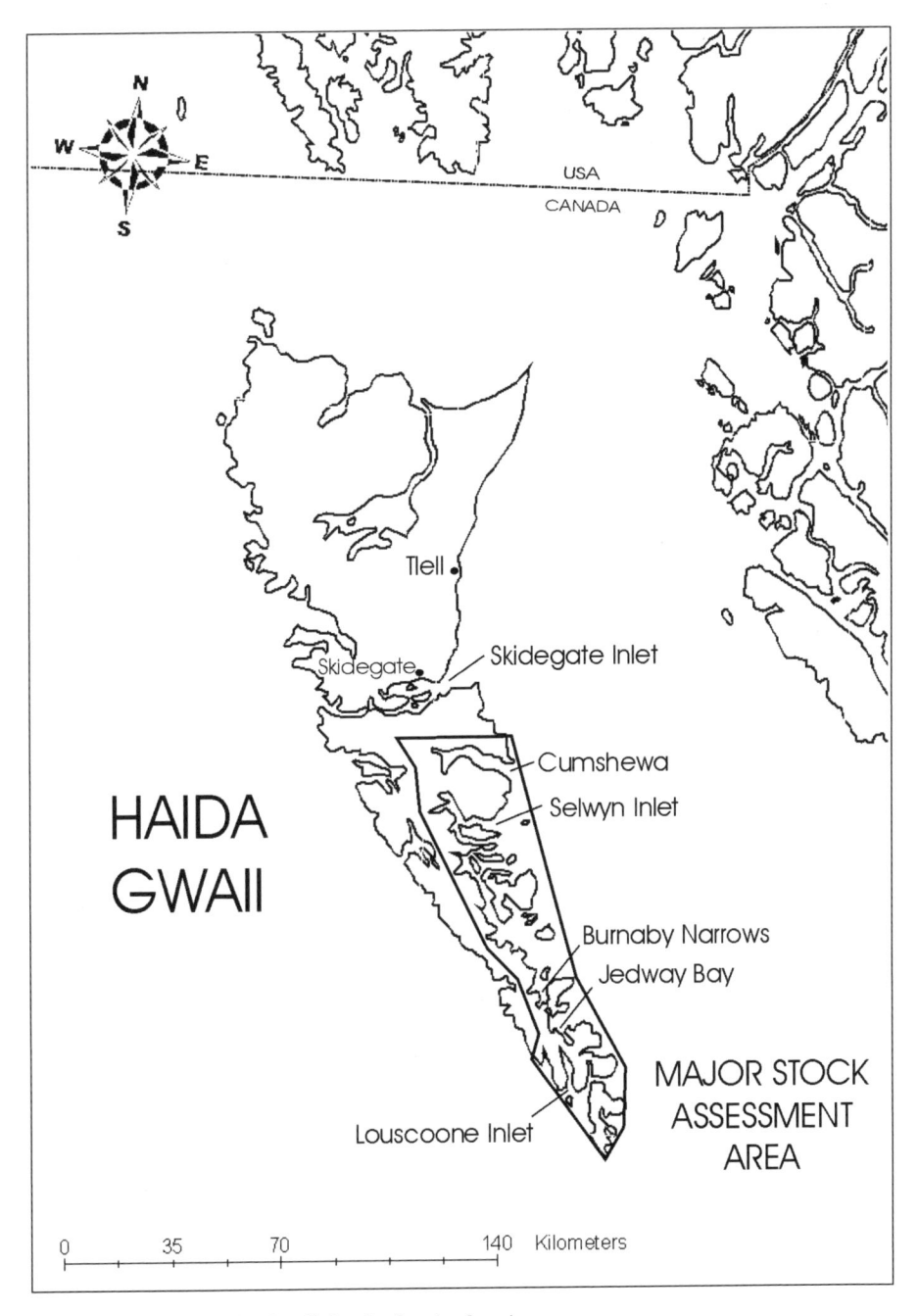

FIGURE 5.1 *Map of Haida Gwaii showing herring locations.*
Redrawn from Jones (2000).

Management of herring fisheries has adopted more precautionary approaches in an effort to avoid coastwide stock collapses such as occurred in the late 1960s (Stocker, 1993; Schweigert and Ware, 1995; Jones, 2000). During the 1950s and 1960s herring had been viewed as inexhaustible, with catches limited only by market demand. Herring-roe fisheries were reopened in 1972, and until 1985 were managed under a fixed escapement policy which aimed to secure a minimum spawning biomass of herring in specified areas throughout the coast (Stocker, 1993). In 1985 a fixed harvest rate strategy has been used in which catches are set before the season begins at a maximum of 20 per cent of the forecast biomass for several major assessment regions. In addition, no commercial roe fishing is allowed when stocks fall below a predetermined cut-off point of 25 per cent of the estimated unfished biomass (Haist et al., 1986). Although this policy is still in force, a recent assessment predicted that Haida Gwaii stocks would continue to be depressed under such a fixed harvest rate policy during low productivity periods (or 'warm' regimes) (Ware and Schweigert, 2002).

METHODS

The oral history described in this paper was gathered as part of an ethical analysis of Haida Gwaii herring fisheries (Jones, 2000). As part of that study in 1998, the author interviewed seven Skidegate Haida men ranging in age from 44 to 91. Individuals were selected for their experience with herring or knowledge of the area. The two eldest had not fished herring commercially, but the others had participated in herring bait, reduction, herring spawn-on-kelp and herring-roe fisheries. A checklist of interview questions was used, but depending on the situation divergences were allowed from this format. All of the interviews were tape-recorded.

Relevant information was selected from the interviews. This was compared to historic fisheries data for two historically important herring spawning areas in Haida Gwaii: Skidegate Inlet and Burnaby Narrows (also known as Dolomite Narrows) and vicinity (Figure 5.1). There are only remnant populations of herring in Skidegate Inlet, while Burnaby Narrows is actively managed for commercial herring spawn-on-kelp and herring-roe fisheries. Historic catch and herring spawn data are shown in Figures 5.2 and 5.3 for Skidegate Inlet and the major herring stock assessment area extending from Louscoone Inlet to Cumshewa Inlet. Burnaby Narrows and the surrounding vicinity is a major spawning area within the major stock assessment area. Spawn length is presented (without intensity or width) since it is likely to be the most unbiased estimate of total spawn, providing comparisons are within the same area (Hay and McCarter, 1999).

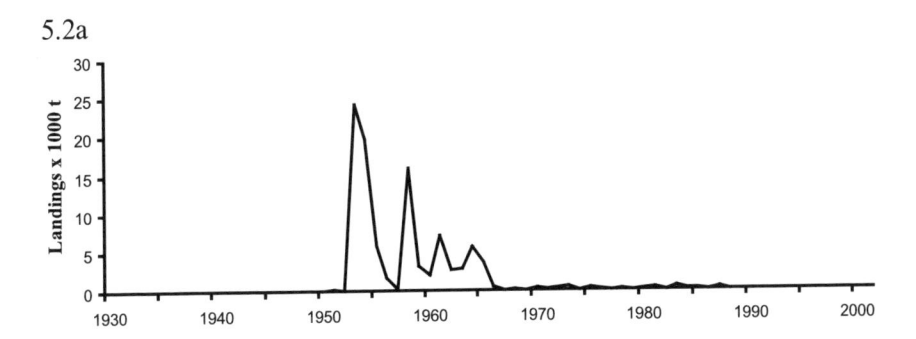

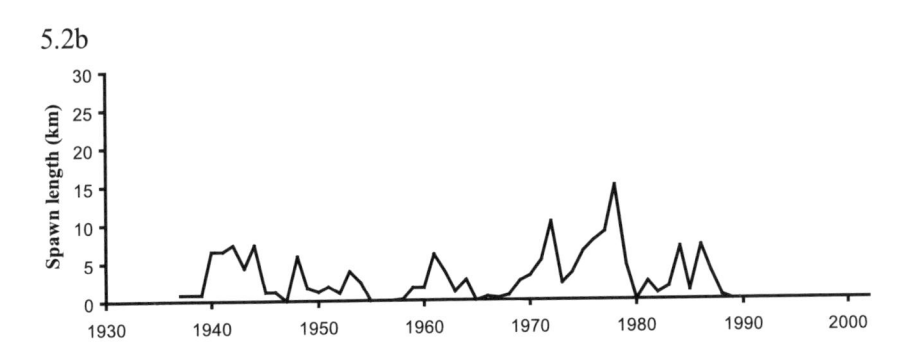

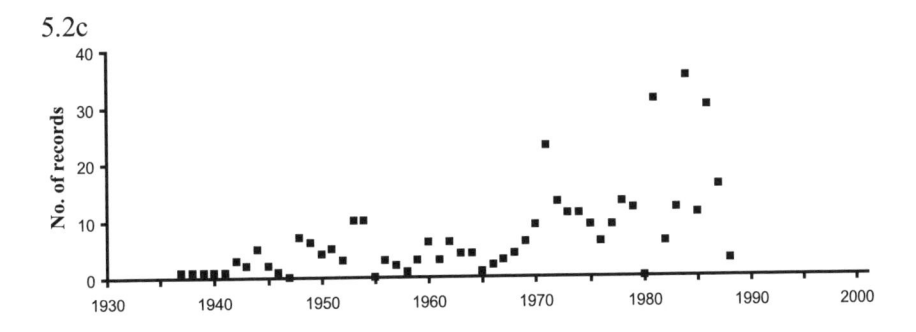

APPLICATION OF HAIDA ORAL HISTORY TO PACIFIC HERRING MANAGEMENT

FIGURE 5.2 *Herring catch (a), spawn length (b) and number of spawn records (c) at Skidegate Inlet, 1930–2001.*
Source: Catch data from 1930–50 from Daniel et al. (2001), 1951–1997 from J. Schweigert, personal communication 1998; spawn information from Hay and McCarter (2001).
Note: No spawn data were available after 1988.

5.3a

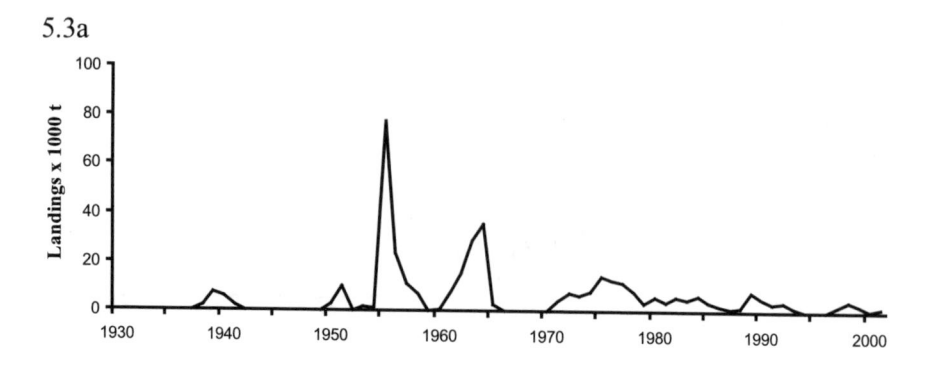

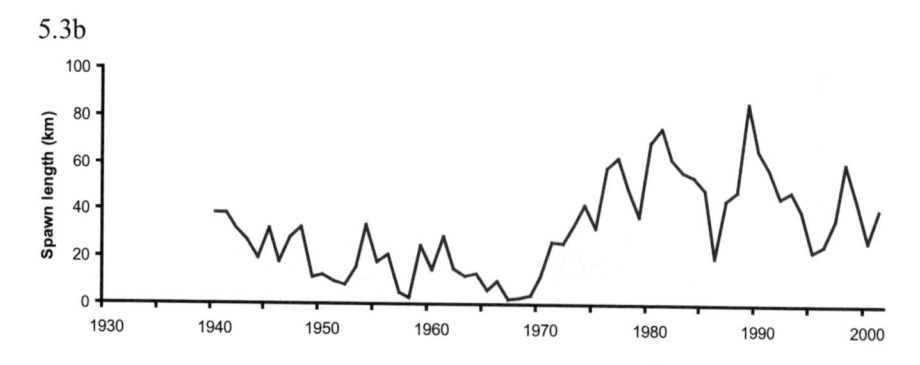

[Graph showing Landings × 1000 t from 1930 to 2000, with a major peak around 1955 reaching ~78, and another peak around 1963 reaching ~33]

5.3b

[Graph showing Spawn length (km) from 1930 to 2000, with values generally low through the 1960s and higher peaks in the 1980s-1990s reaching ~80]

FIGURE 5.3 *Catch (a) and spawn length (b) in Haida Gwaii major stock area (Cumshewa–Louscoone), 1930–2001.*

Source: 1930–1949 catch data from Taylor (1964), 1950–2000 from Schweigert and Fort (1994) and Schweigert (2001); spawn information from Hay and McCarter (2004).

SKIDEGATE INLET: TRADITIONAL KNOWLEDGE AND FISHERY HISTORY

O NE interviewee described an ancient process of oil extraction from herring in Skidegate Inlet:

'I'll start off with before we got oolichan grease.[2] Before they [the Haida] went over to the mainland. Before canoes. Herring was really thick in here. Outside South Bay [in Skidegate Inlet] there was great big herring there. You could use that herring for oil, its oil content, because we had no oolichan grease. You could

---

2. Oil or 'grease' from the eulachon (*Thaleicthys Pacificus*) is an important food, health supplement and trade item for aboriginal peoples in the Pacific north-west. There are no eulachon streams on Haida Gwaii, so it was the subject of trade with other Tribes.

get the oil out of there. ... You just put it up on a stick. You cut it open. I guess you gut it too. They were great big herring the size of humps. Then you put it by the fire and there's a container below it. The head up and the tail down and it just drips right into the container.'

Besides documenting utilization of herring for oil, this account provides important size structure information indicative of a lightly fished stock. It is very rare today to find herring as large as 'humps' or pink salmon (*O. gorbuscha*) (2–3 lb.) anywhere on the BC coast. Big herring have occasionally been reported in unfished populations in other parts of BC, although there is poor documentation of these observations.[3]

Skidegate Inlet is an important area for gathering *k'aaw* because it is near the village of Skidegate. An 1887 photo shows ten skiffs and twenty-four people fishing for herring eggs in Skidegate Inlet.[4] Herring were caught with herring rakes and were also once fished with nets, as indicated in a Haida story recorded about the turn of the century (Enrico, 1995, p. 173).

From 1889 to 1931, catches from Skidegate Inlet for pickled and dry salted markets were low, fluctuating from 10 to 100 tonnes, except from 1912 to 1914 when landings in successive years were 360, 370 and 890 tonnes (Taylor, 1964). The first saltery in northern BC was built in Skidegate Inlet at Alliford Bay in 1912 (Tester, 1935). One of the oldest Haida operated a towboat bringing herring in from the west coast to a plant in Queen Charlotte City in the 1920s. Several Haida interviewees described Skidegate Inlet prior to reduction fisheries in the 1950s:

'It [the spawn in Skidegate Inlet] used to be so thick. Every week it spawned different places. ... It used to start spawning in April, May, June, July, still the odd little spawn. That's how many months we used to get spawn. And damn near every week it spawns in different places. ... There isn't enough herring around now. The whales eat the roe.[5] As soon as it starts spawning, even one day spawning, the whales are in there kicking up the roe and then they siphon it out.

There used to be a lot of sea lion. Even *k'aalw* – cormorants.[6] There used to be lots on both Islands (near Skidegate village). In the evenings you'd hear them. You'd hear them plainly. You don't see them now. No feed for them. No herring. ... Grey cod[7] and tommy cods,[8] they're edible but they're not there now. ... Soles,[9] that's gradually disappearing too.'

---

3. Doug Hay, Fisheries Scientist, Fisheries and Oceans Canada, Pacific Biological Station, Nanaimo, personal communication.
4. (BC Provincial Museum, PN 355 by Newcombe 1897).
5. Grey whales (*Eschrichtius robustus*).
6. Large diving birds (*Phalacrocorax penicillatus*).
7. Pacific cod (*Gadus macrocephalus*).
8. Pacific tomcod (*Microgadus proximus*).
9. Skidegate or butter sole (*Isopsetta isolepis*).

INDIGENOUS PRACTITIONERS AND RESEARCHERS

Further evidence for larger sea lion populations in the past is the name of an historic Haida settlement in Skidegate Inlet – Kay 'llnagaay (Sea Lion village), located at Second Beach between Skidegate and Skidegate Landing. Large grey cod (Pacific cod, *Gadus macrocephalus*) that were once caught in Skidegate Inlet are now absent.

DFO records indicate a very large herring fishery in Skidegate Inlet beginning in 1953/54 with landings of about 24,100 tonnes, dropping to about 1,600 tonnes by the 1957/58 season (Figure 5.2). Taylor et al. (1956) remarked that in the third season of the fishery it was closed on 1 February because of the large proportion of small fish, and that 'there was a marked decline in the catch for the second successive year; no spawning was reported for the third year in succession'. Several of the Haida interviewed recalled the fishery:

> 'We used to get some [*k'aaw*] right in here. Until they had no quota in the inlet, fished it out a couple of times and then – clap – no more fish in the inlet. Just been lacking off and on ever since, after they really cleaned it out. Ah, at night-time, wintertime, you'd see it, like a big city out there, all these big seine boats with their lights. Just tons and tons, taking fish out, taking herrings out of here. ... It never really came back after they cleaned it out. ... [Fishing was] with a big seine, a real deep seine. Big boats, double-deckers. That was what was doing all the fishing. Yeah, there was no limit here at all so it didn't take long for them to clean it [out]. Ever since then it's never been the same.
>
> The first year they moved in on the reduction herring in the inlet, this inlet, they took 75,000 tons right out of here. Then they moved down to Skincuttle and they took 50,000 tons out of there.[10] All in that one year. [They packed them in] those great big barges that hold 400 tons. Nelson Brothers had them, you know. They'd tow them out of here. They lost one down here. Some of the planks from the hatchcovers drifted ashore down there. They lost a couple of loads coming out of here.'

Despite the large catches, stocks recovered and further reduction fisheries took place (Figure 5.2). Herring fisheries closed coastwide in 1968 except for bait and food fisheries. When they reopened in 1972, catches from Skidegate Inlet in bait, roe herring gillnet and herring spawn-on-kelp fisheries were relatively small. Spawn length increased in the 1970s, although this may have been partly due to increased monitoring (Figure 5.2). Spawn records are incomplete after 1988. Small spawns have occurred in some years since then and most but not all have been recorded although they do not yet appear in the DFO database.[11]

---

10. Cumulative catches may have been close to these figures.
11. Robert Russ, Haida Fisheries Program, Skidegate, BC, personal communication.

One Haida described herring behaviour and spawning locations in Skidegate Inlet now and in the past:

'All the islands [in the inlet] used to [get] spawn even part-way up the inlet. Now you are lucky if one or two places spawn. But the herring are smart, they just spawn a little bit along right from Tlell. We used to hunt seals up there, when they're chasing the herring. It used to spawn a bit all along the beach right from Tlell all the way up to the inlet right to Charlotte [Queen Charlotte City]. They still do that, they just spawn a bit on the eelgrass. That's Mother Nature's way of making them survive, I guess. So the Indians don't get any. Spawn a little bit on the eelgrass. They still do it. It happened again this year. I see the birds following them out there. [It spawned up the coast like that before] in the sixties when we used to hunt seals up there. And it still happens that way. You see, I walk the beach all the time and I see the birds moving along as the herring move in. Within a three week or a month period they move along the beach.'

This observation was interesting since I didn't find any DFO records of spawns between Skidegate and Tlell.

BURNABY NARROWS: TRADITIONAL KNOWLEDGE AND FISHERY HISTORY

Collecting *k'aaw* (herring spawn-on-kelp) was a seasonal activity that often involved travel to fishing camps, as another Haida recalled from the 1930s:

'We used to go there [Kiit or Burnaby Narrows] for drying k'aaw in April. There used to be about, oh, I'd say about eight, ten families there, they used to have houses. Oh, April is a poor time to try and dry k'aaw. Don't matter what you are doing, if you hear somebody let out a scream, drop everything, run like heck for the beach where you got it stretched out. Pick them up and bring them to a drying shed. Oh, boy. That was work. ... Everybody helps, nobody is excluded. ... When it's good they did fairly good. Only for 22 cents a pound. Dried up stuff. ... They used to sell it to Japanese right at Jedway. ... I don't think there is even any sign of the houses that used to be there. ... We used to come back in May, just to get ready for going to North Island, for trolling time too. And then from North Island they'd go across to Skeena. ... Dried, take it across and trade with the Indians over there for oolichans and oolichan grease and soapberries.'

The first reduction landings in Haida Gwaii were at a plant at Pacofi that operated from 1938 to 1943 (Tester, 1945). Tester states that, beginning in 1939, 'a newly-discovered run in the vicinity of Burnaby Strait contributed largely to the catch'. The next major landings were in 1951–52. By this time a large mobile fleet was fishing, locating schools with hydro-acoustic depth sounders and using mercury lights to attract fish at night. From February 1955 to March 1956, a catch of 77,650 tonnes was taken in the vicinity of Burnaby Island, the largest catch ever taken from a single area (Taylor et al., 1956).

One Haida described changes in herring spawns and predator abundance at Burnaby Narrows after the main reduction fishery in the 1950s:

> 'There used to be big spawns ... as far as you could see. But being younger and being a kid everything looks big, bigger than it actually is. Like I go to Jedway quite often now and when I was a kid I used to marvel at how far away the dock was, if I got to the dock, holy man what a feat. And here it's just a tiny little bay. And to walk across to the next bay was a big adventure. ... Now it's just a five minute walk. So things amplify when you are younger I guess. ... But I know that there were millions of tons of fish, because when they started moving through Burnaby Narrows it sounded like a big rainfall or something, at night time going through the Narrows. And then the sealions and the killerwhales[12] right with them too. Hear the sealions roaring all night going through the Narrows after the herring. ... When we go looking for *k'aaw* in the spring there's not nearly as much spawn (now). And a few sealions, maybe twenty or thirty sealions passing through. ... But seals getting abundant every year. All along the beach here there are seals everywhere out here when the herring come into the inlet [Skidegate].'

Significant landings continued to be recorded from the Burnaby Island area until the reduction fishery was closed coastwide in 1967–68 (Figure 5.3). A Haida who participated in the reduction fishery in the 1970s, described his concerns:

> 'But there should have been quotas set. Even when I was running boats, they should have closed Jedway [near Burnaby Narrows]. I don't see why they kept opening it, opening it. Because in the reduction days, the amount of herring that's in there [now] is nothing compared to that. When the herring came in there they got wiped out right away. It is a wonder there is anything left down there.'

Spawn in the major stock assessment area that includes Burnaby Island is shown in Figure 5.3.

---

12. *Orcinus orca*.

DISCUSSION

The oral history provides information relevant to evaluating current herring harvest policy. The data collected provide evidence of ecosystem changes over time in two major herring spawning areas in Haida Gwaii. The interviews also indicated that herring abundances in Haida Gwaii prior to the reduction fisheries of the 1950s were probably larger than at present.

In Skidegate Inlet, ecosystem changes included evidence for exceptionally large-sized herring before industrial-scale fishing began; high herring abundance in the 1950s, as shown by the catches; the disappearance of cormorants that once nested on islands near Skidegate; disappearance of grey cod from the inlet; a decrease in the number of sea lions and an increase in grey whales in the inlet. In the Burnaby Narrows area spawns were larger in the 1930s than today, with so many herring passing through the narrows that before they 'sounded like raindrops'; there were more herring in the 1950s and 1960s than at present, as shown by soundings and catches; and the numbers of sea lions and killer whales during herring spawning times have decreased.

The change in herring abundance in Skidegate Inlet from the 1950s to the present is demonstrated by fisheries data. Catches are the strongest evidence for the historic size of the population. In the 1950s Skidegate Inlet was known as one of nine major stocks of herring in BC (Taylor, 1964). There is some uncertainty about the size of herring populations there before the 1950s. No particularly large spawns were recorded or noted before the 1950 fisheries by those interviewed. Spawn data for Skidegate Inlet are better than other areas of the island as it was closer to settlements (Hay et al., 1989). It is thus possible that herring populations may have been at an historic high in the 1950s. Another possibility is that some spawn was not recorded. It is also possible that the schools of herring caught in Skidegate Inlet were fish that would have spawned in other areas, although this is somewhat doubtful as catches took place close to the normal spawning time of other North Coast populations.[13]

Changes in killer whale, sea lion, seal and seabird distribution could be local effects or signs of long-term trends not necessarily related to herring. For instance the decline of grey cod in Skidegate Inlet – major stocks occur in Hecate Strait – may be more related to stock depletion in trawl fisheries than changes in herring abundance. Grey whales have been increasing coastwide since the cessation of whaling, and began appearing in Skidegate Inlet in the 1980s.

---

13. There is biological evidence for this theory as herring frequently aggregate in the winter prior to dispersal to spawning grounds. For instance large schools of herring greatly exceeding the spawning biomass were observed in Juan Perez Sound in winter surveys (November–December) in the late 1980s and early 1990s (see for example McCarter et al., 1994). But it is somewhat doubtful that herring found in Skidegate Inlet would have spawned elsewhere given the timing of the peak 1953–54 fishery – 28,550 tonnes caught in Skidegate Inlet in fourteen days between 25 February and 15 March (Taylor, 1955, p. 153) – and the spawning period of most B.C. North Coast stocks (late March to late April in 1953–54, Hourston et al. 1972, p. 35).

The oral history indicates greater herring abundance at Burnaby Narrows area in the 1930s than at present. Those interviewed described larger spawns and greater predator abundance at Burnaby Narrows prior to the reduction fisheries. It was interesting that one account considered how the perceptions might differ for a child compared with an adult. He considers the roaring of sea lions and the sound of the herring flipping at night as supporting the belief that there were more herring in the early days than now. It shows that those interviewed examined their personal observations carefully before drawing conclusions.

The observations indicate changes in ecosystem conditions, particularly the relative abundance of visible herring predators. Modelling has also suggested changes in the productivity of herring over the period of industrial fisheries (Ware and Schweigert, 2002). Currently the size of the unfished population (its equilibrium unfished biomass) is estimated by simulation modelling and is used to establish a cut-off level for herring-roe fisheries (Haist et al., 1986; Schweigert, 2002). If the unfished populations were actually larger than is presently estimated, then the present cut-off estimate should be higher. An increase in the cut-off estimate could reduce the effects of fishing at low stock levels, because the current harvest policy demands that the commercial herring-roe fishery should close when stocks are assessed at levels below the cut-off. It may also have implications for the harvest rate if current harvest rates do not allow depressed stocks to rebuild to target levels that may also be set according to estimates of the unfished population.

Haida perceptions about current herring stocks are based on direct observations by Haida fishers and Haida Fisheries Program staff. Haida participants in the commercial herring spawn-on-kelp fishery understand that the size of the herring stock changes. They experience annual stock fluctuations at first hand as they travel throughout the area during the spawning season trying to find and catch herring with hydroacoustic sounders and seine nets. Many fishers have participated in this fishery for thirty or more years and have a large body of knowledge about where to find herring. They are also able to interpret soundings because they often set their seine nets on herring that they locate on their sounders. When stocks have been low, herring have been hard to find and catch, and fishers opposed the reopening of roe fisheries in 1998 and 2002. Haida Fisheries Program staff are active on herring spawning grounds throughout the season. Since 1999 the Haida Fisheries Program has conducted herring spawn surveys under contract to the DFO by means of a chartered seiner and dive team. Spawn survey data are used in stock modelling and forecasts.

DFO forecasts, however, also consider other information, such as the age structure of the stocks, and their predictions have tended to overestimate the level of stocks in Haida Gwaii in recent years (Schweigert, 2002, Figure 4.12). The model has recently been changed, weighting the spawn index more heavily than other data to try to get a better overall fit to spawn index data (Schweigert, 2001). Haida fishers did not believe DFO forecasts in 1998 and 2002. In retrospect, considering the most recent results of modelling herring populations, these Haida fishers were correct.

Oral history has limitations. Most of those interviewed did not recall the exact year of their observations. Moreover, observations from specific geographic locations cannot necessarily be extrapolated to a larger area. The first of these problems could be addressed by more in-depth questioning during the interview to determine approximate dates by using related events or career histories (e.g. Ames, this volume; Neis et al., 1999). The second problem may not be a major issue since Skidegate Inlet is a relatively small area close to settlements and Burnaby Narrows is the main spawning area in the major stock assessment area. Both areas have a long documented history of spawning according to both the DFO and the Haida.

DFO data before 1950 and even up to 1970 have limitations as well. But it is informative to look at the trends in spawn length (see Hay and McCarter [1999] for information on how this abundance index is measured) before, during and after the major reduction fisheries that occurred between the mid-1950s and late 1960s (Figures 5.2 and 5.3). DFO spawn data for Skidegate Inlet show considerable annual variation in spawn, with reduced spawn lengths during the major reduction period and similar lengths in the 1940s and 1980s (Figure 5.2). Spawn length peaked in the 1970s. This increase is not consistent with the interview results. One possible explanation is that there were few annual spawn observations in the period prior to the reduction fisheries (Hay and McCarter, 1999). This may tend to bias the spawn length downwards in the early period.[14]

It would thus be difficult to estimate trends from spawn data alone. Catch data for both areas indicate relatively light exploitation prior to about 1955. This would imply that stocks (which were fished to the point of collapse in the late 1960s) should have been higher before 1955 unless there were environmental factors affecting stock productivity. This inference depends on the timing and intensity of herring reduction fisheries in Haida Gwaii and may not hold true for other BC north-coast areas.

This example shows the difficulty of reaching conclusions about long-term trends in abundance even for relatively well-documented fish stocks. It also suggests that oral history can assist with interpretation of historic fisheries data, and has other benefits such as providing details about historic utilization and the early history of fisheries. Further information could be gained by interviewing more people, particularly Haida women who participated in gathering and processing *k'aaw*. It would also be useful to conduct interviews in more depth to more accurately pinpoint the time and frequency of observations.

---

14. Hay and Kronlund (1987) found that increase in the 'number of records' of spawn was due in part to a methodological change to a system of recording short spawns as separate events rather than pooling them. But sporadic records such as in the late 1930s and 1940s are still likely to be underestimates due to the lengthy reported spawning period in Skidegate Inlet (April to June, see Hourston et al., 1972).

REFERENCES

Anon. 2003. Objective based fisheries management and the science based review of Pacific herring management. Discussion Document, 31 January 2003. Nanaimo, BC, Stock Assessment Division, Science Branch, Fisheries and Oceans Canada, 11 pp.

Berkes, F. 1999. *Sacred ecology, traditional ecological knowledge and resource management*. Philadelphia, Taylor and Francis, 209 pp.

Berkes, F.; Mahon, R.; McConney, P.; Pollnac, R.; Pomeroy, R. 2001. *Managing small-scale fisheries: Alternative directions and methods.* Ottawa, International Development Research Center, 320pp. Available online at: http://web.idrc.ca/es/ev-9328-201-1-DO_TOPIC.html (last accessed 26 June 2005).

Daniel, K.S.; McCarter, P.B.; Hay, D.E. 2001. The construction of a database of Pacific herring catches recorded in British Columbia from 1888 to 1950. *Canadian Technical Report Fisheries and Aquatic Science*, Vol. 2368, 108 pp. Available online at: http://www-sci.pac.dfo-mpo.gc.ca/herring/herspawn/hcatch01.htm (last accessed 24 June 2005).

Enrico, J. 1995. *Skidegate Haida myths and histories.* Collected by John R. Swanton, edited and translated by John Enrico. Skidegate, Queen Charlotte Islands Museum Press.

Freeman, M.M.R. 1992. The nature and utility of traditional ecological knowledge. *Northern Perspectives*, Vol. 20, No. 1, pp. 9–12.

Haist, V.; Schweigert, J.F.; Stocker, M. 1986. Stock assessments for British Columbia herring in 1985 and forecasts of potential catch in 1986. *Canadian Management Report Fisheries and Aquatic Science*, Vol. 1889, 48 pp.

Hay, D.E.; Kronlund, A.R. 1987. Factors affecting the distribution, abundance, and measurement of Pacific herring (*Clupea harengus pallasi*) spawn. *Canadian Journal of Fisheries and Aquatic Science*, Vol. 44, pp. 1181–94.

Hay, D.E.; McCarter, P.B. 1999. Distribution and timing of herring spawning in British Columbia. Fisheries and Oceans Canada, Ottawa. *Canadian Stock Assessment Secretariat, Research Document*, 99/14, 44 pp.

——. 2004. Herring spawning areas of British Columbia: A review, geographical analysis and classification. Revised edition, 2004. Available online at: http://www-sci.pac.dfo-mpo.gc.ca/herring/herspawn/pages/project_e.htm (last accessed 24 June 2005).

Hay, D.E.; McCarter, P.B.; Kronlund, R.; Roy, C. 1989. Spawning areas of British Columbia herring: A review, geographical analysis and classification. Volume 1: Queen Charlotte Islands. *Canadian Management Report Fisheries and Aquatic Science*, Vol. 2019, 135 pp.

HOURSTON, A.A. 1980. The decline and recovery of Canada's Pacific herring stocks. Rap. P.-v. Reun. *Conseil International pour l'Exploration de la. Mer*, Vol. 177, pp. 143–53.

HOURSTON, A.S.; OUTRAM, D.N.; NASH, F.W. 1972. Millions of eggs and miles of spawn in British Columbia herring spawnings, 1951 to 1970. *Journal of the Fisheries Research Board of Canada*, Vol. 359, 144 pp.

JOHANNES, R.E. 1993. Integrating traditional ecological knowledge and management with environmental impact assessment. In: J.T. Inglis (ed.), *Traditional ecological knowledge, concepts and cases*. Ottawa, International Development Research Centre, pp. 33–40.

JONES, R.R. 2000. The herring fishery of Haida Gwaii: An ethical analysis. In: H. Coward, R. Ommer and T.J. Pitcher (eds.), *Just fish: Ethics and Canadian marine fisheries*. St Johns, ISER books, pp. 201–24.

MCCARTER, P.B.; HAY, D.E.; WITHLER, P.; KIESER, R. 1994. Hydroacoustic herring survey results from Hecate Strait, 22 November–2 December 1993 W.E. Ricker Cruise 93HER. PSARC H94-05, 40. Nanaimo, BC, Fisheries and Oceans Canada, Biological Science Branch Canada.

NADASDY, P. 1999. The politics of TEK: Power and the 'integration' of knowledge. *Arctic Anthropology*, Vol. 36, No. 1–2, pp. 1–18.

NEIS, B.; SCHNEIDER, D.C.; FELT, L.; HAEDRICH, R.L.; FISCHER, J.; HUTCHINGS, J.A. 1999. Fisheries assessment: What can be learned from interviewing resource users? *Canadian Journal of Fisheries and Aquatic Sciences*, Vol. 56, pp. 1949–63.

SCHWEIGERT, J.F. 2001. *Stock assessment for British Columbia herring in 2001 and forecasts of the potential catch in 2002*. Fisheries and Oceans Canada, Ottawa. Canadian Science Advisory Secretariat Research Document, 2001/140, 83 pp.

——. 2002. *Stock assessment for British Columbia herring in 2002 and forecasts of the potential catch in 2003*. Fisheries and Oceans Canada, Ottawa. Canadian Science Advisory Secretariat Research Document, 2002/110, 90 pp.

SCHWEIGERT, J.F.; FORT, C. 1994. Stock assessment for British Columbia herring in 1993 and forecasts of the potential catch in 1994. *Canadian Technical Report Fisheries and Aquatic Science*, Vol. 1971, 67 pp.

SCHWEIGERT, J.F.; WARE, D. 1995. Review of the biological basis for B.C. herring stock harvest rates and conservation levels. *Pacific Stock Assessment Review Committee Working Paper*, H 95:2. Nanaimo, B.C, Canada, Fisheries and Oceans Canada, Biological Science Branch, Pacific Biological Station, 8 pp.

STOCKER, M. 1993. Recent management of the British Columbia herring fishery. In: L.S. Parsons and W.H. Lear (eds.), *Perspectives on Canadian marine fisheries management. Canadian Bulletin of Fisheries and Aquatic Science*, Vol. 226, pp. 267–93.

———. 2001. Report of the PSARC Pelagic Subcommittee Meeting, August 29–30, 2001. Fisheries and Oceans Canada, Science Branch, *Canadian Science Advisory Secretariat, Proceeding Series,* 2001/25, 64 pp.

TAYLOR, F.H.C. 1955. The status of the major herring stocks in British Columbia in 1954–55. *BC Department of Fisheries, Nanaimo. Report of Provincial Fisheries Department,* 1954, pp. 51–73.

———. 1964. Life history and present status of British Columbia herring stocks. *Bulletin of the Fisheries Research Board of Canada,* Vol. 143, 81 pp.

TAYLOR, F.H.C.; HOURSTON, A.S.; OUTRAM, D.N. 1956. The status of the major herring stocks in British Columbia in 1955–56. BC *Department of Fisheries, Nanaimo, Report of Provincial Fisheries Department,* 1955. pp. 45–77.

TESTER, A.L. 1935. The herring fishery of B.C.: past and present. *Biological board of Canada Bulletin,* Vol. 47, pp. 37.

———. 1945. Catch statistics of the British Columbia herring fishery to 1943–44. *Bulletin of the Fisheries Research Board of Canada,* Vol. 67, 47 pp.

WARE, D.M.; SCHWEIGERT, J.F. 2002. Metapopulation dynamics of British Columbia herring during cool and warm climate regimes. Fisheries and Oceans Canada, Ottawa. *Canadian Science Advisory Secretariat Research Document,* 2002/107, 37 pp.

CHAPTER 6 The use of traditional knowledge in the contemporary management of a Hawaiian community's marine resources
Kelson K. Poepoe, Paul K. Bartram and Alan M. Friedlander

ABSTRACT

It is traditional for Hawaiians to 'consult nature' so that the methods, times and places of fishing are compatible with local marine resource rhythms and biological renewal processes. The Ho'olehua Hawaiian Homestead continues this tradition in and around Mo'omomi Bay on the north-west coast of the island of Moloka'i. This community relies heavily on inshore marine resources (coastal fishes, invertebrates and seaweeds) for subsistence, and consequently has an intimate knowledge of these resources. The shared knowledge, beliefs and values of the community are channelled by Hui Malama o Mo'omomi ('the group caring for Mo'omomi') to promote culturally appropriate fishing behaviour.

The first author of this paper is recognized as the senior caretaker (*kahu*) of the Mo'omomi coastal area because he was trained by *kupuna* (wise elders), has over forty years of local fishing experience and is a teacher (*kumu*) of others (including the second author). His ability to think ecologically and articulate resource knowledge fosters a practical understanding of local inshore resource dynamics, and thus lends credibility to standards for community fishing conduct. This chapter describes this unwritten code of conduct and verifies the effectiveness of some of the 'mental models' that guide local fishing behaviour. A new school curriculum has been built around these models and at least a dozen other coastal communities in Hawai'i are considering how to adapt them.

INTRODUCTION

As is elsewhere throughout the world, coastal fisheries in Hawai'i are facing unprecedented overexploitation and severe depletion (Shomura, 1987; Smith, 1993; Friedlander et al., 2003). The decline in fish abundance and size, particularly around the more populated areas of the state, is probably the cumulative result of years of chronic overfishing (Shomura, 1987; Friedlander and DeMartini, 2002) and habitat degradation (Hunter and Evans, 1995). Fishing pressure on nearshore resources in heavily populated areas of Hawai'i appears to exceed the capacity of these resources to renew themselves (Smith, 1993), and the near-extirpation of apex predators and heavy exploitation of lower trophic levels by intensive fishing pressure have resulted in a stressed ecosystem that does not contain the full complement of species and interrelationships that would normally prevail (Friedlander and DeMartini, 2002).

Factors contributing to the decline of inshore fisheries in Hawai'i include a growing human population, destruction or disturbance to habitat, introduction of new and overly efficient fishing techniques (inexpensive monofilament gillnets, scuba equipment, spear guns, power boats, sonar fish finders) and loss of traditional conservation practices (Brock et al., 1985; Lowe, 1996; Friedlander et al., 2003). Intensive fishing pressure on highly prized and vulnerable species has led to substantial declines in catch as well as size, and has raised concerns about the long-term sustainability of these stocks (Friedlander and Parrish, 1997; Friedlander and DeMartini, 2002; Friedlander and Ziemann, 2003).

Despite the opinion of many fishers that overharvesting is one of the major reasons for the long-term decline in inshore marine resources, there is poor compliance with state fishing laws and regulations (Harman and Katekaru, 1988). The lack of marine-focused enforcement and minimal fines for those few cases that have been prosecuted weaken the incentive to abide by fisheries management regulations. Overfishing by a growing population that no longer recognizes traditional conservation practices has greatly contributed to the decline in inshore fisheries (Lowe, 1996).

Contemporary State of Hawai'i controls on fishing have evolved from Western concepts of resource management that were introduced around the end of the nineteenth century. Fishery management based on the expertise of government resource managers has displaced management based on indigenous knowledge systems throughout most of the world, and Hawai'i is no exception. The Western industrial societies' approach to management asserts that it should be left to 'professionals' and that the users of resources should not also manage them (Berkes, 1999). This view is fundamentally different from traditional Hawaiian marine resource use and conservation, where the resource users were the managers (Johannes, 1978; Jones and others, this volume).[1] The Ho'olehua Hawaiian Homestead continues the Hawaiian tradition in and around Mo'omomi Bay on the north-west coast of the island of Moloka'i (Figure 6.1) Without perpetuation of the 'Mo'omomi system' for community self-management, the local fisheries upon which this community depends for subsistence might be in the same state of decline as elsewhere in the populated Hawaiian islands.

---

1. The term 'Hawaiian' is used throughout to mean the original Polynesian settlers of the Hawaiian Islands and their descendants.

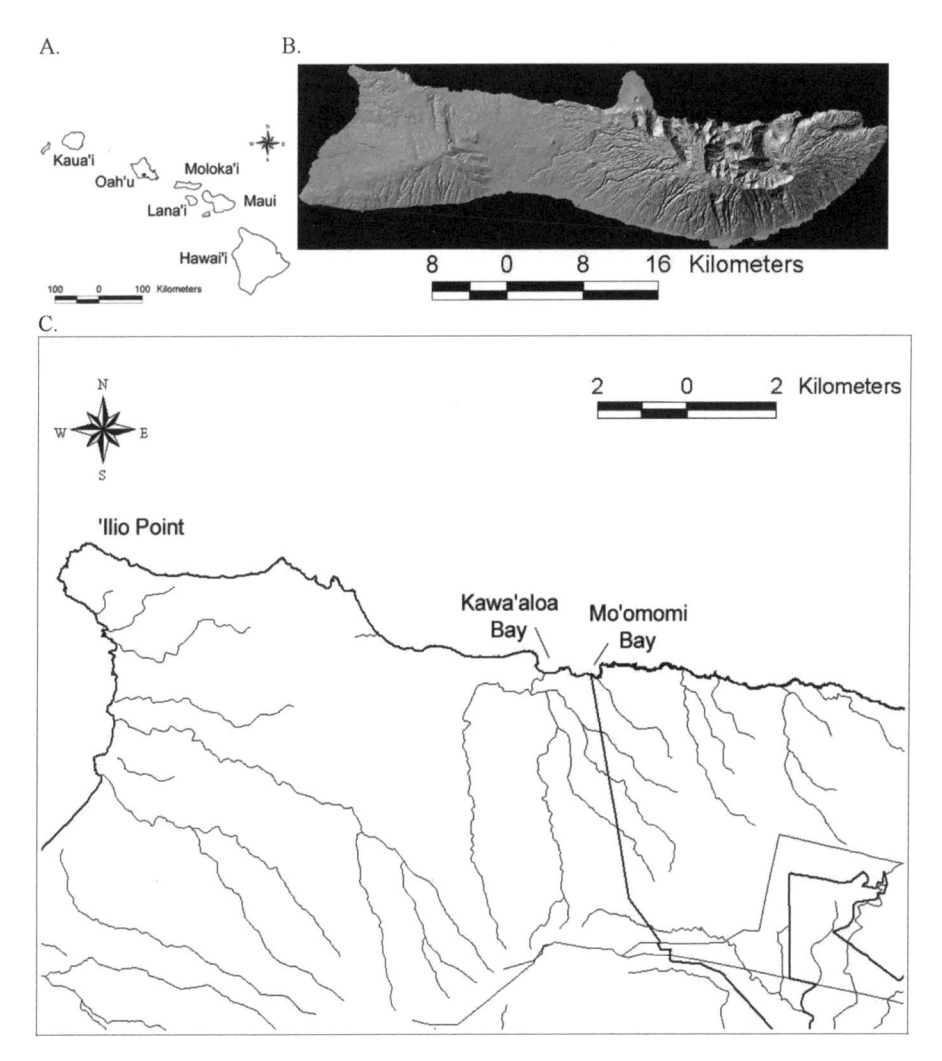

FIGURE 6.1 *(A) Location of the main Hawaiian Islands, (B) island of Moloka'i (Landsat 7 ETM/1G Satellite Imagery) and, (C) Mo'omomi and Kawa'aloa Bays located on the north shore of Moloka'i.*

Long before they had any association with Westerners, Hawaiians depended on fishing for survival. The need to avoid food depletion motivated them to acquire a sophisticated understanding of the factors that cause limitations and fluctuations in marine resources. Through their familiarity with specific places, and much trial and error, Hawaiian communities were able to develop ingenious social and cultural controls on fishing that fostered, in modern terminology, 'sustainable use' of marine resources (Titcomb, 1972; Hui Malama o Mo'omomi, 1995).

Hawaiians no doubt learned about resource limits by exceeding them at some stage of population expansion prior to European contact. Information compiled from *kupuna* (wise elders) refers to some fish species as 'famine food' (Titcomb, 1972), suggesting that food needs were not always met in pre-contact Hawai'i.

It is important to recognize Hawaiian practices not as merely traditional, but as adaptive responses to resource availability and limitations. Kirch's (1989) Hawaiian cultural sequence, when integrated with Cordy's (1981) and Hommon's (1976) sequences for Hawaiian society, suggest that, by the seventeenth century, a growing population and increasing control by the chiefs (*ali'i*) had led to a more effective system of responsibilities for meeting food needs. Fishing activities and catch distribution were strictly disciplined by rules (*kapu*) that prescribed death for severe transgressions (Titcomb, 1972). Overseers (*konohiki*) enforced the *kapu* on behalf of the chiefs (*ali'i*).

Community self-management of inshore fisheries in and around Mo'omomi Bay is a contemporary version of the traditional *konohiki* system ('Opu'ulani Albino, *kupuna*, personal communication). It is an example of 'folk management', as characterized by Dyer and McGoodwin (1994). Moral suasion, education, and family and social pressure have become the means to elicit proper behaviour rather than the harsh punishments of ancient times.

'Traditional' knowledge and practice should not be interpreted as static, rigid or non-changing. 'The culture lives on through its practitioners' (EKF, 1995) and cultural activities have a strong sense of 'place'. Tradition, as it exists in the world of contemporary Ho'olehua homesteaders, is an accumulation of knowledge and behavioural norms that have strong roots in culture, local history and experience, and which are constantly being verified and augmented. It is legitimate in its own right and does not ask to be recast in the idiom of Western industrial society or verified through the methods of contemporary government resource managers. However, the Mo'omomi system, like many other indigenous knowledge systems, does need to be communicated more effectively if it is to be useful to other coastal communities. That is the purpose of this chapter.

OBJECTIVES

- Explain the practices of the Ho'olehua homesteaders as adaptations to a harsh environment with limited resources for subsistence.
- Describe an unwritten code of conduct of Hui Malama o Mo'omomi to guide fishing behaviour in the community.
- Translate 'mental models' and management practices of the Hui Malama o Mo'omomi into written conservation principles for two fish and one seaweed species.

- Verify the effectiveness of these models in maintaining healthy local populations of the three species.

HISTORY OF THE COMMUNITY AND FOUNDATIONS OF KNOWLEDGE

DESPITE its rugged shoreline and windward exposure, the north-west coast of Moloka'i has a long history of use for subsistence fishing and gathering. As early as the eleventh century, Hawaiians from the island's wet, north-east valleys spent summer months at Mo'omomi Bay and nearby coastal areas (Figure 6.1), catching and salt-curing fish to see them through winters when the ocean was too rough for fishing (Summers, 1971).

Marine resources along a 22 km length of Moloka'i's north-west coast, on both sides of Mo'omomi Bay, are now mostly harvested by a community of native Hawaiians who reside in the nearby Ho'olehua Hawaiian Homestead. Opened in 1924, Ho'olehua was the second homestead established after the US Congress passed the Hawaiian Homes Commission Act in 1921 with the intent of returning Hawaiians to the land. The community comprises 5,500 hectares of land supporting a population of about 1,000 native Hawaiians (Hui Malama o Mo'omomi, 1995).

The homestead occupies arid saddle lands between the two volcanoes that formed the island of Moloka'i. The first homesteaders arrived eighty years ago at a time of severe drought, to find the nearby rocky shores mostly inaccessible and exposed to large waves for much of the year. With their survival at stake, they learned to adapt to local environmental and ecological conditions and cycles.[2] The knowledge acquired in the struggle for survival is not merely practical perception and 'know-how' but patterns of thought and understanding, and models of local ecosystem workings (Friedlander et al., 2002).

Succeeding generations of homesteaders have endured despite the harsh land and ocean environment (Governor's Moloka'i Subsistence Task Force, 1994). They choose to live with less on Moloka'i rather than relocate to the more affluent population centre on the island of O'ahu 100 km to the west ('Opu'ulani Albino, *kupuna*, personal communication). The communal identity of Ho'olehua Hawaiian Homestead continues to be defined by a shared cultural heritage and is maintained by a system of interdependence and social reciprocity that is expressed in many ways, including the sharing of seafood gathered through subsistence fishing. This system enables the homesteaders to live well and with confidence in a sometimes difficult environment. The repetition of subsistence activities is one way that knowledge, values and identity are transferred to succeeding generations. Cultural survival is

---

2. 'Opu'ulani Albino, *kupuna*, personal communication, based on readings of the journals written by an original homesteader.

thus entwined with sustainable resource use (Governor's Moloka'i Subsistence Task Force, 1994).

Residents of Ho'olehua Homestead are more dependent than those in other communities in the state on subsistence farming and fishing (which provide a third of the food consumed by the 1,000 residents of this community) (Governor's Moloka'i Subsistence Task Force, 1994). The majority of Ho'olehua households include active fishers, and preliminary estimates suggest that the average household consumes nearly 11 kg of seafood per week, or about ten times as much as on O'ahu. Fishing is intensive when measured by kg harvested per km² of inshore area (Hui Malama o Mo'omomi, 1995). Homesteaders still eat a diet that is heavy on the traditional Hawaiian staples of fish, *limu* (seaweeds) and *poi* (the starch plant taro pounded into a paste) (Abbott, 1984; Hui Malama o Mo'omomi, 1995). The marine species most important for community subsistence include a diversity of shallow-water reef fish, invertebrates and seaweeds, some of which are shown in Table 6.1 (Hui Malama o Mo'omomi, 1995).

TABLE 6.1 *Important seafood resources and methods of harvest for the Ho'olehua homesteaders.*

Hawaiian name	Common name	Scientific name	Fishing method
Fishes			
Moi	Pacific threadfin	*Polydactylus sexfilis*	throw-net, pole and line
Papio/ulua	jacks – small/large	Carangidae	throw-net, pole and line, spear
Uhu	parrotfishes	Scaridae	spear
Kumu[1]	whitesaddle goatfish	*Parupeneus porphyreus*	spear
Enenue	rudderfishes	*Kyphosus* spp.	throw-net, spear
Kole[1]	Goldring surgeonfish	*Ctenochaetus strigosus*	spear
Kala	bluespine unicornfish	*Naso unicornis*	spear
Aholehole[1]	Hawaiian flagtail	*Kuhlia xenura*	throw-net, spear
Invertebrates and algae			
Limu kohu	red seaweed	*Asparagopsis taxiformis*	hand harvest
'opihi	limpet	*Cellana* spp.	hand harvest
'ula	lobster	*Panulirus* spp.	spear, tangle-net
He'e	octopus	*Octopus* spp.	hand harvest, spear
A'ama	rock crab	*Grapsus tenuicrustatus*	hard harvest
Wana	sea urchins	Echinoidea	hand harvest
Ha'uke'uke	helmet urchin	*Colobocentrotus atratus*	hand harvest

1 = endemic species.

Memories of the first homesteaders' (grandparents' generation) teachings about survival and 'sustainable' resource use are relatively fresh in the minds of younger generations of homesteaders. Hence, the Ho'olehua community is one of the few places remaining in the Hawaiian Islands where the traditional Hawaiian system still provides a framework for fishery resource use and conservation ('Opu'ulani Albino, *kupuna*, personal communication).

HISTORY OF COMMUNITY-BASED MANAGEMENT AT MO'OMOMI

NETWORKS of social ties and cooperation generated by the subsistence activities of Ho'olehua community members foster a collective interest in resource conservation and a consensus about the proper conduct of fishing (Hui Malama o Mo'omomi, 1995). In the 1980s, a few fishers from the community began targeting fish species, especially the endemic whitesaddle goatfish known as *kumu* (*Parupeneus poryphyreus*), that have high commercial value in Honolulu. Commercial marketing of their catches brought them into contact with off-island fishers, who also began to fish intensively in and around Mo'omomi Bay for the same species, causing local depletion of *kumu* (K.K. Poepoe, personal observations). This deviation from traditional subsistence fishing practices and resource conservation norms motivated some Ho'olehua homesteaders to form Hui Malama o Mo'omomi (Hui Malama o Mo'omomi, 1995).

In 1994, the Governor's Moloka'i Subsistence Task Force suggested that the Ho'olehua Hawaiian Homestead be allowed to manage shoreline marine resources in nearby areas for subsistence fishing (Governor's Moloka'i Subsistence Task Force, 1994). The 1994 Hawai'i State Legislature established a process for designating community-based subsistence fishing areas. In response to this legislation, the Hui prepared a fishery management plan for the north-west coast of Moloka'i (Hui Malama o Mo'omomi, 1995).

The plan had three major objectives:

- to establish a marine resource monitoring programme that integrates traditional and science-based techniques
- to foster consensus about how fishing should be conducted to restore community values and care-taking
- to revitalize a locally sanctioned code of fishing conduct.

The Hui's long-term goal is to bring fishery management in the coastal areas in and around Mo'omomi Bay down to the level of the users who have the most detailed understanding of the local resources and the greatest long-term interest in their sustainable use (Hui Malama o Mo'omomi, 1995).

In response to the legislation, the State Department of Land and Natural Resources (DLNR) designated Mo'omomi and Kawa'aloa bays, a small portion of the Ho'olehua community's fishing grounds, as a community subsistence fishing area, with fishing gear restrictions and monitoring of resources and fishing activities during a two-year experimental period (DLNR, 1996). After the experiment, the state drafted regulations for permanent government designation of a subsistence fishing area limited to the two bays.

In October 2000, the DLNR held a public hearing on Moloka'i. Community leaders who attended favoured a much larger special area and proposed a traditional *ahupua'a* framework (Smith and Pai, 1992) in which the watershed and adjacent marine areas would be managed as interconnected units. The Hui proposed to manage local fisheries according to mutually agreed standards that would allow the state to evaluate the community's management performance (K.K. Poepoe, personal observation). State officials continue to review these proposals but no regulations have been implemented and no immediate state government action is planned.[3]

CONTEMPORARY SELF-MANAGEMENT THROUGH A CODE OF CONDUCT

THE Hui continues informal management through internal cultural norms and values that guide and instruct the behaviour of the community and that encourage responsible fishing based on individual conscience, social and family pressure, and the training of youth to become 'good marine citizens'. An unwritten 'code of conduct' focuses on how fishing should be practised in and around Mo'omomi Bay to maintain regular biological renewal processes rather than on how much fish should be harvested (Pacific American Foundation/Hui Malama o Mo'omomi, 2001).

The wisdom and insights of leaders and *kupuna* who possess and transmit traditional knowledge and values are crucial in lending credibility to the code of conduct (Pacific American Foundation/Hui Malama o Mo'omomi, 2001). *Kupuna* wisdom is based on cultural protocol (EKF, 1995). Protocol combines knowledge, practice and belief – fundamental characteristics of most traditional systems (see Berkes 1999) – that evolve over time within a specific cultural and ecological context.

Hawaiian protocol is built on an old foundation of responsibilities that link people with their environment (EKF, 1995). These responsibilities define behavioural norms in the Ho'olehua Homestead and other Hawaiian communities on Moloka'i to such an extent that a new educational curriculum has been developed around them for use in public schools (Pacific American Foundation/Hui Malama o Mo'omomi, 2003). The most important of the responsibilities (from Handy et al., 1972; Pukui et al., 1972; Kanahele, 1986; EKF, 1995; Hale, undated) are:

3. W. Puleloa, State aquatic biologist, Moloka'i, personal communication.

- *Concern about the well-being of future generations.* Meet present food needs without compromising the ability of future generations of people to meet their needs. Irresponsible resource use is tantamount to denying future generations their means of survival.
- *Self-restraint.* Take only what one needs for immediate personal and family use and use what one takes carefully and fully without wasting. A good Hawaiian fisher is not the one with the largest catch but the one who can get what he or she needs without disrupting natural processes. An example from the compilation of 'sayings of wisdom' by Pukui (1983) illustrates this conservation ethic, *'E 'ai I kekahi, e kapi kekahi':* Eat some now and save some for another time.
- *Reverence for ancestors and sacred places where ancestors rest.* Hawaiians inherited valuable knowledge from their ancestors. At one time, this knowledge was crucial for physical survival. The 'ancestry of experience' (Holmes, 1996) stored in the memories of living Hawaiians is still largely transmitted through non-written processes. It is taught to succeeding generations by telling stories, creating relationships among people and between people and places, and establishing personal meaning. Ancestors are worshipped because the survival of Hawaiian culture depends on knowledge and skills passed from generation to generation (Holmes, 1996).
- *Malama* ('take care of living things'). The Hawaiian perspective is holistic, emphasizing relationships and affiliations with other living things. Accountability, nurturing and respect, important for good human relationships, are also beneficial in relationships with marine life.
- *Pono* ('proper, righteous') behaviour. Hawaiians are expected to act properly and virtuously in relationships with past, present and future generations and with the food sources that sustain them.

MONITORING AND EDUCATION PROGRAMME

THE local code of fishing conduct is reinforced through continual feedback based on community resource monitoring, education and peer pressure. Local resource monitors, supervised by the community leaders and *kupuna*, acquire an intimate knowledge of the local marine environment though daily observation. Emphasis is placed on acquiring and applying knowledge about the habitats, natural resource rhythms, and spawning and feeding patterns of shoreline and inshore food species. These observations, recorded in daily journals or held in memory, become the raw material to help develop mental models of resource rhythms and processes. As resource monitors develop a sophisticated understanding of local resources, they look for anomalies, such as resources expected to be present that are not, and potential reasons for their absence.

INDIGENOUS PRACTITIONERS AND RESEARCHERS

Many natural processes that affect fish distribution are monitored by the community, but the most important of these are seasons and moon phases. The moon was as essential in scheduling the activities of the ancient Hawaiians as clocks are to modern man. The moon calendar is a predictive tool based on awareness of natural cycles and their relationship to fishing and farming success. Its wisdom reflects lifetimes of observations and experiences by many generations of Hawaiians in their quest for survival (EKF, 1995). Modern-day people of Hawai'i still refer to the calendar to plan fishing and planting activities, and a popular form of the calendar is published annually.

The moon calendar emphasizes natural processes that repeat at different time scales: seasonal, monthly and daily. Distinctions are made between two general seasons (*ka'u* or dry; *ho'oilo* or wet) and three general phases of the moon after the new moon: *ho'onui* (nights of enlarging moon); *poepoe* (nights of full moon); and *emi* (nights of diminishing moon). In addition to diagramming seasons and moon phases, Figure 6.2 also gives the Hawaiian names for the twelve months of the year.

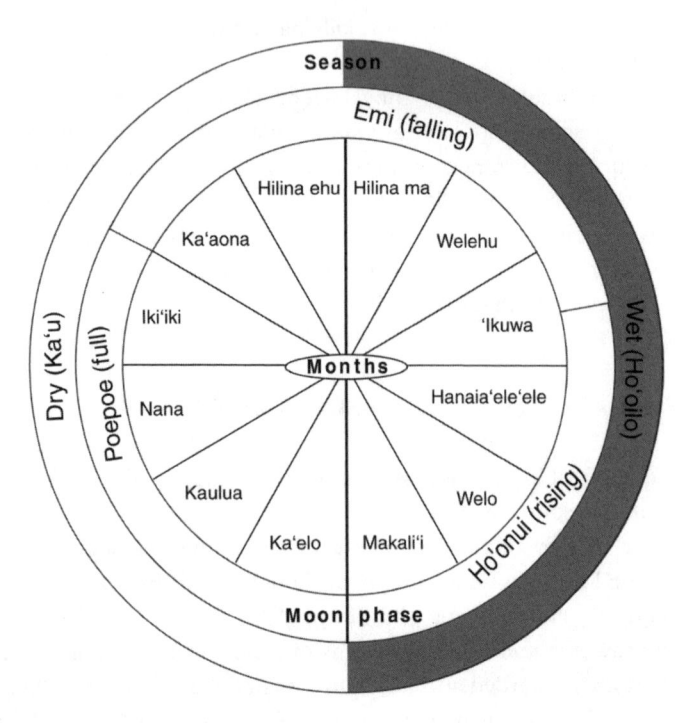

FIGURE 6.2 *Hawaiian moon calendar showing months, seasons and moon phases that are used to guide fishing activities. Names used for months in this calendar are specific to the island of Moloka'i.*

Adapted from Friedlander et al. (2002).

Specific names are also given to each night of the Hawaiian lunar cycle (Figure 6.3). Prohibitions (*kapu*) occurred on a monthly basis. The exact details of many of these prohibitions have been lost over time but their timing is associated with periods of the lunar cycle when spawning and movement of some important resource species usually occur. These prohibitions may have served to limit overall fishing effort on a monthly basis, in particular during periods when certain species might be more susceptible to harvest or when their capture could disturb reproduction or other important activities.

By observing spawning behaviour and sampling fish size and reproductive state, community monitors can construct a calendar identifying the spawning periods of major food fish species. The year 2000 calendar (Table 6.2) shows that peak spawning for *ulua* (jacks, *Caranx ignobilis*), *moi* (Pacific threadfin, *Polydactylus sexfilis*), *uhu* (parrotfish, Scaridae) and '*a'awa* (Hawaiian hogfish, *Bodianus bilunulatus albotaeniatus*, an endemic subspecies) occurred during the summer months. Late winter/early spring spawning was observed for *aholehole* (Hawaiian flagtail, *Kuhlia xenura*, a Hawaiian endemic) and *kumu*. Surgeonfishes (Acanthuridae) typically spawned in late winter, as well as in early spring. By identifying peak spawning periods for important resource species, traditional closures or *kapu* can be applied so as not to disturb the natural rhythms of these species.

MENTAL MODELS

THE traditional Hawaiian resource-use system involved measuring and evaluating natural processes to produce representations of the workings of ecosystems, as in Western scientific methods.[4] Thus, theoretical constructs of Hawaiian scientific thought are *mental models* that recognize different states or 'frames' (after Starfield et al., 1993), capturing the essential aspects of dynamics that may apply to the same ecosystem at different times. However, Hawaiian knowledge relies on memory and does not incorporate the rigorous quantitative estimates or written records of Western science. There was no written Hawaiian language prior to the early nineteenth century (Kuykendall, 1938), so traditional knowledge was orally transmitted from generation to generation through chants, stories and demonstration.

Due to the local importance of *aholehole*, *moi* and the red seaweed *limu kohu* (*Asparagopsis taxiformis*) as food items in the Ho'olehua community, these species were examined closely, and written conservation principles were derived from the first authors' and other community monitors' 'mental models' of resource dynamics. Pertinent biological life history information is included to provide background for readers unfamiliar with the species. The models are presented using both scientific and traditional Hawaiian terminology because of the dual audiences for this information.

---

4. Isabella I. Abbott, *kupuna* and PhD in botany, University of Hawai'i, personal communication.

INDIGENOUS PRACTITIONERS AND RESEARCHERS

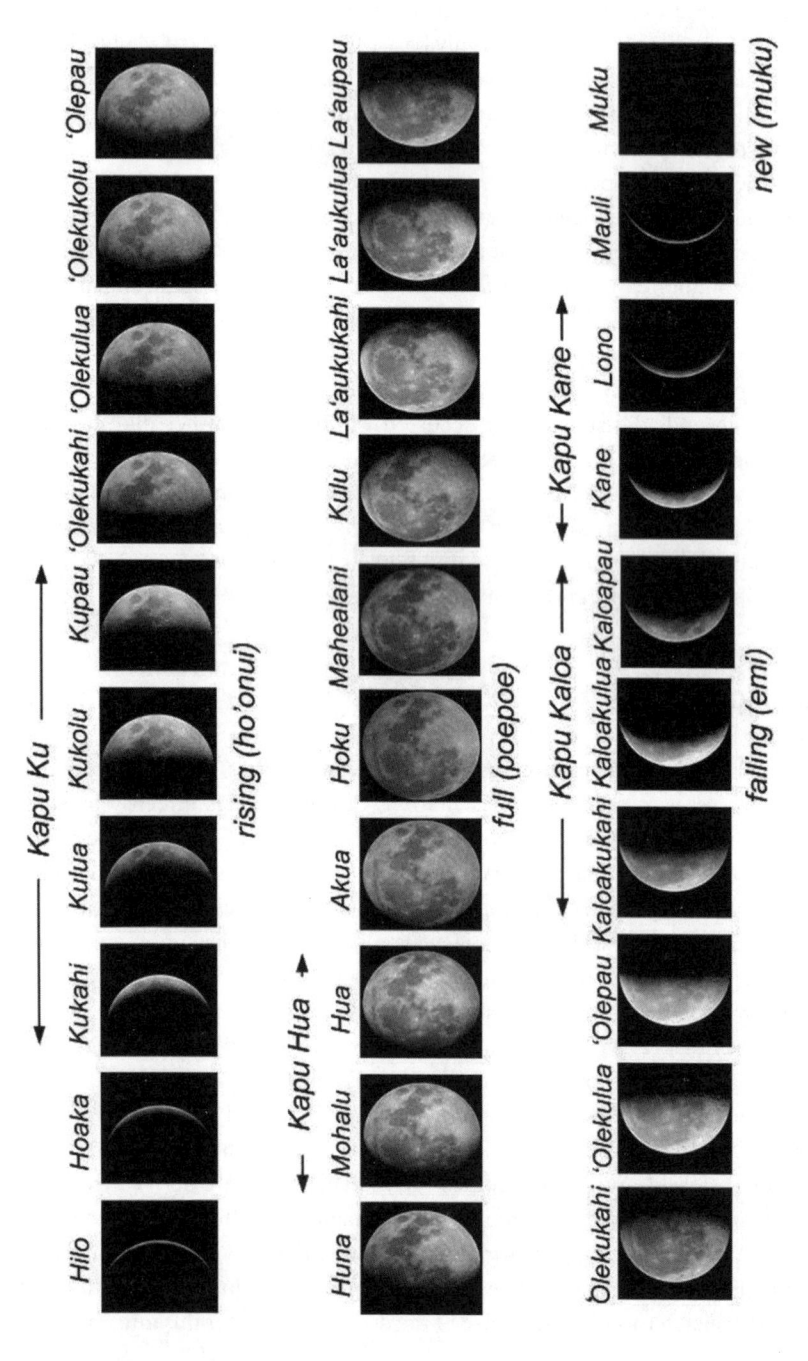

FIGURE 6.3 *Hawaiian names for nights of the rising (ho'onui), full (poepoe) and falling (emi) moon phases and prohibition (kapu) periods.*
Source: Adapted from Prince Kuhio Civic Club (2001).

TABLE 6.2 *Mo'omomi Bay fish spawning calendar for the year 2000 for key resource species. Black boxes indicate months of peak spawning. Grey boxes indicate other months when spawning was observed.*

Species	Jan.	Feb.	Mar.	Apr.	May	Jun.	Jul.	Aug.	Sep.	Oct.	Nov.	Dec.
ulua (Caranx ignobilis)						■	■					
aholehole[1] (Kuhlia xenura)			■	■	■							
moi (Polydactylus sexfilis)					▒	■	■	■				
'u'u (Myripristis species)							▒	■	■	■		
kumu[1] (Parupeneus porphyreus)				▒	■							
'aweoweo (Priacanthus species)								▒	■			
ta'ape (Lutjanus kasmira)						■	■					
'a'awa[2] (Bodianus bilunulatus)					■	■	■					
enenue (Kyphosus species)			■	▒	▒	■						
uhu (Scaridae)						■	■	■				
uhu palukaluka (Scarus rubroviolaceus)						■	■					
ponuhunuhu (Calotomus carolinus)						■	■					
pualu (Acanthurus xanthopterus)				▒	■							
palani (Acanthurus dussumieri)				▒	■							
kala (Naso unicornis)				■								
kole (Ctenochaetus strigosus)				■								
manini[2] (Acanthurus triostegus sandvicensis)				▒	■							

1 = endemic species.
2 = endemic subspecies.
Source: Adapted from Friedlander et al., 2002.

Conservation principles for *aholehole*

Aholehole are endemic to the Hawaiian Islands (Randall, 1996). The young live in shallow water along the shoreline and may be found in tide pools, streams and estuaries. They feed mainly on planktonic crustaceans but also on polychaete worms, insects and algae. Their length at maturity is about 18 cm, with spawning occurring year-round, although mainly during the winter and spring months. The *aholehole* was used in sacrifices in ancient Hawai'i to keep away evil spirits when a white fish or pig was needed (Titcomb, 1972).

At Mo'omomi Bay, *aholehole* spawn during the wet season, typically in late winter/early spring. Much of the distribution of *aholehole* is based on the movement of sand in and out of nearshore habitats (Table 6.3). During the winter months, large swells re-suspend sand, providing ample space inside reef holes (*puka*) along the shore for *aholehole* to school. During the summer months, tradewind swells from the east transport sand inshore; the reef *puka* are thus filled in and the *aholehole* have to move offshore. This change in habitat between seasons coincides with, and may possibly be a cue for, the onset of spawning. The conservation principles developed by the Hui Malama o Mo'omomi for harvesting *aholehole* discourage both catching sub-reproductive individuals and harvesting during times of peak spawning in the late winter and early spring.

TABLE 6.3 *Season movement patterns of aholehole (Kuhlia xenura) in relation to changes in habitat at Mo'omomi Bay.*

Season	Sand movement	Reef holes *(puka)*	**Aholehole** distribution
Winter	Offshore	Exposed	Inshore
Summer	Inshore	Filled	Offshore

Conservation principles for *moi*

The Pacific threadfin (*Polydactylus sexfilis*) or *moi* is a very popular and much sought-after sport and food fish in Hawai'i, which also supports a small subsistence fishery (Friedlander and Ziemann, 2003). In ancient Hawaiian culture, *moi* were reserved for the ruling chiefs and forbidden to commoners (Titcomb, 1972). Hawaiians developed a number of traditional strategies to manage *moi* for sustainable use. *Kapu*, or closures, were placed on *moi* during the spawning season (typically from June through August) so as not to disrupt spawning behaviour.

Moi are protandric hermaphrodites, initially maturing as males of about 20–25 cm after a year, with most individuals then undergoing a sex reversal, passing through a hermaphroditic stage and becoming functional females of between 30 and 40 cm fork length at about three years of age (Santerre et al., 1979). Spawning occurs inshore, and eggs are dispersed and hatch offshore (Lowell, 1971). Larvae and juveniles are pelagic until the juveniles attain a fork length of about 6 cm, whereupon they enter inshore habitats, including surf zones, reefs and stream entrances (Santerre and May, 1977; Santerre et al., 1979). Newly settled young *moi*, locally called *moi-li'i*, appear in shallow waters in summer and autumn, when they are the dominant member of the nearshore surf zone fish assemblage.

Moi typically spawn in *moi holes* west of Mo'omomi Bay (Figure 6.1). They usually come inshore to spawn from June through August. Sand movement is very important in determining when and where *moi* spawn, because shelter is an important controlling factor in reducing the risk of predation during the spawning period. In the west end of Kawa'aloa Bay, for example, *moi* move inshore to spawn in an interval when sand has stopped moving but before too much sand has filled the *puka* in the reef. Stable sand leads to higher numbers of *moi* prey (shrimp and crabs). Observation of sand movement and the height of sand waves can give a good indication of when *moi* will move inshore to spawn. As sand waves flatten out, the sand becomes more stable, while steep sand waves indicate the movement of sand.

A mental model of the life history of *moi* is used by Hui community monitors. Conservation principles and management practices were derived from this model by integrating seasonal movement, spawning aggregation behaviour and the relationship of different life history phases to these behaviour patterns. Table 6.4 is an attempt to construct a written representation of the knowledge concerning the behaviour of *moi* and how it relates to Hui Malama o Mo'omomi's conservation principles. These include restrictions on harvest of *pala moi* (hermaphrodites) *or moi* (females), depending on population structure, and restrictions on harvest during the spawning season. Minimizing the disturbance to spawning and nursery habitats is another important conservation practice.

TABLE 6.4 *Seasonal movement and aggregation of moi around the island of Moloka'i.*

Fish size	Dispersed	Aggregated	Aggregated and spawning
Adults (*mana moi, pala moi, moi*)	Autumn through winter	Spring, in reef holes prior to spawning	June, July, and August; one spawning per month cued by moon phase
Juveniles (*moi li'i*)	Leave for adult habitat once grown	In autumn, as new recruits feeding in sand bottom areas with nearby rocky shelter	N/A

Moi have a readily identifiable aspect of their life history (sex reversal) that has contributed to their decline in Hawai'i: continued overfishing results in relatively few females left in the population around heavily fished areas of the state (Friedlander and Ziemann, 2003). Awareness of the need to protect both immature *moi* and the female breeding stock from overharvest is an example of how Hawaiian resource knowledge can validate Western science, which has 'discovered' this method of conservation and named it as 'slot limits.'

Conservation principles for seaweed (*limu kohu*)

Hawai'i is rich in edible seaweed (*limu*) owing to the high volcanic islands and associated rainfall, which provides nutrients for the growth of *limu*. While the use of seaweeds among other Polynesian peoples was infrequent in the past or has been curtailed today (Abbott, 1984), a wide variety of seaweeds are consumed by Hawaiians even to this day. One of the most prized species is *limu kohu* (the supreme *limu* in Hawaiian) (A*sparagopsis taxiformis*) (Abbott, 1984).

Fronting Mo'omomi and Kawa'aloa Bays, *limu kohu* grows in areas of strong surge from the splash zone on intertidal benches (*papa*) to boulder and flat limestone bottoms as deep as 12 m. This seaweed is well adapted to the shallow-water habitats off Mo'omomi, which are wave washed almost year round. There are, however, marked seasonal changes in the distribution of *limu kohu* (Table 6.5). During *ho'oilo* (wet season), the tides rotate in the opposite pattern to *ka'u* (dry season), when the highest tides occur during the day and the lowest tides at night. During the wet season, the coast is exposed to intense wave action generated by North Pacific swells and strong trade winds. Under these conditions, *limu kohu* is able to attach itself to and flourish on long stretches of *papa* that experience more water movement than during the dry season.

From January 2000 to January 2001, seasonal changes in the distribution, abundance and reproductive condition of *limu kohu* were studied at the major harvest site. Information collected during this twelve-month period of detailed observation is summarized in Table 6.5. The survey period began during the latter half of the 2000 wet season (January to April), through the dry season (June to October) followed by the start of another wet season (November to January 2001). These data were collected by the authors and community resource monitors. Severe drought conditions later in 2001 severely retarded the growth of *limu kohu* on the *papa* over this time period.

A number of environmental factors affect the growth of *limu kohu* on intertidal benches and sub-tidal areas (Table 6.5). The change of seasons from *ho'olio* (wet) to *ka'u* (dry) exposes growths of *limu kohu* on the intertidal benches to dehydration, UV damage and eventual death. Patterns observed in the relative abundance and heights of plants (Table 6.5) indicate that the wet season provides the best growing conditions on shallow (0–1 m) benches, or *papa*. Marked changes in bench cover by this seaweed occurred during the wet season or after rainfall, with young stands of *limu kohu* growing to between 2 and 5 cm high during one cycle of the moon.

Limu kohu reproduces by spores. The observations during the wet season indicate that shallow-water plants bear spores after they have grown to a height of 7 cm, and sporing continues until full growth of 10 to 13 cm is completed (Table 6.5). As they grow taller, shallow-water stands of *limu kohu* are torn by high wave energy, starting with the fronds, until eventually the main stems weaken and break off.

Observations during *ka'u* (the dry season) indicate that daylight exposure during minus tides, long days and reduced water movement make the shallow *papa* an inhospitable environment for *limu kohu* (Table 6.6). However, the longer days stimulate lush growths and sporing of this seaweed on subtidal areas of boulders and limestone flats to a depth of about 6 m. At greater depths, growth is sparser because of limited sunlight (Table 6.6).

The continued availability of *limu kohu* at Mo'omomi Bay depends on the recruitment and growth of new plants. Successes in reproducing (through sporing) and in attaching to local substrata are key processes that sustain the supply of this seaweed. The critical conservation principle derived from the mental model for *limu kohu* is to retain spores so they are more likely to settle out on local substrata. That

TABLE 6.5 *Observations of the edible seaweed limu kohu (Asparagopsis taxiformis) at the major shallow-water (0–1 m) harvest site, January 2000–January 2001.*

Time of observations	Condition of shallow plants	Height of shallow plants (cm)	Condition of reproductive spores	Other information
Wet Season (Ho'oilo)				
Jan. 2000	Abundant	7–10 cm	Attached	
Feb. 2000	Long plants breaking off, dying back, losing red colour	7–10 cm	Large numbers attached, some being released	Wave action breaking off plants
March 2000	Shorter, sparse and pale in colour	7 cm	Large number being released from shallow plants; evident on deep plants (7 m)	
April 2000	Still abundant but long plants have broken off; pale colour	5–7 cm on bench; 7–10 cm in pools	Same as March	
Dry Season (Ka'u)				
May 2000	Pale colour; what long plants remain are overgrown with epiphytes and dying back; some plants very close to shore	5 cm	Few spores attached to shallow plants; increasing number on deep plants (7 m)	Time of peak harvest; collecting may spread spores for regrowth
June 2000	Sparse and short growths	5 cm	Not evident on shallow plants; abundant on deepwater plants	Lack of rainfall
July 2000	Plants getting longer	7 cm	Sparse on shallow plants; abundant on deepwater plants	Less than 0.25 cm rainfall in month
August 2000	Abundant	7–10 cm	Sparse on shallow plants; abundant on deepwater plants	0.6 cm rainfall on 25 August
Sept. 2000	Sparse	6 cm	Not evident	0.8 cm rainfall in month; wave action breaking off plants
Oct. 2000	Abundant	7 cm	Sparse	
Wet Season (Ho'oilo)				
Nov. 2000	Abundant	7 cm	Increasing on longer plants	2 cm rainfall in month
Dec. 2000	Scattered, red colour	7 cm on bench; 7–10 cm in pools	Increasing on longer plants	0.3 cm rainfall in month
Jan. 2001	Abundant, dark purple colour	7–10 cm	Abundant on shallow plants	0.81 cm rainfall in month

is why *limu kohu* gatherers are encouraged to rub plants against a rough surface (such as the collector's bag) as they are harvested. Many spores are trapped within the basal mass, and leaving this mass in the water increases the chance that spores will attach and grow near the original harvest location. Observations during the peak harvest period in May 2000 (see Table 6.6) suggest that *limu kohu* may replant in shallow inshore areas of the *papa* as a result of this conservation practice.

TABLE 6.6 *Environmental factors affecting the distribution and growth of the edible seaweed limu kohu (Asparagopsis taxiformis) in and around Mo'omomi Bay.*

Season	Limu kohu habitat	
	Shallow (0–1 m depth)	Deep (1–10 m)
Wet (*Ho'bilo*)	Growth favoured by winter rainfall (introducing nutrients), minus tides at night, short days, ocean turbulence dispersing reproductive spores	Growth favoured by water motion dispersing reproductive spores but inhibited by short days
Dry (*Ka'u*)	Growth inhibited by lack of rainfall, 'sunburn' during minus tides, long days	Growth favoured by long days
Sediment movement		
Abrasion	Grinds off old seaweed, opening surfaces for new seaweed growth	
Grain size	Reproductive spores probably attach successfully to particular grain sizes; grains that are too big may smother attached spores, whereas small grains may not settle out on the bottom	

EFFECTIVENESS OF HUI MANAGEMENT PRACTICES

THE research that formed the basis for this chapter allowed for a comparison among various locations around the state as well as for reference points for policy makers and others to evaluate the success of community management practices. As fisheries resources at Mo'omomi are very healthy compared with most areas around the state, management of these resources can focus more on wise sustainable use rather than rebuilding of stocks.

Quantitative science-based visual transect methods showed that the fish abundance in Mo'omomi Bay and adjacent coastal waters was between three and five times higher than similar north shore locations around Hawai'i (Figure 6.4). Important resource species in Mo'omomi Bay such as *moi* and *aholehole* were larger and more abundant than in other areas of the state (Figures 6.5A and B).

Moi populations at Mo'omomi appear to be in much better condition than the *moi* populations around O'ahu. A sample of 104 *moi* obtained from Mo'omomi (mean = 27 cm) showed significantly larger fish ($P < 0.001$) than those taken by the fishery on O'ahu in 1999 (mean = 23.9 cm), with more females in the population than on O'ahu (Figure 6.5A). Tagging of *moi* conducted by community resource monitors resulted in a number of recaptures at spawning holes after one or more years at large. This fidelity to spawning sites reinforced the practice of caring for these locations during the spawning season.

Size frequency for *aholehole* from Mo'omomi Bay was compared with size frequency information from Hilo Bay, Hawai'i after a net ban went into effect in the late 1980s (Figure 6.5B). Despite this ban on gillnets, the size frequency distribution in Hilo Bay was still less than at Mo'omomi Bay. This again confirms the health of the fish populations at Mo'omomi Bay compared to other fish populations around the state. Tagging studies at Mo'omomi revealed that *aholehole* did not move much along the shoreline, with all recaptures occurring at or near release sites.

DISCUSSION

THE residents of Ho'olehua Hawaiian Homestead tend to care deeply about what becomes of their subsistence resources, not only as a source of food for themselves and future generations, but also because their way of life and identity are at stake. Fishing behaviour by community members continues to evolve through

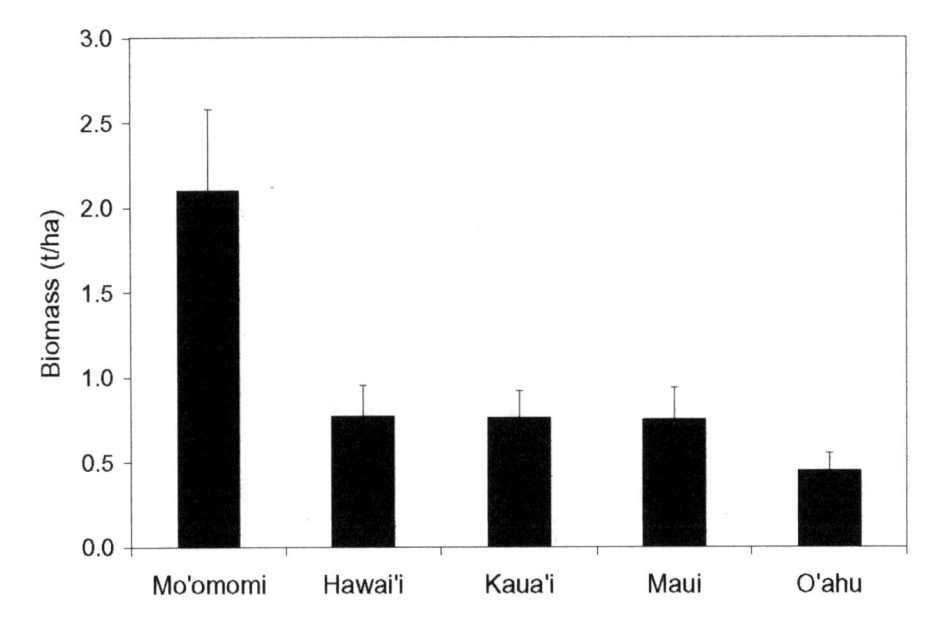

FIGURE 6.4 *Comparison of fish biomass (t/ha) at Mo'omomi Bay and similar exposed north shore locations around the main Hawaiian Islands. Error bars are standard error of the mean.*
Adapted from Friedlander et al. (2002).

INDIGENOUS PRACTITIONERS AND RESEARCHERS

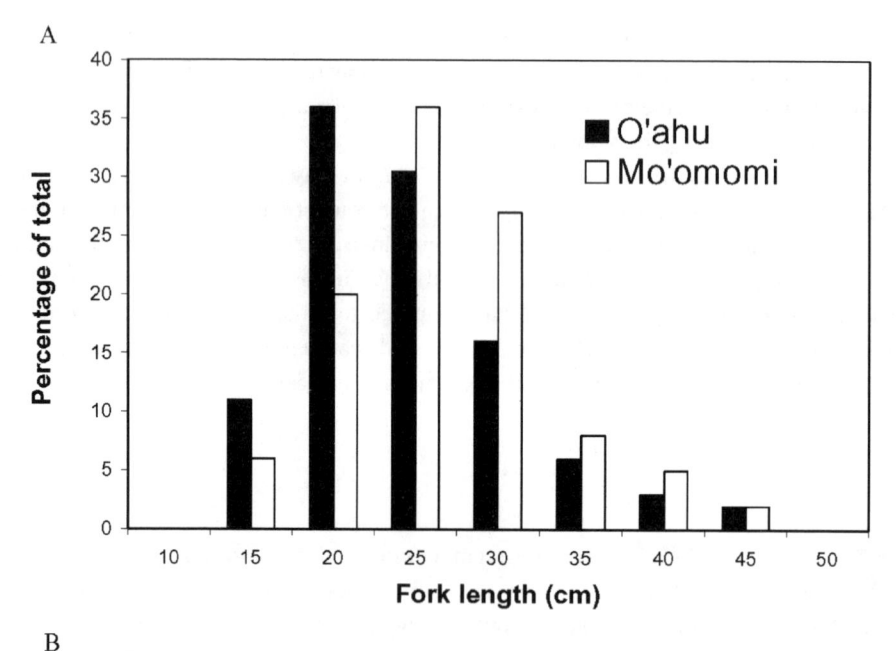

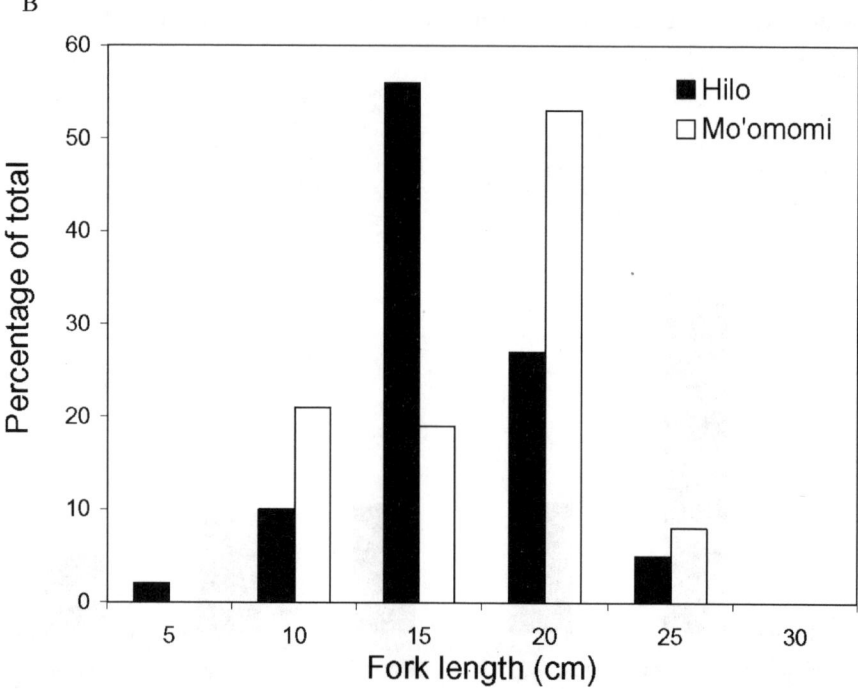

FIGURE 6.5 *(A) Fork length (cm) of moi (Polydactylus sexfilis) harvest along windward O'ahu and in Mo'omomi Bay in 1999. (B). Length frequency distributions for aholehole (Kuhlia xenura) caught at Hilo Bay after gillnet ban and at Mo'omomi Bay in 1999.*

social learning (by oral transmission, imitation and demonstration) of *pono*, or proper, behaviour. Despite substantial deterioration of Hawaiian ancestral marine resource knowledge in general (Titcomb, 1972), Hui Malama o Mo'omomi has found that it remains dynamic, capable of being verified, regenerated and even expanded in some localities by new generations of Hawaiians (Pacific American Foundation/Hui Malama o Mo'omomi, 2001).

Is the 'Mo'omomi model' simply a *kipuka* (a high spot around which lava flowed), removed from mainstream contemporary life in the State of Hawai'i, or can other community-based organizations find ways to adapt the lessons of Mo'omomi? In 1997, Hui Malama o Mo'omomi recognized the need to synthesize and document local marine resource management knowledge based on memory and the spoken word so that this information could be used in educational programmes focused within the community. Growing interest in the Mo'omomi model from outside the community has generated a wider audience requesting written documentation.

Directly transferring Mo'omomi's traditional knowledge to other places risks weakening its vitality, and increases the chances of dislocation and misapplication outside the restricted context in which the detailed knowledge evolved and is effective. Furthermore, not all coastal communities have the fundamental conditions required for a 'folk management system' to arise (Pinkerton, 1994). At least a dozen other coastal communities in Hawai'i are considering how to adapt some of the management techniques of the Hui Malama o Mo'omomi, with modification of specific details. The Hui proposes the following guiding principles when conditions are suitable for community-based fisheries management of other coastal areas in Hawai'i:

- Fishers have responsibilities in the use of marine resources.
- Fundamental tenets of traditional resource management include understanding basic processes of renewal and conducting harvest practices so as not to disrupt these processes.
- To be effective, fishery conservation must function within a specific local context. Communities and their individual members must exercise control over local inshore marine resource use and be accountable for the health and productivity of local resources.
- The emphasis should be on how fishing is conducted, not the quantity of fish harvested.
- The time dimension of 'sustainable use' should be intergenerational, not the four-year time cycle between political elections and agendas.
- Sustainable yield does not mean maintaining resource abundance at a fixed level or an unexploited level. Fishing should be modulated in response to changing rhythms of resource abundance and productivity.

Much more could be done to explore the ways in which the traditional knowledge of the Hui Malama o Mo'omomi might be integrated with contemporary fishery management, and to consider whether such integration is even desirable. Synergy between these knowledge systems could add to biological and ecological understanding of marine resources (Weeks, 1995). Berkes (1999) cautions that the use of indigenous knowledge is political because it threatens to change power relations between indigenous groups and the dominant society. The example of Ho'olehua Hawaiian Homestead may, nevertheless, inspire new approaches and suggest more participatory and locally based alternatives to top-down, centralized resource management. Ruddle's (1994) summary of the contemporary importance of local knowledge can be said to apply to the Ho'olehua community:

- It has practical usefulness for the specific resources and areas harvested.
- It has academic interest to visiting scientists.
- It is an instrument of empowerment for the Hui Malama o Mo'omomi and possibly for other coastal community organizations in the State of Hawai'i.

ACKNOWLEDGEMENTS

THE US Department of Commerce, US Administration for Native Americans and US Department of Education have provided funding support. The Pacific American Foundation provided administrative and management support. The Hawai'i Community Foundation's Natural Resources Conservation Program assisted Hui Malama o Mo'omomi in acquiring a small vessel that has greatly extended the range of community resource monitoring. The Oceanic Institute contributed portions of Dr Alan Friedlander's time. The authors have benefited from the wisdom and advice of Dr Isabella Aiona Abbott.

REFERENCES

ABBOTT, I.A. 1984. *Limu: An ethnobotanical study of some Hawaiian seaweeds.* Lawai, Kauai, Hawaii, Pacific Tropical Botanical Garden, 35 pp.

BERKES, F. 1999. *Sacred ecology: Traditional ecological knowledge and resource management.* Philadelphia, Pa., Taylor and Francis, 209 pp.

BROCK, R.E.; BUCKLEY, R.M.; GRACE, R.A. 1985. An artificial reef enhancement program for nearshore Hawaiian waters. In: F.M. D'Itri (ed.), *Artificial Reefs: Marine and Freshwater Applications.* Chelsea, Mich., Lewis Publications, pp. 317–36.

Cordy, R. 1981. *A study of prehistoric social change: The development of complex societies in the Hawaiian Islands.* Academic Press, New York.

Department of Land and Natural Resources (DLNR) 1996. Status report to the 19th legislature regular session of 1997 on the subsistence fishing pilot demonstration project, Molokai. In response to Act 271, Session Laws of Hawaii 1994. Honolulu, Hawaii, DLNR.

Dyer, C.L.; McGoodwin, J.R. 1994. Introduction. In: C.L. Dyer and J.R. McGoodwin (eds.), *Folk Management in the World's Fisheries.* Niwot, Colo., University Press of Colorado, pp. 1–15.

Edith Kanaka'ole Foundation (EKF) 1995. Draft Ke Kalai Maoli Ola No Kanaloa, Kaho'olawe Cultural Use Plan. Consultants to the Kaho'olawe Island Reserve Commission, 28 pp.

Friedlander A.M.; Brown, E.K.; Jokiel, P.L.; Smith, W.R.; Rodgers, K.S. 2003. Effects of habitat, wave exposure, and marine protected area status on coral reef fish assemblages in the Hawaiian archipelago. *Coral Reefs,* No. 22, pp. 291–305.

Friedlander, A.M.; DeMartini, E.E. 2002. Contrasts in density, size, and biomass of reef fishes between the northwestern and the main Hawaiian Islands: The effects of fishing down apex predators. *Marine Ecology Progress Series,* Vol. 230, pp. 253–64.

Friedlander, A.M.; Parrish, J.D. 1997. Fisheries harvest and standing stock in a Hawaiian Bay. *Fisheries Research,* Vol. 32, No. 1, pp. 33–50.

Friedlander, A.; Poepoe, K.; Helm, K.; Bartram, P.; Maragos, J.; Abbott, I. 2002. Application of Hawaiian traditions to community-based fishery management. *Proceedings of the 9th International. Coral Reef Symposium,* Vol. 2. Jakarta, Ministry of Environment, pp. 813–18.

Friedlander, A.M.; Ziemann, D.A. 2003. Impact of hatchery releases on the recreational fishery for Pacific threadfin in Hawaii. *Fishery Bulletin,* Vol. 101, pp. 32–43.

Governor's Moloka'i Subsistence Taskforce. 1994. Kelson Poepoe and Donna Hanaiki, co-chairs. Department of Land and Natural Resources, State of Hawaii.

Hale, E.K. (undated) *Na Waiwai Hawaii: Straight from the heart, an introduction to Hawaiian cultural values.* Honolulu, Hawaii, Hawaiian Studies Institute, Kamehameha Schools, 12 pp.

Handy, E.; Craighill, S.; Pukui, M.K. 1972. *The Polynesian family system in Ka'u, Hawaii.* Rutland, VT, Charles E. Tuttle, 206 p.

Harman, R.F.; Katekaru, A.Z. 1988. *1987 Hawaii commercial fishing survey.* Hawaii Department of Land and Natural Resource, Division of Aquatic Resources, 71 pp.

Holmes, L. 1996. *Elders' knowledge and the ancestry of experience in Hawaii.* University of Toronto, PhD dissertation.

HOMMON, R.J. 1976. *The formation of primitive states in pre-contact Hawaii.* PhD dissertation. University of Arizona.

HUI MALAMA O MO'OMOMI. 1995. Proposal to designate Mo'omomi community-based subsistence fishing area, northwest coast of Moloka'i. Prepared for the Department of Land and Natural Resources, State of Hawaii, 37 pp. + appendices.

HUNTER, C.L.; EVANS, C.W. 1995. Coral reefs in Kaneohe Bay, Hawaii: Two centuries of western influence and two decades of data. *Bulletin of Marine Science,* Vol. 57, No. 2, pp. 501–15.

JOHANNES, R.E. 1978. Traditional marine conservation methods in Oceania and their demise. *Annual Review of Ecological Systems,* Vol. 9, pp. 349–64.

KANAHELE, G.H.S. 1986. *Ku kanaka, stand tall: A search for Hawaiian values.* Honolulu, Hawaii, University of Hawaii Press, 530 pp.

KIRCH, P.V. 1989. *The Evolution of the Polynesian chiefdoms.* Cambridge, UK, Cambridge University Press, 328 pp.

KUYKENDALL, R.S. 1938. *The Hawaiian kingdom. Vol. 1, Foundation and transformation, 1778–1854.* Honolulu, Hawaii, University of Hawaii Press, 462 pp.

LOWE, M.K. 1996. Protecting the future of small-scale fisheries in an economy dominated by tourism and coastal development, based on the results of the main Hawaiian Islands marine resources investigation (MHI-MRI). In: S. Nagata (ed.), *Ocean resources: Development of marine tourism, fisheries, and coastal management in the Pacific Islands area.* Proceedings of the Sixth Pacific Islands Area Seminar, Tokai University, Honolulu, Hawaii, pp. 137–42.

LOWELL, N. 1971. *Some aspects of the life history and spawning of the moi (Polydactylus sexfilis).* University of Hawaii, MA thesis, 45 pp.

PACIFIC AMERICAN FOUNDATION AND HUI MALAMA O MO'OMOMI. 2001. *Application of Hawaiian traditions to community-based fishery management.* Completion report for two-year demonstration project in and around Mo'omomi Bay, Moloka'i, Hawaii. Prepared for Administration for Native Americans, US Department of Health and Human Services, 53 pp.

PACIFIC AMERICAN FOUNDATION AND HUI MALAMA O MO'OMOMI. 2003. *'Imi'Iki* [Search for knowledge]. Completion Report for Native Hawaiian Education Program, US Department of Education.

PINKERTON, E.W. 1994. Summary and conclusions. In: C.L. Dyer and J.R. McGoodwin, (eds.), *Folk management in the world's fisheries.* Niwot, Colo., University Press of Colorado, pp. 317–37.

PRINCE KUHIO HAWAIIAN CIVIC CLUB. 2001. Ancient Hawaiian moon calendar related to fishing and farming. Honolulu, Hawaii.

PUKUI, M.K. 1983. *Olelo noeau:* Hawaiian proverbs and poetical sayings. *Bishop Museum Special Publication,* No. 71. Honolulu, Hawaii, Bishop Museum Press.

Pukui, M.K.; Haertig, E.W.; Lee, C.A. 1972. *Nana I ke kumu* [Look to the source], Vols. 1. and 2. Honolulu, Hawaii, Hui Hanai, Queen Liliuokalani Children's Center, 333 pp.

Randall, J.E. 1996. *Shore fishes of Hawaii.* Vida, Oreg., NaturalWorld Press, 216 pp.

Ruddle, K. 1994. Local knowledge in the folk management of fisheries and coastal marine environments. In: C.L. Dyer and J.R. McGoodwin (eds.), *Folk management in the world's fisheries.* Niwot, Colo., University Press of Colorado, pp. 161–203.

Santerre, M.J.; Akiyama, G.S.; May, R.C. 1979. Lunar spawning of the threadfin, *Polydactylus sexfilis*, in Hawaii. *Fisheries Bulletin,* Vol. 76, pp. 900–4.

Santerre, M.J.; May, R.C. 1977. Some effects of temperature and salinity on laboratory-reared eggs and larvae of *Polydactylus sexfilis* (Pisces: *Polynemidae*). *Aquaculture,* Vol. 10, pp. 341–51.

Shomura, R. 1987. *Hawaii's marine fishery resources: Yesterday (1900) and today (1986).* Honolulu, US Dept. Commerce/NOAA, National Marine Fisheries Service Southwest Fisheries Science Center Administrative Report H-87-21, 14 pp.

Smith, M.K. 1993. An ecological perspective on inshore fisheries in the main Hawaiian Islands. *Marine Fisheries Review,* Vol. 55, pp. 31–46.

Smith, M.K.; Pai, M. 1992. The *ahupua'a* concept: relearning coastal resource management from ancient Hawaiians. *NAGA,* Vol. 15, No. 2, pp. 11–13.

Starfield, A.M.; Cumming, D.H.M.; Taylor, R.D.; Quadling, M.S. 1993. A frame-based paradigm for dynamic ecosystem models. *Artificial Intelligence Applications,* Vol. 7, No. 2–3, pp. 1–13.

Summers, C.C. 1971. Molokai: A site survey. Pacific Anthropological Records No. 14. Honolulu, Hawaii, Department of Anthropology, Bernice P. Bishop Museum, pp. 40–1.

Titcomb, M. 1972. *Native use of fish in Hawaii.* Honolulu, Hawaii, University of Hawaii Press, 175 pp.

Weeks, P. 1995. Fisher scientists: The reconstruction of scientific discourse. *Human Organization,* Vol. 54, No. 4, pp. 429–36.

PART II: INDIGENOUS AND ARTISANAL FISHERIES

CHAPTER 7   Traditional marine resource management in Vanuatu
Worldviews in transformation

Francis R. Hickey

ABSTRACT

Much of the marine-related traditional knowledge held by fishers in Vanuatu relates to increasing catches while managing resources of cultural, social and subsistence value. Traditional beliefs and practices associated with fisheries and their management follow natural cycles of resource abundance, accessibility and respect for customary rules enshrined in oral traditions. Many management-related rules that control fishers' behaviours are associated with the fabrication and deployment of traditional fishing gear. A number of traditional beliefs, including totemic affiliations and the temporal separation of agricultural and fishing practices, serve to manage marine resources. Spatial-temporal 'refugia' and areas of symbolic significance create extensive networks of protected freshwater, terrestrial and marine areas.

The arrival of Europeans initiated a process of erosion and transformation of traditional cosmologies and practices related to marine resource management. More recently, the forces of development and globalization have emerged to continue this process. The trend from a primarily culturally motivated regime of marine resource management to a more commercially motivated system is apparent, with the implementation and sanctioning of taboos becoming increasingly less reliant on traditional beliefs and practices. This chapter reviews a number of traditional marine resource management beliefs and practices formerly found in Vanuatu, some of which remain extant today, and documents the transformation of these systems in adapting to contemporary circumstances.

INTRODUCTION

Vanuatu is a Y-shaped archipelago, roughly 1,000 km long, located in the western South Pacific (Figure 7.1). Vanuatu means 'Our Land', the name adopted at independence from the joint colonial rule of England and France in 1980 (it was previously known as the New Hebrides). There are eighty-two islands, mostly volcanic in origin, seventy of which are inhabited. Most are surrounded by narrow, highly productive fringing reefs, which are relatively small due to the steep nature of volcanic islands. There are only a limited number of other highly productive aquatic ecosystems such as mangroves, estuaries and lagoons (Cillaurren et al., 2001).

There is great linguistic and cultural diversity among Vanuatu's lush tropical high islands, with approximately 113 Austronesian languages spoken at present (Tryon, 1996). There are a number of Polynesian outliers, including the islands of Futuna, Aniwa, Mele, the villages of Mele on Efate Island, as well as some villages on Emae Island that have Polynesian-influenced populations. Many other islands exhibit varying degrees of Polynesian influences (Spriggs, 1997).

A number of factors affect food security on the islands. Volcanic eruptions, cyclones, tsunamis, earthquakes, landslides, storm surges, floods and droughts all affect crops and reefs. Various mitigating strategies were traditionally employed, including the creation of the complex network of refugia and other fishery management strategies described in this chapter.

To ensure a successful communal harvest of fish, a taboo would be placed on the area to be fished prior to harvesting. Such taboos could forbid anyone to swim or even walk by on the shore. This would both maintain the sanctity of the taboo and make the fish less wary of entering the area.

While the season for a communal harvest was clearly prescribed by local custom (which in turn was based on seasonal resource abundance and/or annual tidal cycles and therefore accessibility), specialists determined the actual timing of the harvest. Optimal tidal conditions, clearly recognized to coincide with lunar phases, were carefully chosen for fish to migrate shoreward over the tidal reef flats from the deeper waters beyond the reef edge. The optimal reef-gleaning season was also determined by annual tidal cycles whereby reefs were fully exposed during daylight hours.

Methods of overcoming food shortages included storing fermented fruits and utilizing alternative foods (like wild yams and cycad fruits) not normally eaten. Another strategy was to create 'giant-clam gardens', with fishers gathering giant clams (Tridacnidae) into discrete areas on reef flats for their exclusive use in times of need. This increased reproductive success by maintaining a close proximity of a breeding population dependant on external fertilization. Thus, it may also be considered a management strategy.

Starting in the early 1800s, diseases introduced by Europeans reduced the population from an estimated half million or more in the pre-contact period to 45,000 by the 1940s (Bedford, 1989). By 1999, the population had rebounded to 189,000 (National Statistics Office, 2000). Christianity, primarily the Presbyterian, Anglican and Catholic faiths, was introduced some 150 years ago and overlaid and influenced island traditions to varying degrees. The diversity of traditions, coupled with extensive migration from inland to coastal areas, the introduction of modern fishing gear and the commercialization of resources, often makes it difficult to

TRADITIONAL MARINE RESOURCE MANAGEMENT IN VANUATU

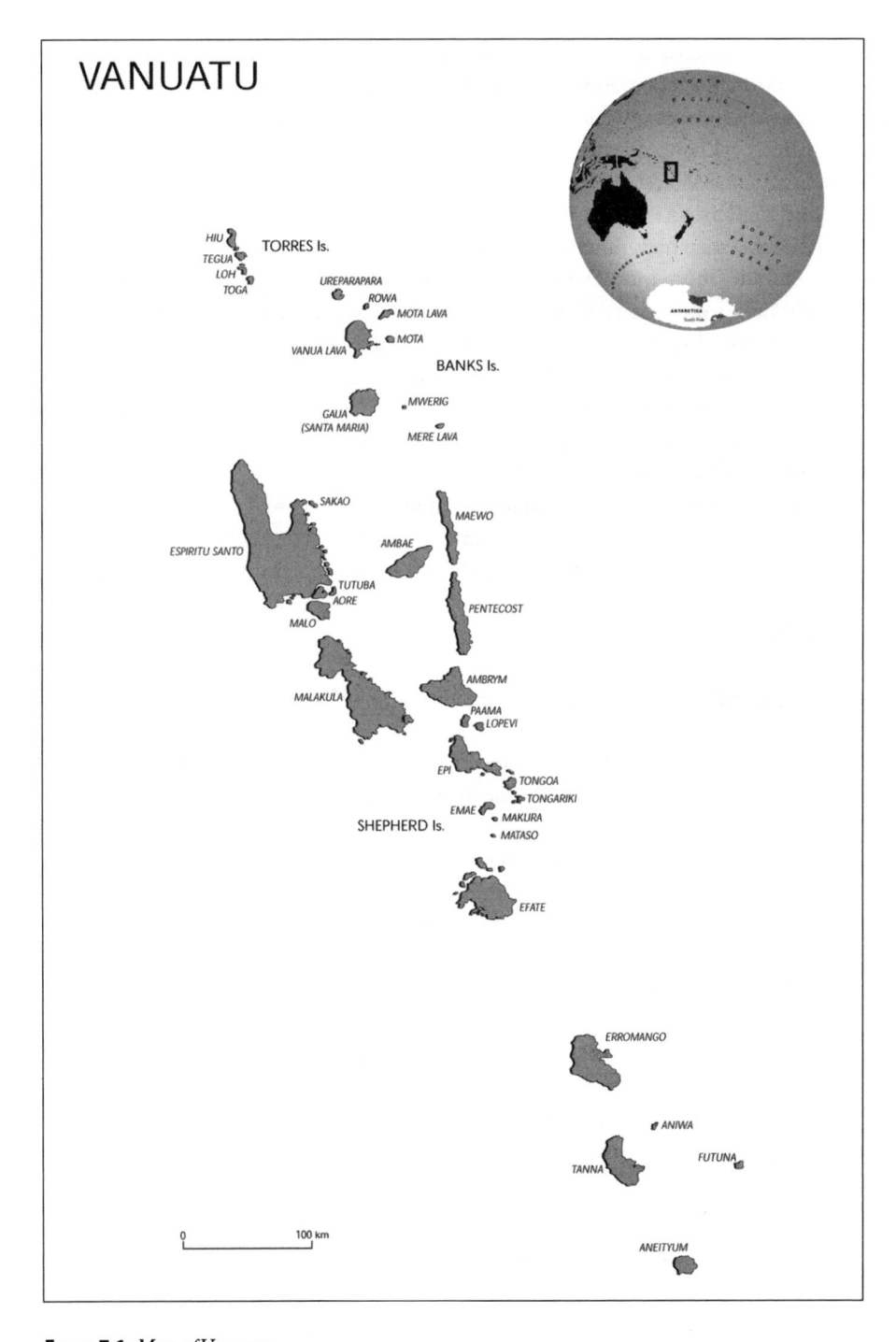

FIGURE 7.1 *Map of Vanuatu.*

generalize about customary fishing beliefs and practices. Clearly though, despite the impacts of the colonial period, Vanuatu has a strong cultural heritage of traditional resource management.[1] While some traditions have been severely undermined and transformed by contact with Europeans, others are still extant and much cultural knowledge remains in living memory.

The Vanuatu Fisheries Department emphasizes the fundamental role of traditional management practices, while also introducing some national regulations; these include measures such as setting size limits for some commercialized invertebrates, protecting turtle nests and eggs, and banning the harvesting of berried spiny lobsters. However, the monitoring and enforcement of these regulations in rural areas remains extremely difficult and cost prohibitive,[2] and the regulations are rarely enforced outside urban areas due to logistical and financial constraints. Their main value is to control the export of commercial fisheries products like trochus from the two urban centres.

The increasing population, concentrated in coastal regions, and the global market pressure for Western-style economic development make the strengthening of traditional management of marine resources critically important.

Traditional fisheries

Traditional fishing methods vary somewhat among islands and cultural groups. Most traditional harvesting, however, is focused on nearshore reefs. Reef gleaning for various fish and shellfish, crab, octopus, sea urchins, spiny lobsters and numerous other invertebrates provides a significant portion of the catch. Women and children's contribution in providing sustenance through reef gleaning is significant and often under-acknowledged. Other methods, including fish poisoning, spearing and shooting fish with bow and arrow from reef edges, hook and lining, netting and fish trapping, and communal harvesting methods like coconut leaf-sweeping, fish driving, and weir fishing are commonly used in different areas. However, hooks and lines were apparently not used everywhere in former times.

There are also fisheries for marine turtles and, in the past, for dugongs (*Dugong dugon*), as well as the annual harvesting of the palolo seaworm (*Polycheata*). In some areas, there are traditional offshore fisheries for deepwater Etiline snappers, breams (Lethrinidae) and groupers (Serranidae), as well as for flying fish, tuna and tuna-like species, although the latter were fished mainly in areas of Polynesian influence. All of these fishing methods are based on extensive traditional ecological knowledge (TEK) of the various resources so as to optimize catches, and encompass a significant corpus

---

1. The term traditional here is meant to refer to practices, beliefs and knowledge considered to have a foundation in the past, particularly before European arrival.
2. The Department of Fisheries no longer employs an Enforcement Officer, due to budget constraints and the observation that it fostered community dependency on the state to solve local resource management problems.

of traditional beliefs and practices, including numerous prohibitions controlling fisher behaviour.

Most of these fishing practices are still in use today. However, their modern counterparts have largely replaced traditional nets and hook and lines. Newer methods, such as the use of snorkelling gear, spearguns, underwater torches and long gillnets have become increasingly common. Outboard-motor boats are now widely used for pelagic and deepwater fishing and inter-island transport. However, the outrigger canoe, with styles varying among islands, still serves most coastal villages for nearshore fishing and transport (Hickey, 1999).

TRADITIONAL RESOURCE MANAGEMENT

Cosmology

Marine resource management was never formerly compartmentalized. The knowledge, beliefs and practices that contributed to the management of resources pervaded all facets of life. Numerous beliefs, practices and protocols governed much of the activities and behaviours, not only of fishers but of all clan members engaged in any of the traditional activities of life. Arts such as weaving baskets and mats, making ceremonial carvings and headdresses, preparing traditional medicines or making canoes all involved following strictly prescribed protocols based on area-specific cosmologies.

These protocols, encoded and enshrined in oral traditions, were often derived from island deities/cultural heroes and sanctioned by the ancestors as 'the way'. 'The way' was orally transmitted to subsequent generations as a holistic approach to life on the islands, including the synergistic management of resources. Following the way specified by island deities led to a fruitful life on the islands, where people were also ritually part of that sanctified world and were symbolically one with the gods and ancestral spirits (Eliade, 1957).

Consequently, it is important to consider the context in which management measures, as well as harvesting techniques, were practised: that is, within the framework of the cosmology or belief system held in ancient times. Life in the islands of Vanuatu had, and still largely has, an inherent sanctity stemming from the animistic cosmological belief that 'all things have a spirit' and that all things and events, are inherently connected through this spiritual medium (Lucas, this volume). By extension of that belief, people may hope to influence natural forces otherwise beyond their control by the use of sanctified rituals, and so enhance survival in the face of various threats to food security.

Many practices stemming from this underlying cosmology are highly ritualized and are undertaken by specialists who received this knowledge from elders. Most involve the use of sacred stones and leaves, often used synergistically, along with other rituals whose secrets are closely guarded. In many cases, the power of the omnipresent ancestral spirits that live 'on the other side' is evoked to achieve the desired influence over nature and worldly events. Communication with these spirits was often ritually enhanced through the use of a narcotic drink prepared from kava (*Piper methysticum*) (Lebot et al., 1992).

Evoking the power of the ancestral spirits or island deities to increase resource abundance was an integral part of any traditional taboo on resources. Reef taboos were never formerly static, but were always accompanied by ritualized practices underpinned by cosmological beliefs invoking ancestors/deities to increase resources. Today the abandonment of these practices is sometimes cited as the reason for resource depletion.

Environmental knowledge and indicators

Tidal patterns are important, since much of the nearshore marine resources come from reef gleaning or communal fishing activities requiring good low tides. The overall maximum tidal amplitude in Vanuatu is roughly 1.5 m. The annual lows, often zero or negative tides, occur during the austral winter months of June and July. Extreme low water of the winter spring tides occurs at midday, and the reefs are optimally exposed for gleaning during daylight hours around the new and full moons. During the summer months, extreme low water occurs at midnight during the new and full moons, but these low tides never get as low as those of the winter months.

Winter is also the optimal season for communal fish harvesting methods such as the traditional leaf sweep, fish drives, use of fish weirs and fish poisons in tide pools. These techniques all depend on spring tides that are high enough to allow fish to come inshore over reef flats to feed, yet low enough to strand fish in pools behind natural barriers or those created by these methods. The winter season also coincides with the period when most nearshore resources are believed to be not spawning.

The flowering of *waelken* (*Miscanthus sp.*) in late summer is the environmental cue that indicates the seasonal spawning of many reef fish in the southern islands of Vanuatu (where seasonal temperature variations are more pronounced). As their flowers swell up in maturation, so do fish swell up with their eggs. When the flowers burst in late February and March, the fish release their eggs. The post-spawning period is considered good for hook and line fishing, as reef fish feed hungrily to recharge fat reserves depleted through egg production, and are quick to take bait. Other species, such as Siganids, have spawning peaks earlier in summer, from October to January, indicated by the onset of flowering of the coastal tree *Excoecaria agallocha*.

In the colder winter months, when the reefs are optimally exposed for harvesting through gleaning, many of the nearshore resources charge their fat reserves, and are thus preferred for their taste. A commonly cited environmental cue is the flowering of *narara (Erythrina variegata)*, when reef fish, crab and lobster are said to be full of fat. This time is also known to be best for catching octopus, which are said to come out of their holes to see the bright red *narara* flowers. The appearance of the constellation Pleiades on the western horizon after sunset (in April) is used on most islands to herald the New Yam season, and the return of the seasonal low tides.

Some islands (like Ambrym) cite Orion's Belt, which follows about a month later in the same position, as symbolic of a fisher returning from the exposed reefs with baskets of shellfish to be prepared with yam puddings in this season, while for other islands it is symbolic of people returning from the gardens with baskets full of yams. The annual cycle of tidal patterns that determine optimal reef gleaning and communal fishing methods is thus synchronized with the annual agricultural cycle of yam production.[3]

Communal fish harvests in the winter months capable of producing large catches were thus part of an annual cycle of ceremonial feasts or ritualized exchanges with inland villages in return for resources such as yams or fruits from island interiors. These practices served to redistribute a seasonal abundance of resources between different island biomes, while strengthening alliances and maintaining peaceful trade relations between kinship groups.

Seasonal cycles

Seasonal abundance – the occurrence of spawning migrations and aggregations – in addition to harvesting method constraints such as tidal patterns also determined which species were targeted at particular times. Nearly every marine resource had a discrete season when it was targeted, often encoded by an environmental cue such as a flowering plant (Narcisse, this volume) or the appearance of a star or other environmental cue, as in the examples above. This is expressed by village elders who say, 'Everything has its own time.' Many species would, for example, primarily be targeted when their fat reserves were at a maximum and they thus tasted better. As this was also generally the time preceding spawning peaks, this cycle minimized fishing pressure during reproductive periods.

Nearly every month of the year, different resources would be considered ripe or become abundant; an example is the annual seasonal appearance of the marine Palolo worm. In later months, sharks would come ashore to bear live young. Shark

---

3. Communal fishing methods such as the coconut leaf sweep are still ritually practised on some islands, but the introduction of long monofilament gillnets now allows for large catches with much less communal effort. However, the optimal tidal pattern for large catches of reef fish on diurnal migrations from reef flats to drop-offs using modern nets largely remains as described for communal harvesting methods.

pups remain inshore for some time and are easily harvested using hand spears. In the early summer months, with the return of the rains, terrestrial crabs (*Cardisoma* spp.) would intensify their foraging activity near the coast, fattening up prior to aggregating to specific coastal areas to release their eggs in the sea, making them easily harvestable. Summer months would also see flying fish (Exocoetidae) and their predators, the tunas, come inshore where they could be harvested. Later, the pelagic scads (*Selar* spp.) and small mackerel (*Rastrelliger* and *Scombrus* spp.) would mature, forming large schools in inshore lagoons and bays. Sardines (*Sardinella* spp.) would also form large shoals inshore where they could be easily harvested, and rabbit fish (Siganidae) would migrate to a known location to aggregate for spawning purposes. All of these smaller species would, in turn, attract larger predators like jacks and trevallys (Carangidae) and barracuda (Sphyraenidae) that could also be harvested.

This annual cycle of different resources becoming plentiful at different times clearly indicated the season to target them. In this way, fishing pressure on various resources was concentrated on a given resource for only a brief period of the year on a rotational basis. Even if some were harvested during a spawning migration or aggregation, there would be only minimal pressure on the population in the remainder of the year.

TRADITIONAL MANAGEMENT IN TRANSITION

Except for a few high-value benthic species, tropical, small-scale, multi-species fisheries in places like Vanuatu are prohibitively expensive and notoriously difficult to manage using Western models that require extensive data collection (Johannes, 1998a; Johannes and Neis, this volume). Johannes (1998b) suggested that unrealistic emphasis on quantitative management ideals like optimum or maximum sustainable yields for these fisheries could justifiably give way to a new paradigm that he called 'data-less marine resource management', emphasizing that it is not management in the absence of information. The use of reproductive and lifecycle information, coupled with TEK of resources and traditional marine tenure, is invaluable for achieving management objectives.

Traditional marine tenure

The fundamental management strategy for nearshore reefs is based on traditional marine tenure (TMT) and the accompanying traditional beliefs and practices that prohibit or restrict the harvest and consumption of marine resources. The principle underlying TMT is the ability of families, clans, chiefs and/or communities to claim exclusive rights to fishing areas, exclude outsiders, and regulate activities in these areas. The benefits of their restraint may then be realized at a later date, thus providing the motivation to protect resources. The systems of TMT in the Pacific have been relatively well documented by Johannes and MacFarlane (1991), Ruddle et al. (1992) and Hviding (1996), among others. The well-entrenched heritage of TMT is legally recognized in Vanuatu and continues to provide an ideal framework for a decentralized system of village-based management of marine resources.

TMT effectively devolves responsibility for marine resource management to traditional leaders, communities, clans or families; that is, to those with the most intimate knowledge of the resources and the greatest motivation to manage well. Devolution of management responsibility is possible, as the government of Vanuatu recognizes and supports TMT in the constitution of the republic, and traditional leaders and resource custodians continue to see the introduction of village-based prohibitions as their traditional right and responsibility.[4]

However, a growing concern is that contemporary taboos tend to be less firmly rooted in tradition, and consequently command less respect than traditional ones. The ancient traditional taboos, as outlined below, were associated with elaborate traditional practices and ritual underpinned by traditional cosmologies and primarily sanctioned through supernatural forces (Satria, this volume). Contemporary taboos tend to be less ritualized and therefore less steeped in tradition, with a consequent decrease in reliance on supernatural sanctioning. The influence of the church, particularly the notion that traditional beliefs are 'heathen and uncivilized', makes this ritualization and reliance on supernatural sanctioning less acceptable in some communities.

Bans and taboos

The earliest transformations of traditional marine management systems stemmed from the introduction in the late 1800s of an export trade in nearshore resources for dried sea cucumbers *(Holothuroidea)* and later included trochus *(Trochus niloticus)* and green snail *(Turbo marmoratus)*. Traditionally derived taboos began to be regularly placed on these resources in response to commercial pressure. This trend in protecting commercially harvested resources through the use of taboos has continued into the present, as more resources are targeted for commercial purposes for export to urban centres and overseas.

Contemporary village-based management prohibitions, often referred to by villagers today as 'bans' to distinguish them from ancient traditional taboos, continue to be locally monitored and enforced by village leaders. These bans are enforced through the traditional institution of the village court which, although not legally

[4]. More recently, the forces of development and globalization are increasingly eroding government recognition and support for TMT as new legislation is introduced affecting land tenure and more land titles (which affect reef accessibility) are transferred to foreigners for development purposes.

recognized, continues to adjudicate effectively on most offences occurring in rural areas, as it has for centuries.

Fishers recognize that fishes often retreat into areas under taboo when being pursued. Taboo areas, even when they are not particularly large, if widely distributed act as a mosaic of refugia or sanctuaries for mobile marine life. Turtles are found to become accustomed to the presence of divers observing them in areas where hunting is taboo for sufficiently long periods (personal observation). Dugongs (*Dugong dugon*), protected from hunting for some years now, have even been tamed to swim with humans, and along with unwary turtles are used to attract tourists to generate revenue. The knowledge that fishes and other marine life increase in abundance and lose their wariness in areas under taboo is put to good use by the regular placement of closures.

TRADITIONAL MARINE RESOURCE MANAGEMENT PRACTICES OF VANUATU

The categories of traditional marine resource management practices vary significantly among cultural groups in the archipelago, reflecting their cultural diversity. Some of these practices are extant today, while others survive only through oral histories. Many of the marine management strategies described are also applied to freshwater and terrestrial resources. Reefs were viewed as extensions to the land, and their custodianship was generally, but not always, the responsibility of the adjacent landholder (Satria, this volume). The information below summarizes research conducted by the author in collaboration with the Vanuatu Department of Fisheries and Vanuatu Cultural Centre over the preceding decade. Virtually all of the traditional strategies described below have direct parallels in modern resource management strategies founded on Western scientific principles, but long predate them. The Western classification terms are used below to highlight these parallels.

Privileged-user rights

The right of reef custodians to control and restrict fishing and other activities is fundamental to the principle of TMT and is reflected in the modern management strategy of 'limited access'. Under TMT, there may also be complex tiers of user rights for different groups, based on historical connections with reef areas. Groups arriving later in an area may be accepted, but only with secondary rights, by the original founding group who retain primary rights. Also, neighbouring, often inland, groups may retain tenure over original canoe-landing sites or may in the past have bartered for user rights to defined reef areas, and these rights may remain in effect for ensuing generations. Respect for TMT is said to have been universally very high in the past, and transgressions would be dealt with harshly, as well as through supernatural intervention. While remaining flexible through consultation, the system thus controlled and limited fishing effort within nearshore areas.

Species-specific prohibitions

In most areas it was taboo to eat turtle or turtle eggs if one planned to go to the yam garden in the next couple of days. It is said that to do so would result in one's yams being stunted like the fins or eggs of a turtle. In some areas, equivalent prohibitions applied to octopus, lobsters, giant clams, certain species of fish and other foods, including oily fruits and nuts. These prohibitions also applied to working in water taro (*Colocasia esculenta*) and other types of gardens, such as those for bananas. In some areas, it was taboo to go to gardens if one's leg had made contact with the sea, as doing so would risk damaging crops.

Food prohibitions could sometimes be overcome by making a small 'devil's garden', distant from the main one, after consuming one of the prohibited foods.[5] The yams from the devil's garden would then be offered to the spirit responsible for stunting the yams, and the yams from the main garden would thus be spared.

Various informants suggest that these prohibitions may relate to the negative effect of introducing oily substances from turtles and other foods to gardens, as these could attract wild pigs or insects to food gardens. Making small devil's gardens prior to working in the main yam gardens would result in most of the oils being deposited in the devil's garden, though it would require additional time and energy. Salt also harms many garden crops, which may explain the negative association between seawater exposure and gardening. These effects apparently led to a temporal separation of gardening and fishing activities throughout many areas of Vanuatu and are elaborated on below in the section on seasonal closures.

Another species-specific prohibition is the practice of showing respect to the memory of recently deceased clan members by tabooing their favourite food or the last food they ate. For example, a certain type of fish, spiny lobster, octopus, a type of shellfish or fruit may be tabooed in honour of a deceased clan member for a year or more. The time period is generally commensurate with the respect shown to their memory. This relieves fishing pressure on that resource within the clan's area during that period.

Additional species-specific restrictions include prohibitions against children and pregnant women eating turtles, as this was found to result in children developing sores. In some areas, those with asthma were also prohibited from eating turtle as it aggravated their condition. In other areas it was taboo for young girls to consume

---

5. The term 'devil', introduced by early missionaries, is commonly used today to refer to various manifestations from the spirit world.

giant clams (Tridacnidae) until after their first menstruation, while young boys were forbidden to consume many species of large angelfish (*Pomacanthus* spp.) until they were circumcised. These prohibitions stemmed from area-specific cosmological beliefs and resulted in reduced fishing pressure on these resources.

In some areas it was also considered taboo to collect small gastropods (e.g. *Turbo* spp.) that had no encrusting growth on them, to avoid taking immature ones.

Seasonal closures

During the summer months when yams were being cultivated and many reef resources were restricted by gardening taboos, as well as by the tidal cycles outlined above, a range of fruit and nut trees ripened to provide alternative sources of nutrition. When new yam gardens were prepared, there was much labour involved in clearing garden plots and planting the tubers. With the coming of the spring rains, weeding and training the vines required frequent trips to the garden. The production of yams was a central aspect of food production and featured prominently in the customs of most areas of Vanuatu. Cultivating yams was thus treated as a serious endeavour. Given the importance of agricultural production in Vanuatu (Weightman, 1989), it is apparent that gardening restrictions which limited fishing activities also served to reduce fishing pressure on nearshore reefs during the months of yam production. As noted above, the tides of this season are also less suitable for reef-gleaning activities, and thus reef gleaning and communal harvesting methods were further separated temporally from gardening activities by tidal cycles.

The yam production period, starting as early as August/September, and extending until April/May, covers the entire hot season. This period includes the turtle-nesting season, the time when turtles are most vulnerable to exploitation by humans. It is also thought to be the season when many nearshore reef species are at their spawning peaks. Fishing prohibitions during the main agricultural season thus have highly significant management value because they reduce fishing pressure during peak reproductive periods. The yam production season also encompasses the period when trade winds collapse and winds become light and variable. Johannes (1978) highlights the advantage for fish of spawning during periods when prevailing winds and currents are at their weakest, which will reduce the transport of larvae far from their point of origin.

In areas like Futuna, Tanna, Paama and Ambrym Islands, most nearshore resources are considered to be taboo from the time the yam gardens are started until the New Yam Ceremony some six months later. This would ensure a good harvest of seafood for New Yam celebrations as well as during the subsequent months of the yam harvest season. As this closure coincides with the time when most nearshore fish and invertebrates are believed to be at their spawning peaks, the annual half-year taboo also serves to protect resources during this vulnerable period.

An additional incentive to limit consumption of nearshore reef fish during summer months is that they are more frequently found to be ciguatoxic during this period when new coral growth is observed to be highest. Ciguatera fish poisoning in humans results from eating fish that has consumed toxic species of dinoflagellate.

Food avoidance

Many cultural groups in Vanuatu are associated with different totems that include specific types of fish, octopus, giant clams, turtles, sharks and moray eels as well as various terrestrial resources. The practice of not consuming ones' ancestral totems out of respect and reverence for them serves as a management strategy by reducing or controlling fishing pressure on those resources. In some areas, highly controlled, ritualized harvests of totemic species are undertaken for exchange to other areas, thereby limiting fishing pressure.

Protected areas

In virtually all parts of Vanuatu there were formerly numerous coastal protected areas, known locally as 'taboo places', that had spiritual significance for which people had the greatest reverence and which they would respectfully avoid. These taboo places were also common in terrestrial and freshwater areas and were often associated with areas of high biodiversity. Examples of such areas include burial places, and places where spirits resided or island deities were based. Volcanic lakes on Ambae and Gaua Islands are two such large inland freshwater areas high in biodiversity. Many rivers and creeks were also considered taboo areas and were thus protected, as they were considered to be paths of spirits travelling between the sea and inland areas.

These permanently taboo areas were commonly found along coasts, as well as at offshore islands and reefs. Access to them was restricted or controlled at all times, unlike spatial-temporal refugia. The taboo areas formed a network of marine and terrestrial protected areas whose management benefits included the production of larger, more abundant marine organisms that export larvae (and marine plant propagules) as well as spillover effects. By protecting a number of different habitat types colonized by species unique to them, taboos also preserved and enhanced biodiversity.

These areas were, by their very nature, protected and sanctioned by the spirits residing there. Compliance was thus very high, as the enforcement of these areas was endogenously based on the belief system of supernatural sanctioning. This is unlike the Western counterpart of marine protected areas that is sometimes promoted in Vanuatu, which relies increasingly on state sanctioning. While many taboo places are no longer respected, primarily due to the influence of Christianity, Western education and development pressure, others continue to protect resources in areas where respect for them remains.

Behavioural prohibitions

The numerous customary protocols associated with the fabrication and deployment of traditional fishing gear and techniques were integral to the traditional resource management regime. Once certain fisheries were initiated with the fabrication of, for example, a spiny lobster trap, a fisher's behaviour became regulated by protocols associated with that fishery. Taboos could vary among cultural groups and depended on the fishery type.

A widely known example of a behavioural prohibition is the requirement for sexual abstinence before engaging in fishing activities as well as during the fabrication of fishing devices. This reduces fishing pressure within a clan's area while providing additional benefits relating to birth control. Other examples of behavioural prohibitions that reduce fishing pressure follow:

- In some areas, it is taboo to swim or remain on the shore during sunset, as certain spirits are known to be active then. As spawning aggregations are known to occur at sunset, this prohibition protects them (Johannes, 1978).
- Fishers cannot be seen departing, or at least others must not know they are joining a fishing expedition, as this brings 'bad luck' and so the trip may be aborted. Also, it is taboo to call out or make noise when embarking on a fishing trip.
- If a visitor arrives and spends the night, then it is taboo to go fishing the next day.
- It is taboo to eat certain foods or drink kava or go to certain places when one is involved in the fabrication or deployment of certain fishing devices.
- Pregnant or menstruating women, and men with pregnant wives, are automatically excluded from most fishing activities. This taboo relates to the belief that the spirit of an unborn child has a negative effect on fishing activity.

Thus, there is an extensive and complex web of taboos associated with fishing that act synergistically with other traditional management measures to reduce fishing effort. A fisher who is unable to respect any behavioural taboos must refrain from fishing for the following day or two, thereby reducing fishing effort in a given area. As there are ways to find out who has not followed the rules, this puts shame on any offenders, affecting their reputations as fishers, and is thus avoided.

Spatial-temporal refugia

Some of the cultural practices that created spatial-temporal refugia throughout Vanuatu are outlined below. These refugia allowed for an increase in abundance and diversity and provided spillover benefits, decreasing the wariness of resources while also potentially protecting spawning activities and increasing biodiversity. Events associated with such spatial-temporal refugia are described below. These areas would be open to fishing once the taboo has been removed so as to make use of resource abundance in line with Pacific peoples' strong subsistence, social and cultural links with resources.

> Death of a traditional leader
In some areas, such as the Banks Islands, the death of a traditional leader ('chief' or highly ranked member of a hierarchical society) would be honoured by tabooing the reef of the leader's clan. Depending on the degree of respect, this total closure could last for many years. This taboo is associated with the enactment of many rituals. When the reef is re-opened, a final communal feast is held to honour the deceased, using in part the resources harvested from the closed area.

> Death of any clan member
The death of any individual of a clan – man, woman or child – may mean that the clan's area of reef is put under taboo, or closed to all harvesting for one to three years, as is the case on northern Epi.

> Grade-taking
In areas of north and central Vanuatu, the all-important rituals of grade-taking by men, and in some areas women, as part of ascending a social and spiritual hierarchy (Layard, 1942; Bonnemaison, 1996) are accompanied by taboos being put on terrestrial, freshwater or reef resources from one to four years, and often for as long as six years in the case of marine taboos. These practices include multiple pig killings, kava drinking, dancing, singing, feasting and other rituals.

> Passing on of a hereditary chief's title
In the Shepherd Islands of Central Vanuatu, the practice of hereditary chiefs' passing title to their progeny is associated with a reef taboo. The taboo duration may be the time taken for a young pig to develop a full circle tusk, some six to seven years. Offerings to ancestors are also traditionally made to evoke their assistance in monitoring and enforcement. The tusked pig will be sacrificed to remove the taboo, and the marine resources harvested from the taboo area are used as part of the ordination feast.

> Yam season
As outlined above, in some areas of Vanuatu, most nearshore reef resources are annually closed to harvesting during the summer months from around the time of yam planting until the New Yam celebrations approximately six months later. In other areas the taboos are species specific, but nearly all areas protect turtles. These

agricultural taboos are now less commonly respected, while some areas continue to limit fishing during this period because of the management value of doing so during spawning periods. It is also generally acknowledged that yams produced these days are much smaller than in former times due to the loss of respect for traditional practices.

> *Circumcision*
Cleansing rites that are part of circumcision rituals are sometimes associated with reef taboos, which are generally of a short duration, sometimes one month. These short closures are particularly effective in conserving resources if their timing coincides with spawning migrations or aggregations.

> *In preparation for specific feasts or other traditions*
In most areas, specific feasts or other traditions, such as the harvest and exchange of marine resources to inland villages, are preceded by a reef taboo. Ritual specialists then evoke the ancestors to increase resources and ensure a good catch. Inland villages would later reciprocate with foodstuffs from their areas. This highly ritualized system of exchange effectively controlled fishing pressure on resources both spatially and temporally while redistributing resources during periods of seasonal abundance and strengthening trade and peaceful relations. These taboos are still found in some areas and are sometimes integrated into Christian rituals.

Marine resource management through a mosaic of spatial-temporal refugia

The variety of traditional area closures ensured a number of areas were closed at any one time. When visiting north Pentecost in northern central Vanuatu in 1998, the author was informed of a total of eleven marine closures associated with grade-taking ceremonies. These closures formed a mosaic of spatial-temporal refugia across the top end of this relatively small area, protecting various marine habitats. In 2005, the number of areas closed due to these rituals had increased due to the strong adherence to traditional practices in this area.

Consequences of violating traditional taboos

The consequences of taboo violation included supernatural retribution from island deities and ancestors. Traditional leaders, under the auspices of the village court, also imposed fines of pigs, kava, woven mats and other traditional wealth as an additional deterrent and means of removing the 'wrong' in the eyes of ancestors and other clan members. Typically, ancestors would punish transgressors, or their family members, by making them ill, sometimes terminally so. Some were capable of assuming various forms, including sharks or barracuda that could directly enforce a marine taboo. Practices to ensure the participation of ancestors in enforcement included placing culturally specific taboo leaves in the area to symbolically monitor and enforce the taboo (Johannes and Hickey, 2004). Communication with the spirit world was often enhanced by ritualized kava drinking.

Ancestral icons may also be concealed in the area to symbolically invoke their participation. The killing of pigs at the initiation of the taboo also serves as a symbolic sacrifice to ancestors for their part in monitoring and enforcing the taboo. The killing of another pig is thus required within some cultural groups to remove the taboo and make it safe to harvest again in the area. In other areas, additional culturally significant gifts (such as pigs, kava, yams or white fowl) were offered, sometimes set adrift on a raft, to ensure the ancestor's role in monitoring and enforcing the taboo. This system of sanctioning was considered highly effective in the past, and remains so in numerous areas where traditional belief systems remain strong.

DISCUSSION

TRADITIONAL leaders and reef custodians in Vanuatu increasingly use TMT to put resources, fishing areas or fishing methods under taboo for varying periods of time (Johannes, 1998a; Hickey and Johannes, 2002; Johannes and Hickey, 2004). Some of these taboos are extant versions of traditional practices. Many taboos imposed today, however, are contemporary expressions of earlier ones. The Vanuatu Cultural Centre, Fisheries Department and Environment Unit have supported these traditionally derived contemporary taboos (as well as ancient traditional practices) along with a programme of cooperative management. An awareness programme was initiated by the Fisheries Department Research Section in the early 1990s, initially targeting trochus resources (Amos, 1993). It provides relevant biological knowledge to communities for use in conjunction with traditional knowledge in the management of nearshore reef resources. These cooperative management efforts quickly spread to include other commercially important resources as well as those important for subsistence. This programme was later introduced to the Department's Extension Services by providing appropriate training to rural-based Extension Officers.

Part of this process included raising awareness among rural communities of Department regulations about size limits and other state prohibitions on resources. Once villagers were aware of the regulations and understood the rationale behind them, they generally adopted the regulations as part of their village-based management regime (Johannes, 1998a; Meeuwig et al., this volume). Chiefs and villagers then took over monitoring and informally enforcing these regulations on behalf of the government.

The knowledge gained of the management value of traditional practices, including area and species closures and other prohibitions on harvesting marine resources, has thus been adapted and applied in the expression of contemporary taboos. If the taboo was of sufficient duration, resources were observed to become larger, more abundant and less wary, leading to increased catches after the taboo is lifted. Also, taboos placed during periods of spawning activities assisted recruitment processes.

Another aspect influencing respect for taboos is that the benefits of traditional taboos were generally distributed to the entire community through communal feasts and distribution of resources. Today, however, individual reef owners often expect to prosper from the sale of trochus and other resources. Thus, there is often less incentive for the entire community to respect the taboo. In former times, the paramount traditional leader of an area would have the right, through consensus, to put large reef areas controlled through different clans' tenure under taboo for traditional purposes. In this way, management of large areas was harmonized for communal benefit. Some reef custodians recognize the relationship between respect for taboos and communal benefit sharing, and allow reef access to the entire community to promote widespread respect for taboos placed on individual clans' reefs.

Many communities recognize that the decrease in respect for contemporary taboos is exacerbated by a general decline in respect for traditional authority among youth influenced by Western education or individualistic ideals learned in urban centres. Disputes over land or reef tenure as well as village leadership are also found to weaken respect for village-based taboos (Hickey and Johannes, 2002; Johannes and Hickey, 2004). In response to these factors, some communities attempt to strengthen and revitalize traditional beliefs about resource management by including more traditional practices in their implementation. Others, in areas where traditional beliefs are more influenced by introduced cosmologies, choose also to integrate Christian beliefs and practices in implementation, and this is often effective in assisting with management; still others look increasingly toward the state for assistance in sanctioning village-based taboos.

The trend towards greater state sanctioning, as well as Western notions of conservation that ignore traditional links to resources, have been assisted by foreign donors, regional and volunteer organizations and NGOs. These often have limited appreciation of traditional resource management systems, and are primarily familiar with Western models from their own countries. Some outside groups take village-based taboos and repackage them in Western forms such as 'conservation areas and MPAs', thus tacitly undermining and eroding traditional systems. Government policy makers and bureaucrats, often educated in industrialized countries and increasingly isolated from rural communities, sometimes acquiesce in these introductions, following the locally entrenched notion that 'the West knows best'.

This recent trend towards Western repackaging of traditional practices is of concern as it implies that Western models are superior, when in fact parallels to Western science-based resource management strategies already exist in Vanuatu's traditional systems, as documented above. Reliance on state sanctioning of village-based resource management also has significant limitations, as government capacity to perform this role is severely limited in an archipelago with so many coastal villages. It also raises community expectations and fosters a mentality of depending on the state to solve rural community problems, which rarely respond well to legislation. The application of Western law in villages is seen as divisive, with a win/lose outcome that further erodes the social cohesion necessary for cooperation in village-based management (Johannes and Hickey, 2004). Recognizing and supporting the existing strong cultural heritage of decentralized village-based resource management and strengthening efforts to adapt it to contemporary needs would be much more effective. This could be facilitated by continuing to build the capacity of traditional leaders and communities, and promoting consultation with all stakeholder groups to increase understanding, consensus and compliance prior to implementation of taboos.

In cases where enforcement remains problematic, legal recognition of traditional village court systems, where village-based transgressions including those related to resource management are adjudicated, would be an effective means to assist with enforcement. Legislation to empower traditional leaders and communities to manage resources under traditional tenure would be more effective and economical than creating a parallel system that transfers that power to the state and serves to undermine traditional authority. Fa'asili and Kelokolo (1999) report that legal empowerment of the Chief's Council in Samoa has been successful in supporting the community-based management of resources while reinforcing traditional authority.

CONCLUSION

Vanuatu has a strong cultural heritage of traditional resource management, and a well-entrenched and legally recognized system of TMT to draw upon in continuing to adapt its indigenous system of resource management to contemporary needs. Many elements of traditional systems and authority remain extant and are well respected by the majority of the rural population. Some community elders still retain a large corpus of TEK that is useful for resource management, but their numbers are now dwindling rapidly. Culturally appropriate awareness and education programmes directed towards traditional leaders, fishers and communities have been shown to be highly effective in facilitating the adaptation of traditional systems to contemporary needs (Amos, 1993; Johannes, 1998a; Hickey and Johannes, 2002; Johannes and Hickey, 2004).

Further support is needed for efforts to continue to develop the capacity of traditional leaders and communities in the decentralized management of resources under their tenure through the strengthening of traditional leadership, village-based consultation and consensus-building mechanisms and providing culturally appropriate awareness concerning resources. It is particularly important that young people be made more aware of the practical value and modern-day relevance of traditional management systems and TEK held by elders. This can best be achieved by the active involvement of elders in curriculum development and formal education, and the inclusion of traditional activities as part of the school curriculum (Haggan et al., 2002; Poepoe et al., this volume) as well as through informal educational channels. This will enhance the intergenerational transfer of knowledge (Narcisse, this volume) and promote greater appreciation, pride, self-reliance and transmission of such knowledge and practices. Mobilizing TEK for use in resource management helps to empower communities with the use of their own knowledge while fostering a stronger sense of ownership of a resource management initiative. Both factors have been observed to enhance the sustainability of village-based resource management initiatives in Vanuatu.

Government policy makers, foreign donors, NGOs, and volunteer and regional organizations working in the environment sector could all benefit from greater awareness of the value and efficacy of supporting and building on traditional management systems and the risks of blindly introducing foreign conservation methods originating in industrialized countries without TMT. The trend of some organizations towards devolving resource management authority from TMT to the state risks raising community expectations while fostering dependency on governments that lack the capacity (both human and financial) to deliver (Johannes and Neis, this volume). The repackaging of existing village-based taboos as Western conservation models, often for the edification of tourists and development agencies, tacitly denigrates and further erodes Vanuatu's remaining traditional resource management practices. Attempts at introducing the Western ethos of conservation also ignore the strong social and cultural links of Pacific Islanders with their resources and the efficacy of existing traditional systems of management. Eroding traditional rights of communities of autonomy over their land and resources is not likely to solve problems in Melanesia, but is sure to create them.

ACKNOWLEDGEMENTS

The author gratefully acknowledges the cooperation and support of friends and colleagues at the Vanuatu Cultural Centre, Department of Fisheries and the Environment Unit, particularly members of the Cultural Centre fieldworker network who provided endless hours of discussion and demonstrations on their traditions.

May your knowledge and wisdom continue to guide your children's children's futures. The author would like to emphasize that the traditional knowledge that forms the basis of this paper remains the intellectual property of the respective providers of that knowledge. I would also like to humbly dedicate this chapter to the memory of Bob Johannes, who took the time to provide much encouragement and inspiration.

REFERENCES

AMOS, M. 1993. Traditionally based management systems in Vanuatu. SPC Traditional Marine Resource Management and Knowledge Information Bulletin, Vol. 2, pp. 14–17. Available online at www.spc.int/coastfish/News/Trad/trad.htm (last accessed on 27 June 2005).

BEDFORD, R. (ed.) 1989. *The population of Vanuatu*. Population Monograph 2. Noumea, New Caledonia, South Pacific Commission, 126 pp.

BONNEMAISON, J. 1996. Graded societies and societies based on title: Forms and rites of traditional power in Vanuatu. In: J. Bonnemaison, C. Kaufmann, K. Huffman, K.; D. Tryon (eds.), *Arts of Vanuatu*. NSW, Australia, Crawford House Publishing, pp. 200–16.

CILLAURREN, E.; DAVID, G.; GRANDPERRIN, R. 2001. *Coastal fisheries atlas of Vanuatu: A 10-year development assessment*. Paris, IRD editions, 256 pp.

ELIADE, M. 1957. *The sacred and the profane: The nature of religion*. New York and London, Harcourt Brace, 256 pp.

FA'ASILI, U.; KELOKOLO, I. 1999. The use of village by-laws in marine conservation and fisheries management. *SPC Traditional Marine Resource Management and Knowledge Information Bulletin*, Vol. 11, pp. 7–10. Available online at www.spc.int/coastfish/News/Trad/trad.htm (last accessed on 3 September 2004).

HAGGAN, N.; BRIGNALL, C.; PEACOCK, B.; DANIEL, R. (eds.) 2002. *Education for aboriginal fisheries science and ecosystem management*. Fisheries Centre Research Reports Vol. 10, No. 6. Vancouver, University of British Columbia, 49 pp. Available online at http://www.fisheries.ubc.ca/publications/reports/report10_6.php (last accessed on 27 June 2005).

HICKEY, F.R. 1999. Canoes of Vanuatu. In: H. Deiter Bader and P. McCurdy (eds.), *Proceedings of the Waka Symposium*. Auckland New Zealand Maritime Museum/ Te Huiteananui-a-Tangaroa, pp. 217–58.

HICKEY, F.R.; JOHANNES, R.E. 2002. Recent evolution of village-based marine resource management in Vanuatu. *SPC Traditional Marine Resource Management and Knowledge Information Bulletin*, No. 14, pp. 8–21. Available online at www.spc.int/coastfish/News/Trad/trad.htm (last accessed on 30 June 2005).

HVIDING, E. 1996. *Guardians of Morovo Lagoon*. Hawaii, University of Hawaii Press, 473 pp.

JOHANNES, R.E. 1978. Reproductive strategies of coastal marine fishes in the tropics. *Environmental Biology of Fishes,* Vol. 3, No. 1, pp. 65–84.

——. 1998a. Government supported, village-based management of marine resources in Vanuatu. *Ocean and Coastal Management Journal,* Vol. 40, pp. 165–86.

——. 1998b. The case for data-less marine resources management: Examples from tropical nearshore fisheries. *Trends in Ecology and Evolution,* Vol. 13, No. 6, pp. 243–6.

JOHANNES, R.E.; HICKEY, F.R. 2004. *Evolution of village-based marine resource management in Vanuatu between 1993 and 2001.* Coastal region and small island papers 15. Paris, UNESCO, 48 pp. Available online at http://www.unesco.org/csi/wise/indigenous/vanuatu1.htm (last accessed on 27 June 2005).

JOHANNES R.E.; MACFARLANE, J.W. 1991. Traditional fishing in the Torres Strait Islands. Hobart, Tasmania CSIRO, 22268 pp.

LAYARD, J. 1942. *Stone men of Malekula.* London, Chattus and Windus, 816 pp.

LEBOT, V.; MERLIN, M.; LINDSTROM, L. 1992. *Kava: The Pacific drug.* New Haven Yale University Press, 255 pp.

NATIONAL STATISTICS OFFICE 2000. *The 1999 Vanuatu National Population and Housing Census, Main Report.* Vanuatu Government of the Republic of Vanuatu, 231 pp.

RUDDLE, K.; HVIDING, E.; JOHANNES, R.E. 1992. Marine resource management in the context of customary tenure. *Marine Resource Economics,* Vol. 7, pp. 249–73.

SPRIGGS, M. 1997. *The island Melanesians,* Oxford, UK, Blackwell, 326 pp.

TRYON, D. 1996. Dialect chaining and the use of geographical space. In: J. Bonnemaison, C. Kaufmann, K. Huffman,; D. Tryon, D. (eds.), *Arts of Vanuatu.* Bathurst, Australia, Crawford House Publishing, pp. 170–81.

WEIGHTMAN, B. 1989. *Agriculture in Vanuatu, a historical review.* Portsmouth, UK, Grosvenor Press, 320 pp.

CHAPTER 8   Tropical fish aggregations
in an indigenous environment
in northern Australia
Successful outcomes through collaborative research
Michael J. Phelan

ABSTRACT

To examine the status of an indigenous subsistence fishery where the main catch was the sciaenid *Protonibea diacanthus*, fishers' knowledge was coupled with catch data and biological information to highlight trends in the harvest. The fishers of the Injinoo Aboriginal Community presented information of a rapid change in the landings of aggregations of this fish. Whereas the fishery was historically based on sexually mature fish, recent catches were composed almost exclusively of juveniles.

In response to the research findings of the project described in this chapter, the Injinoo community undertook a self-imposed two-year ban on the taking of *P. diacanthus*. With much consultation, this initiative has developed into a regional agreement with comprehensive support across the region. The outcome appears unique among Australian fisheries, being the only example in the modern context in which indigenous communities have initiated a long-term ban on harvest of a fish species.

INTRODUCTION

AUSTRALIAN fisheries have a history of being studied and managed in cooperation with commercial and recreational fishing groups, a process which has until recently largely neglected the values intrinsic to indigenous subsistence fishers. Fortunately, an expanding realm of cooperative arrangements is starting to ensure that contemporary environmental management is more inclusive of indigenous interests. However, while the value of a holistic approach to resource management has increasingly been recognized (see reviews by White et al., 1994; Alder, 1996), the value of collaborative partnerships in fisheries research has often been ignored.

INDIGENOUS AND ARTISANAL FISHERIES

This chapter focuses on an ongoing research project that commenced in 1998 and is centred on the close involvement of the Injinoo Aboriginal Community. Injinoo is situated 40 km from the northernmost point of the Australian continent (see Figure 8.1). The community lies over 1,000 km from the nearest city (Cairns), though there are a number of small indigenous communities nearby. Injinoo shares the Northern Peninsula Area (NPA) of Cape York (north of the twelfth parallel) with the indigenous communities of Umagico, New Mapoon, Bamaga and Seisia.

The Injinoo Community was founded in the early 1900s when the remnants of the five clans – the Atambaya, Wuthathi, Yadhaigana, Gudang and Anggamuthi – whose customary lands occupy the NPA came together on their own accord to settle at the former meeting place (Sharp, 1992). The current population of Injinoo is under 400 people, of whom typically fewer than ten people living in the community are of non-indigenous descent. The overall population of the region of the NPA is approaching 2,600.

One-third of Australia's 400,000 indigenous people currently live within 20 km of the coastline (Australian Bureau of Statistics, 2001). Many of the coastal clans of Australia's Aboriginal Nations identify themselves as 'saltwater people', and their traditional estates typically extend beyond the coastal zone and into the seas. In

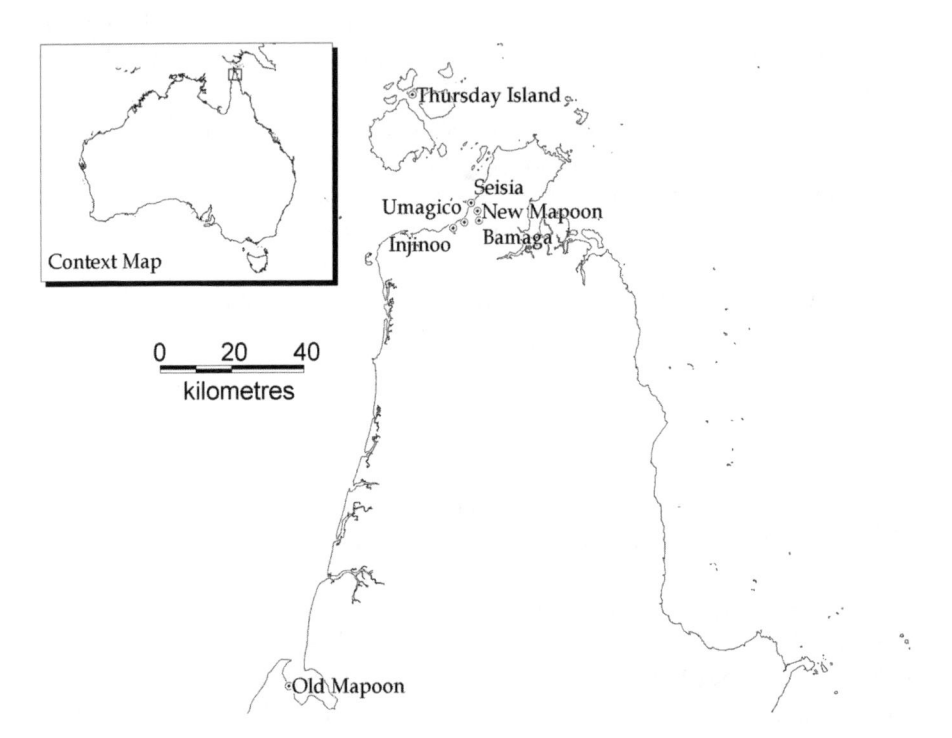

FIGURE 8.1 *Map of the north end of Cape York Peninsula.*

general, these coastal people view the sea as a cultural landscape: an extension of the land with similar inherent responsibilities (Tanna, 1996; Hickey, this volume; Poepoe et al., this volume; Satria, this volume). The term, 'saltwater people' is also used by First Nations on the west coast of Canada who have a similar concept of territory (Elliott, 1990).

In Australia, recognition of the importance of 'land' to Aboriginal cultures is a relatively new concept. Little more than a decade has passed since the Australian High Court decision (*Mabo v Queensland*, 1992) which acknowledged the Native Title rights of indigenous Australians. Acknowledgement of the legal validity of Aboriginal 'sea estates' is even more recent, having been recognized only in 1999 (*Mary Yarmirr and Others v the Northern Territory of Australia and Others*, 1999).

Following these High Court decisions, the inherent rights and responsibilities of indigenous people under customary law are now recognized under Australia's common law (Crisp and Talbot, 1999). As a consequence, the rights of indigenous peoples to their traditional marine resources, and their role in the management of their customary estates, are of increasing relevance to coastal and marine resource administration in Australia.

In all there are about 100 coastal communities, mostly in northern Australia, occupying land under some form of Aboriginal or islander leasehold or title (Smyth, 1993). Indigenous members of the northern communities are largely exempt from commonwealth and state regulations in regard to the harvest of marine resources for traditional or subsistence use. However, there is presently a deficiency of datasets on the contribution of indigenous fishing to the total annual catch.

While indigenous people currently comprise less than 2 per cent of Australia's population, this figure is nonetheless growing rapidly. The number of people who identify themselves as Indigenous Australians increased by 45 per cent in the ten years between 1991 and 2001 (Australian Bureau of Statistics, 2001). The indigenous population of Australia has a much younger age profile than the non-indigenous population, a reflection in part of higher fertility rates. Exemplifying this, at Injinoo 49 per cent of the population is less than 18 years old (Queensland Aboriginal Coordinating Council, 2000).

It follows, then, that in the immediate future there is the potential for a rapid increase in fishing pressure on local resources. This appears more evident when one considers the improving economic situation among many of Australia's indigenous communities. At Injinoo, for example, the community vessel register indicates that there were only five powered vessels in the community in 1990, while ten years later the number had increased to forty-two (at that time there were forty-eight houses in the community).

The research project discussed in this chapter focused on the biology and harvest of Australia's largest tropical sciaenid, *P. diacanthus* (see Figure 8.2). Sciaenids are widely distributed in tropical and subtropical waters (Trewavas, 1977; Sasaki, 2001).

They commonly dominate epibenthic fish assemblages of nearshore waters of both regions (Rhodes, 1998; Blaber et al., 1990), and often form the basis of commercial, recreational and indigenous fisheries (Mohan, 1991; Apparao et al., 1992; De Bruin et al., 1994; Williams, 1997).

Figure 8.2 *An adult* Protonibea diacanthus.

In the NPA, *P. diacanthus* are harvested primarily by local indigenous fishers, but also by local and tourist recreational fishers. *P. diacanthus* may reach sizes greater than 150 cm in length and can exceed 45 kg in weight (Grant, 1999). Aggregations of the fish form annually in the inshore waters of the NPA, and have also been reported at a number of northern Australian locations extending from Central Queensland (Bowtell, 1995) to northern Western Australia (Newman, 1995).

Aggregations of fish, be they formed for the purpose of feeding, spawning or migrations, are renowned as vulnerable fishery targets (Turnbull and Samoilys, 1997; Johannes et al., 1999). The largest member of the family Sciaenidae, *Totoaba macdonaldi*, is a relevant example. *T. macdonaldi* is considered to be critically endangered and is now listed on the IUCN Red List of Threatened Animals as a consequence of overfishing during the annual aggregation period (True et al., 1997).

There is extensive global evidence that target fishing of aggregations can rapidly undermine fishery production. Chronic effects of aggregation fishing include the truncation of size and age structure (Beets and Friedlander, 1992), deterioration of the stock's reproductive capacity (Eklund et al., 2000) and altered genetic composition (Smith et al., 1991). Acute effects include the total loss of aggregations (Sadovy, 1994).

Exemplifying the vulnerability of *P. diacanthus*, the once flourishing commercial fishery along the north-west coast of India has become 'non-existent' (James, 1992). Anecdotal evidence suggests intensive fishing has also severely affected several annual aggregations along the east coast of Queensland (Bowtell, 1998). Yet despite this, there is a dearth of information on the species and the demands made upon those stocks by the various fishery sectors. In particular, the biological purpose and importance of these aggregations has yet to be demonstrated.

This research project was initiated to address concerns of the traditional owners about an apparent increase of local fishing pressure that was targeting *P. diacanthus*. The project is managed by Balkanu Cape York Development Corporation (an indigenous organization representing the people of Cape York) in partnership with the Queensland Department of Primary Industry (the State's fisheries research agency). Funding for the project has been provided by the Fisheries Research Development Corporation (FRDC).

METHODS

INSOFAR as possible, community members were involved in the design and implementation of the project at all stages through to the interpretation of results and the development of the project outcomes. The act of working together on all aspects of the project greatly enhanced the communities' trust in the project staff, and hence their willingness to participate and own the research. At all stages the project adhered to the protocols established by Balkanu for conducting research in indigenous environments (Balkanu Cape York Development Corporation, n.d.). These were designed to allow individual communities to participate in scientific research in a manner that community members deemed culturally appropriate.

Prior to the commencement of sampling, project staff had made a substantial commitment of time in meeting the community residents and promoting the objectives of the project. A key challenge in persuading fishers of the importance of research is that they may perceive that the advancement of such knowledge may 'backfire' and ultimately diminish their rights. From feedback generated at later stages, this initial consultation was deemed critical to the success of the project. Although seemingly unproductive in terms of annotated results, this period was essential to gaining the understanding of community members.

By living in the community, the project's biologist was able to achieve a stronger personal and working relationship with its residents. Over time, this generated a much greater understanding. This was not only from the perspective of the communities' understanding of the research and results, but also of the researchers' understanding of the community. Adopting this approach helps build a bridge between the skills of biologists and those necessary to understand the ethnobiological information critical for the integration of contemporary and traditional practices (Johannes, 1981; Küyük et al. and others, this volume).

As there were no existing catch data on the fishery, oral accounts of traditional owners and long-term residents were collated in order to develop an historical profile. It is in such circumstances, when data is otherwise not available, that oral history proves an invaluable tool in establishing a retrospective analysis of resource use. Nonetheless, the acquisition of such information requires the same critical scrutiny that is applied to any other data set. Only data verified by more than one source were used in this investigation.

The continual involvement of the local fishers has proven integral to the success of the project. Not only did they provide the information on the spatial and temporal scale of the fishery which was necessary to developing the monitoring and sampling programmes, but they have also assisted in the collection of the catch data and biological samples. The assistance provided by the fishers has ensured the project's resources have been used in an efficient manner.

Between January 1999 and December 2000, the harvest of 4,031 individual *P. diacanthus* was recorded, of which almost 15 per cent were measured and weighed. Many of these were sampled to examine the reproductive (n = 270), diet (n = 270) and genetic (n = 109) traits of the fish. A further 114 fish were tagged and released. Standard techniques were adopted in the processing of the samples and the analysis of the data. The project did however adopt the novel DNA fingerprinting technique called amplified fragment length polymorphisms (AFLP).

In order to maintain the high level of community ownership of the project, the community was consulted throughout all stages, and the results were released as soon as they became final. The project staff liaised directly with the community's Council Clerk, who also represented the community by serving on the project's steering committee. This was comprised of elected representatives of each of the stakeholder groups linked to the fishery, and guided the progress and direction of the project. The committee also ensured the transmission of the results to all stakeholder groups.

RESULTS

VERY detailed information was gained by the oral accounts of the traditional owners and the long-term residents of the area. For example, elders recall that the first *P. diacanthus* caught by the traditional owners was taken at Muttee Head in 1946 by the late Daniel Ropeyarn. The collection of accounts has provided a record of the fishery since its inception, and presented evidence of changes in the demographics of the fishery, the harvest and stock condition.

The accounts present anecdotal evidence of:

- the gradual expansion of the fishery in terms of the number of fishers targeting *P. diacanthus* and the number of aggregations being fished
- a corresponding reduction over a five-decade period in the size of the fish caught and the duration of time the fish were present at the aggregation sites
- changes in the fishing practices and locality of fishing as aggregations become less productive.

The indigenous fishers displayed a fine understanding of the spatial and temporal attributes of the aggregating behaviour of the fish stock, with the seasonal, lunar and tidal patterns of the fish movements being common knowledge. Catch and effort data compiled in 1999 and 2000 suggest that this understanding may facilitate the harvest of this species. Almost all of the recorded catch in 1999 (3.9 tonnes) and 2000 (4.5 tonnes) occurred during the aggregation period described by fishers. Recorded catches typically exceeded fifty fish per boat, with catches of over a hundred fish not uncommon. Catch per unit effort (CPUE) ranged up to 225 kg per hour/boat.

When the historical accounts are compared with data gained from the 1999 and 2000 catches, it appears that the state of the fishery has declined within the last decade. Oral records suggest that specimens close to their maximum size (>150 cm) were caught up until 1994, whereas 1999 catch records reveal the fishery was dominated by 3-year-old fish (75–80 cm), while the 2000 catch was dominated by 2-year-olds (60–65 cm) (see Figure 8.3).

The results of the biological sampling support the notion that the fishery has undergone a recent change. The indigenous people of the Injinoo, who eat the eggs of *P. diacanthus*, state that ripe eggs were readily available during previous aggregations. However, sexually mature fish comprised less than 1 per cent of the catch examined in a sampling programme biased towards the largest individuals available. Among the fish showing evidence of sexual maturity, the development of their gonads coincided

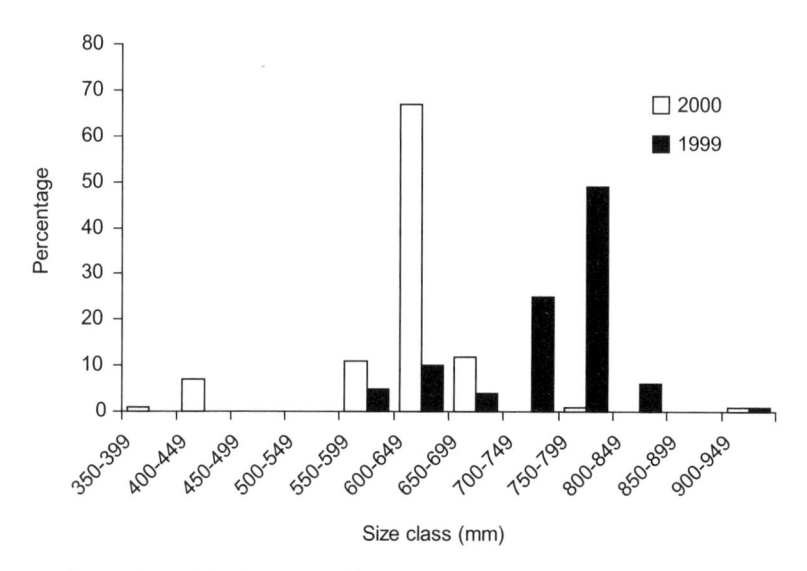

FIGURE 8.3 *Composition of the size classes of P. diacanthus harvested in the Northern Peninsula Area in 1999 and 2000.*

with the aggregation season. However, no hydrated or spent gonads were observed, so the exact timing and location of spawning could not be determined.

Examination of the gonad samples also revealed a decrease in the age of first maturity among females. In a previous study in the adjacent Gulf of Carpentaria, first maturity in females was observed at 4 years of age (McPherson, 1997). No 4-year-old fish were present in the 1999 catch, and among the 3-year-olds, no evidence of sexual development was observed in that year. However, in the following year, even though the 3-year-old stock was greatly reduced, some of these displayed evidence of sexual maturity.

Food items observed in the analysis of the diet of the fish included a variety of teleosts and invertebrates. The range of animal taxa represented in the prey items support the description of an 'opportunistic predator' presented by Rao (1963). The data presented no evidence to support the notion that the seasonal migration of *P. diacanthus* was related to the increased availability of prey items in the inshore waters, as is suggested by Thomas and Kunja (1981).

The tag and release component of the present project provided limited data on the movement patterns of *P. diacanthus* in the NPA waters. Tag returns prove that some of the fish remain at, or return to, the aggregation site at least into the following day. The recaptures also revealed the movement of an individual fish between two distinct aggregation sites. This was supported by DNA fingerprinting. No significant genetic variation was found among fish sampled from the adjacent aggregation sites. As both sites are fished, their participation in multiple aggregations may increase their susceptibility to capture.

THE MANAGEMENT OUTCOMES

FOLLOWING the analysis of the data gathered by this project, the results of the study were presented to the NPA community. In addition to the final report, a series of meetings was held. In accordance with the local custom, the meetings commenced with the elders of the traditional owner groups, and were expanded under their instruction to the wider community. At this stage, Queensland Fisheries Service visited the NPA to further their knowledge of the local fish stock and the public's expectations.

In response to the research findings, the Injinoo Land Trust (representing the traditional landowner groups of the Anggamuthi, Atambaya, Gudang and Yadhaykenu Aboriginal people), in cooperation with the Injinoo Community Council, self-imposed upon its people a two-year ban on the taking of *P. diacanthus*. The area of closure incorporates the inshore waters of the NPA north of the southern boundaries of Crab Island on the west coast and Albany Island on the east coast (see Figure 8.4). The aim of the ban was to allow local stocks to reach a mature size in order to improve the reproductive capacity.

Following extensive consultation led by Injinoo residents, this community initiative developed into a regional agreement with comprehensive support across the NPA. Representing each of the communities of the NPA, the community councils of Umagico, Bamaga, New Mapoon and Seisia undertook to participate in the two-year prohibition on the take of *P. diacanthus*. Furthermore, Torres Shire and the Kaurareg Nation of the adjacent Torres Strait region were also signatories to the ban. Proprietors and operators of all tourist accommodation and fishing charter boats operating in the NPA also pledged their full cooperation with the initiative.

Parties to the regional agreement recognized that the two-year closure might not provide adequate time for the complete recovery of the proportion of the adult fish in the population. All parties requested extension of the research so that decision makers would have sufficient information to review management needs at the conclusion of the two-year period. Further funding was provided by FRDC to continue the stock assessment. The current project built on the length-frequency and reproductive studies undertaken during the initial phase of the project, and drew on the experience and assistance of local fishers.

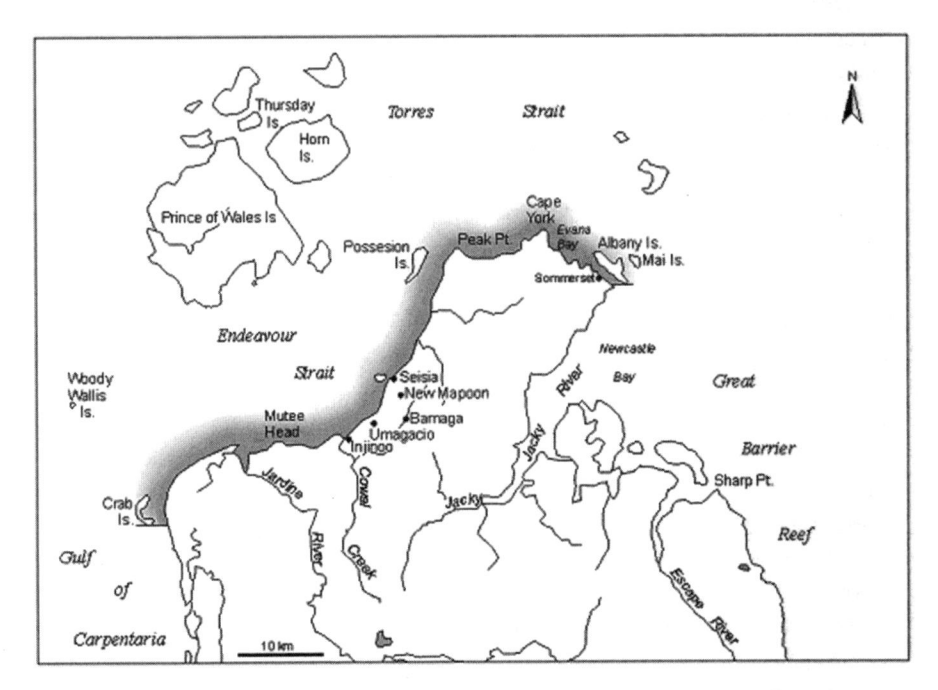

FIGURE 8.4 *Location of the area within the Northern Peninsula Area closed to the harvest of* P. diacanthus *under the regional agreement.*

CONCLUSIONS

THE practice of aggregating is one of the most widespread behavioural mechanisms used by marine fish to reduce natural predation (Die and Ellis, 1999). Yet it is this behaviour that often promotes increased fishing effort and higher catches, as concentrations of fish are both easier to detect and more efficient to harvest (Turnbull and Samoilys, 1997). Information gained by this project suggests that the widespread knowledge of the spatial and temporal attributes of the aggregating behaviour of *P. diacanthus* may facilitate the increased catch of this species.

While the geographical setting of the project was within Queensland, the results should have widespread application to fisheries for *P. diacanthus* and other aggregating fish species. The benefits of collaborative research have been advocated as they apply in Australia, yet they apply almost universally to indigenous fisheries in developed nations, and perhaps also to other fishery sectors. Undoubtedly, the continual cooperation of indigenous fishers greatly enhanced the outcomes of the project.

The comprehensive consultation process conducted throughout the lifetime of the project ensured that the implications of the research were recognized by both management authorities and the communities of the NPA. The implementation of the community-developed two-year closure exceeded all expectations and sets a precedent for similar work. It is believed that this outcome was a product of the communities' understanding, participation in and ownership of the research process.

The outcomes are unique among Australian fisheries, being the only example we know of in the modern context in which indigenous communities have initiated a long-term ban on harvest of a fish species. This outcome serves to demonstrate that, provided with the appropriate opportunities and information; mutually beneficial relationships may be developed between indigenous communities and scientific researchers. This partnership between government institutions and resource users may serve to further enhance prospects of achieving sustainable use of resources.

REFERENCES

ALDER, J. 1996. Have tropical marine protected areas worked? An initial analysis of their success. *Coastal Management*, Vol. 24, pp. 97–114.

APPARAO, T.; LAL MOHAN, R.S.; CHAKRABORTY, S.K.; SRIRAMACHANDRA MURTY, K.V.; SOMASHEKHARAN NAIR, K.V.; VIVEKANANDAN, E.; RAJE, S.G. 1992. Stock assessment of sciaenid resources of India. *Indian Journal of Fisheries*, Vol. 39, pp. 85–103.

AUSTRALIAN BUREAU OF STATISTICS. 1901. Population distribution, Aboriginal and Torres Strait Islander Australians, 100 pp. Available online at http://www.

abs.gov.au/Ausstats/abs per cent40.nsf/ca79f63026ec2e9cca256886001514d7/ 7243c2de7b43332aca2568a900143cbd!OpenDocument (last accessed on 24 June 2005).

BALKANU CAPE YORK DEVELOPMENT CORPORATION. n.d. Draft statement of principles regarding biophysical research in the Aboriginal lands, islands and waters of Cape York Peninsula. Available online at http://www.balkanu.com.au/business/ policy-dev/draftstatement.htm (last accessed on 24 June 2005).

BEETS, J.; FRIEDLANDER, A. 1992. Stock analysis and management strategies for red hind, *Epinephelus striatus* in U.S. Virgin Islands. *Proceedings of the Gulf Caribbean Fisheries Institute,* Vol. 42, pp. 66–80.

BLABER, S.J.M.; BREWER, D.T.; SALINI, J.P.; KERR, J. 1990. Biomasses, catch rates and abundances of demersal fishes, particularly predators of prawns, in a tropical bay in the Gulf of Carpentaria, Australia. *Marine Biology,* Vol. 107, pp. 397–408.

BOWTELL, B. 1995. Heed jewfish warnings. *Fish and Boat,* July, Townsville, Australia, p. 4.

——. 1998. Huge schools of black jew about. *Fish and Boat,* July, Townsville, Australia, p. 11.

CRISP, R.; TALBOT, L. 1999. Indigenous social profile report for SEQ RFA. Final report for FAIRA Aboriginal Corporation, Gurnag Land Council Aboriginal Corporation, and Goolburri Aboriginal Corporation Land Council, 86 pp.

DE BRUIN, G.H.P.; RUSSELL, B.C.; BOGUSCH, A. 1994. *The marine fishery resources of Sri Lanka.* Rome, Food and Agriculture Organization of the United Nations, 315 pp.

DIE, D.J.; ELLIS, N. 1999. Aggregation dynamics in penaeid prawn fisheries: Banana prawns (*Penaeus merguiensis*) in the Australian northern prawn fishery. *Marine and Freshwater Research,* Vol. 50, pp. 667–75.

EKLUND, A.; MCCLELLAN, D.B.; HARPER, D.E. 1900. Black grouper aggregations in relation to protected areas within the Florida Keys National Marine Sanctuary. *Bulletin of Marine Science,* Vol. 66, pp. 721–8.

ELLIOTT, D. 1990. Saltwater people: A resource book for the Saanich native studies program (J. Poth, ed.). Saanichton, BC, School District 63.

GRANT, E.M. 1999. *Grant's guide to fish.* E.M. Queensland, Australia, Grant PTY, 880 pp.

JAMES, P.S.B.R. 1992. Endangered, vulnerable and rare marine fishes and animals. In: P.V. Dehadrai, P. Das; S.R. Verma (eds.), *Threatened fishes of India.* Proceedings of the National Seminar on Endangered Fishes of India. Allahabad, National Bureau of Fish Genetic Resources, pp. 271–95.

JOHANNES, R.E. 1981. *Words of the lagoon: Fishing and marine lore in the Palau district of Micronesia.* Berkeley, Calif., University of California Press, 245 pp.

JOHANNES, R.E.; SQUIRE, L.; GRANAM, T.; SADOVY, Y.; RENGUUL, H. 1999. *Spawning aggregations of groupers (Serranidae) in Palau*. Marine Conservation Research Series Publication 1. The Nature Conservation Agency, 144 pp.

MABO AND OTHERS V. QUEENSLAND (No. 2). 1992. 175 CLR 1 F.C. 92/014. Available online at http://austlii.edu.au/~graham/Slides/London/top20.html (last accessed on 27 June 2005).

MCPHERSON, G.R. 1997. Reproductive biology of five target fish species in the Gulf of Carpentaria inshore gillnet fishery. In: R.N. Garret (ed.), *Biology and harvest of tropical fishes in the Queensland Gulf of Carpentaria gillnet fishery*. Cairns, Department of Primary Industries, pp. 87–103.

MARY YARMIRR AND OTHERS V THE NORTHERN TERRITORY OF AUSTRALIA AND OTHERS [1998] 771 FCA. Available online at http://138.25.65.50/au/cases/cth/federal_ct/1998/771.html (last accessed on 24 June 2005).

MOHAN, L.R.S. 1991. A review of the sciaenid fishery resources of the Indian Ocean. *Journal of the Marine Biology Association of India*, Vol. 33, pp. 134–45.

NEWMAN, J. 1995. Kimberly chaos. *Fishing World* (June). NSW, Australia, Yaffa Publishing Group, 19 pp.

QUEENSLAND ABORIGINAL COORDINATING COUNCIL. 2000. Community house crowding survey 1998/2000. Final Report for the Queensland Aboriginal Coordinating Council. Cairns, Queensland Aboriginal Coordinating Council.

RAO, K.V. 1963. Some aspects on the biology of 'ghol', *Pseudosciaena diacanthus*. *Indian Journal of Fisheries*, Vol. 10, pp. 413–59.

RHODES, K.L. 1998. Seasonal trends in epi-benthic fish assemblages in the nearshore waters of the Western Yellow Sea, Qingdao, People's Republic of China. *Estuarine, Coastal and Shelf Science*, Vol. 4, pp. 629–43.

SADOVY, Y. 1994. Grouper stocks of the Western Central Atlantic: The need for management and management needs. *Proceedings of the 43rd Gulf and Carribean Fisheries Institute*, pp. 43–63.

SASAKI, K. 1901. Sciaenidae: Croakers (drums). FAO species identification guide for fishery purposes. In: K.E. Carpenter; V.H. Niem (eds.), *The living marine resources of the Western Central Pacific*. Vol. 5. Bony fishes part 3 (Menidae to Pomacentridae). Rome, FAO, Rome, pp. 3117–74.

SHARP, N. 1992. *Footprints along the Cape York sandbeaches*. Canberra, Aboriginal Studies Press, 251 pp.

SMITH, P.J.; FRANCIS, R.I.C.C.; MCVEAGH, M. 1991. Loss of genetic diversity due to fishing pressure. *Fisheries Research*, Vol. 10, pp. 309–16.

SMYTH, D.M. 1993. *A voice in all places: Aboriginal and Torres Strait Islander interests in Australia's coastal zone*. Consultant's report to the Coastal Zone Inquiry. Canberra, Resource Assessment Commission, 258 pp.

TANNA, A. 1996. Traditional Priorities for traditional country. Aboriginal perspective's in the Cape York country. *Proceedings of the Australian Coastal Management Conference*, Glenelg, South Australia. pp. 50–56.

THOMAS, P.A.; KUNJA, M.M. 1981. On the unusual catch of ghol *Pseudosciaena diacanthus* off Goa. *Indian Journal of Fisheries*, No. 25, pp. 266–8.

TREWAVAS, E. 1977. The sciaenid fishes (croaker and drums) of the Indo-West-Pacific. *Transactions of the Zoological Society of London*, Vol. 33, Part 4.

TRUE, C. D.; LOERA, A. S.; CASTRO, N.C. 1997. Acquisition of broodstock of *Totoaba macdonaldi*: Field handling, decompression, and prophylaxis of an endangered species. *The Progressive Fish-Culturist*, Vol. 59, pp. 246–8.

TURNBULL, C.T.; SAMOILYS, M.A. 1997. Effectiveness of the spawning closures in managing the line fishery on the Great Barrier Reef. Report to the Reef Fish Management Advisory Committee of the Queensland Fisheries Management Authority. Townsville, Queensland Department of Primary Industries, 28 pp.

WHITE, A.T.; HALE, L.Z.; RENARD, V.; CORTESI, L. 1994. The need for community-based coral reefs management. In: A.T. White; L.Z. Hale; V. Renard; L. Cortesi (eds.), *The need for collaborative and community-based management of coral reefs: Lessons from experience*. Connecticut, Kumarian Press.

WILLIAMS, L.E. 1997. Gulf set net fishery: Monitoring and assessment. In: R.N. Garret (ed.), *Biology and harvest of tropical fishes in the Queensland Gulf of Carpentaria gillnet fishery*. Cairns, Department of Primary Industries, pp. 5–28.

CHAPTER 9   Sustaining a small-boat fishery
*Recent developments and future prospects
for Torres Strait Islanders, Northern Australia*
Monica E. Mulrennan

ABSTRACT

THIS chapter examines strategies for indigenous control of marine territories and fisheries management. Focusing on the tropical marine environment of Torres Strait, northern Australia, it outlines the historical, cultural and ecological basis for demands by the Torres Strait Islanders for a more prominent role in the contemporary management and development of their small-boat fishery. The recent success of Islanders in forging new relationships with government policy makers, managers and scientific researchers is attributed to the various political and legal actions deployed by Islanders to control and manage fisheries resources within the region. It also reflects the wider international and domestic interest being given to small-boat fisheries approaches – particularly those based on traditional resource management systems and local knowledge inputs – as a more sustainable alternative to large-scale industrial fisheries and conventional fisheries management approaches. It remains to be seen whether true power-sharing between Islanders and state authorities, and substantive changes in policy necessary for the survival of small-boat fisheries, can be achieved.

INTRODUCTION

IN recent decades the combined pressures of increased demand for fish, the increased power of fishing fleets, and the limitations of conventional fisheries management approaches have led to overexploitation of many of the world's fisheries, and the marginalization of small-scale (traditional, artisanal, subsistence) fishers through government policies that favour large-scale, industrial fisheries (Berkes, 2003). Growing competition, and associated tension and conflict, have been an inevitable outcome. The 'David and Goliath' scale of the imbalance of power that generates and marks such conflicts is being moderated in a growing number of cases of support for alternative approaches to fisheries management (Dyer and

McGoodwin, 1994; Pomeroy, 1995; Wingard, 2000). A feature of these approaches is the special recognition being given to the potential of more decentralized community-based management systems associated with small-scale fisheries. Indeed this is the only fishery sub-sector specifically mentioned in the UN Code of Conduct for Responsible Fisheries (FAO, 1995; Mathew, 2001).

This chapter examines the case of the Torres Strait islands of northern Australia (see Figure 9.1), where indigenous fishers and hunters retain strong ties to the sea and assert their ongoing right to control and manage the marine resources of their region despite a long history of dispossession and disruption (Beckett, 1987). While the issue of local fisheries knowledge is not explicitly addressed, the chapter speaks directly about the creation of conditions under which Islander knowledge has found a voice in the management and control of local fisheries. Frustration with their marginal status in decision-making, and increasing concerns about the sustainability of government-sponsored approaches to fisheries management, have led Islanders in recent years to adopt a variety of strategies to enhance their role in the region. Focusing on the promotion of an Islander small-boat fishery, Islanders argue that the maintenance and protection of their small-scale fishing activities – based on local knowledge and traditional resource management practices – offer a more biologically sustainable and economically viable alternative to conventional industrial fishing, while also furthering social equity and cultural continuity in the Torres Strait. The success of Islanders in deploying various political and legal strategies to support their case reflects an era in which increasing networks of support and resources are available to help indigenous and local fishing communities assert their rights and claim a voice in decisions relevant to the management and protection of their homelands and seas.

The case study presented here is based on literature survey, archival research, and fieldwork conducted in the eastern part of the Torres Strait over several years. The field research involved participant observation, attendance at fisheries meetings, and semi-structured interviews with elders and active Islander fishers involved in local and regional efforts to bring about regime changes. Inputs from the Islander leadership and fisheries management officials, based on interviews with the author, press statements and other documents, are also included.

CULTURAL AND HISTORICAL CONTEXT OF MARINE RESOURCE MANAGEMENT IN TORRES STRAIT

Recent demands by Torres Strait Islanders for a primary role in the control and management of fishery resources within their traditional marine territories are culturally and historically rooted. Questions of whether or not an indigenous society conserved or managed certain resources, and under what circumstances, are

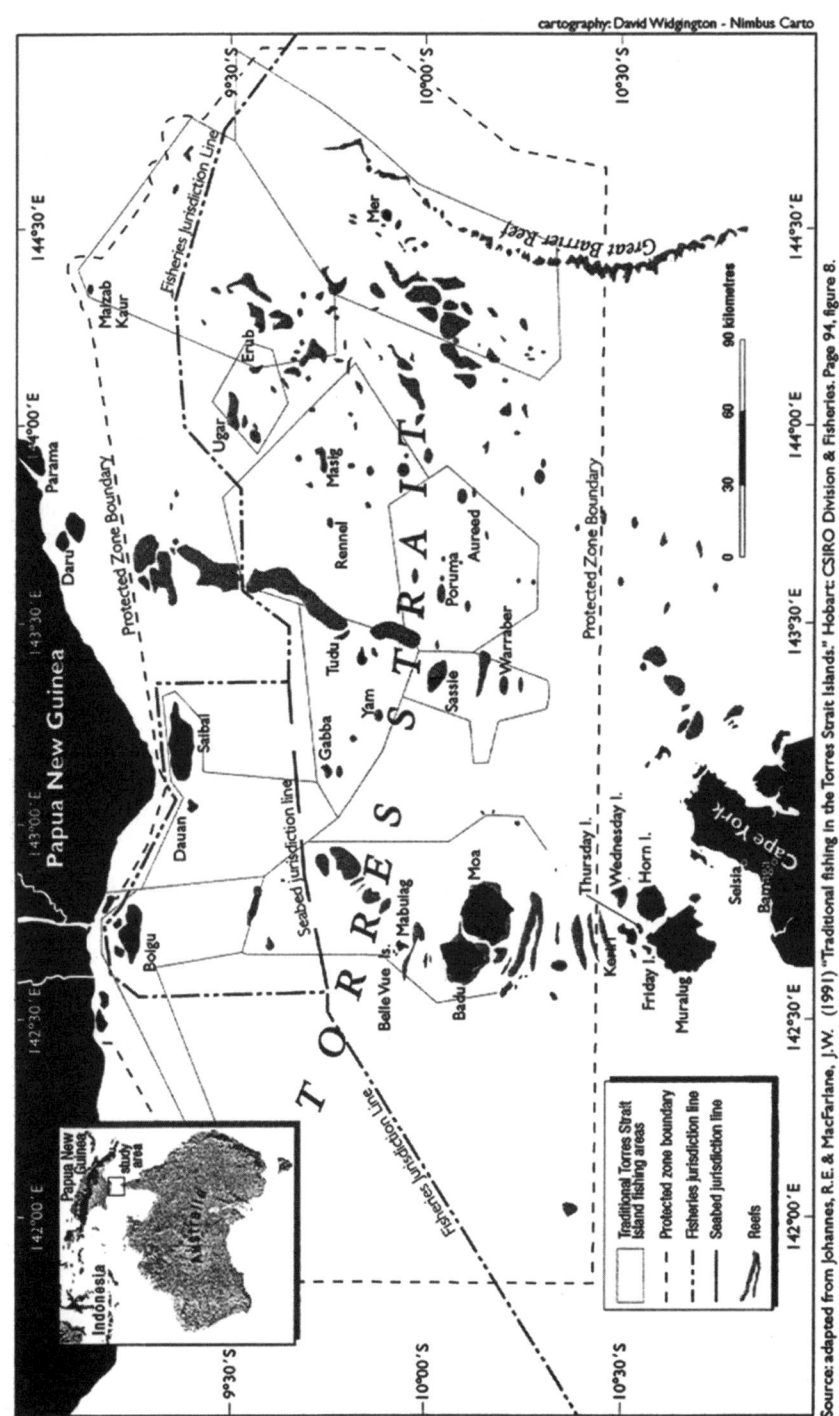

FIGURE 9.1 *Torres Strait, Northern Australia.*

often revealing. So too are questions of how resource management systems evolve, specifically how an indigenous society that is confronted with resource crises, either internally or externally driven, internalizes the experience and makes long-term changes and adjustments consistent with a crisis-and-subsequent-learning model (Berkes and Folke, 1998). The Torres Strait presents an interesting context in which to examine such questions. Located between Papua New Guinea (PNG) and the Cape York Peninsula of northern Queensland, the reef-strewn passage of Torres Strait is home to a group of Melanesian Islanders who have an intimate and long-standing connection to the small islands, extensive reefs and tropical waters of their traditional territory (Figure 9.1).

For the period prior to colonization by Europeans, there is little direct evidence of the resource use and management strategies of Islanders. There was a suggestion, however, in the 1898 research of the Cambridge Anthropological Expedition to Torres Strait that Islander culture involved collective awareness of ecological constraints, and of the possibility of human action directed toward sustainability. Stern taboos regulated human population size (see for example Haddon, 1908). The cultural ideal was two children per nuclear family, and it was against tribal law to have more than three. Infanticide or adopting-out of a fourth child to a family with fewer children was the rule. Although Haddon had nothing to say on the motivation for this taboo, the limiting factor for human population size was surely not seafood supplies; it was almost certainly fresh water, which is in short supply on most Torres Strait islands.

Certain other renewable resources were in short supply, according to Islander oral history. There was no surplus of garden lands, and indeed the relatively barren sand cays of the central Strait depended on trade for vegetable produce from the more fertile Eastern volcanic islands and from PNG coastal communities to the north. In the Eastern islands, seabird manure was used to boost garden production according to local informants today, but local cays and islets did not accumulate guano at a sufficient rate to meet the need. Hence, according to Eastern Islanders, guano collection was among their ancestors' motives for journeying considerable distances to the outer limits of their sea territories, either northward to Bramble Cay near the Fly River estuary of PNG, or long distances southward along the Barrier Reef to Raine Island, where large cays support substantial seabird concentrations.

Sand cays, as sanctuaries for nesting sea turtles and seabirds, are sacred places for Eastern Islanders. Mythology surrounding the creation of Bramble Cay, in the marine estate of Erub (Darnley Island), emphasizes the possibility of marine resource depletion, and human responsibility to protect resources (Scott, 2004). Legendary ancestors used their magic to create the cay because nesting seabirds and turtles had been victims of human overexploitation nearer the home island. In response, ground was taken by clan leaders from the home island to create the cay, which was set far enough away to afford these important resources some protection but close enough to be of use, with the comings and goings of visitors to the cay overseen by clan elders.

A central feature of traditional marine resource management was the ability to control access to resources. As with customary marine tenure systems elsewhere in the Pacific (Ruddle et al., 1992; Hyndman, 1993; Hviding, 1996; Novaczek et al., 2001), this system provided for the restriction and regulation of access and the rotational use of marine resources for management purposes (Nietschmann, 1984; Johannes and MacFarlane, 1991; Johannes, 1988; Sharp, 2002). The establishment of boundaries between neighbouring island communities was a complex matter, with particular boundaries fixed on the basis of stories of creation linked to a home island, histories of military domination, proximity, and equitable sharing in relation to a home island's human needs. Despite its obvious contribution to resource management, according to Johannes (1978), 'the value of marine tenure [in Oceania] was not generally appreciated by Western colonizers. It not only ran counter to the tradition of freedom of the seas which they assumed to have universal validity, but it interfered with their desire to exploit the islands' marine resources – a right they tended to take for granted as soon as they planted their flags.'

From the 1860s to the 1960s, Islanders were involved in a range of industrial fisheries as seamen and diver – for bêche-de-mer (*Holothuria* spp.), pearl shell (*Pinctada* spp.) and trochus shell (Beckett, 1987; Ganter, 1994). This experience, particularly diving-related activities, 'expanded their understanding and description of sea conditions and sea life' (Nietschmann, 1989) and provided object lessons in the exhaustibility of resources that would not have been depleted under pre-contact conditions. Islanders witnessed first-hand the depletion of wild pearl shell, as well as trochus shell, to the point that a crew might dive all day for what a man might formerly have easily gathered in half an hour. The patterns of commercial exploitation, and of management policy to the extent it existed, were, however, out of the hands of Islanders. Islanders were maritime workers for the most part, not owners of vessels. Even the small number of Islander-skippered commercial vessels was strictly under the thumb of the colonial protector until the 1970s.

The 1970s saw further crises, one in relation to giant clams (*Tridacna gigas*), which under aboriginal conditions were an exhaustible resource and seen as such; the other in relation to sardines, which under aboriginal conditions were effectively inexhaustible. Giant clams can easily be overexploited. Their meat is highly savoured and involves limited harvesting effort. Yet giant clams are present in significant numbers even on home reef areas. Food practices limit the consumption of giant clams to infrequent occasions, as a means of varying the diet or during those periods when access to other sources of seafood is limited by unfavourable fishing conditions (see also Hickey, this volume). Giant clams are key symbols in Islander attitudes toward conservation. Their shells should be turned upside down once the flesh has been harvested to serve as a refuge for other life forms. Individuals who fail to observe this practice are labeled *'meme kurup'*, uncouth and uncultured people (Scott, 2004). A crisis occurred when a Taiwanese mother ship engaged in illegal clamming. Although careful to anchor beyond the visual horizons of inhabited islands, the Taiwanese were eventually discovered by Islanders and apprehended. Giant clams had, however, been harvested in such large numbers over an extensive area of reefs that, according to local informants, it took more than twenty years for the clams to re-establish their former size and abundance.

From the early 1970s to the early 1980s, a sea-turtle farming programme, focused on raising green turtles (*Chelonia mydas*), was initiated in the Torres Strait by a foreign biologist. Large numbers of eggs were collected from nesting areas such as Bramble Cay and Raine Island, and brought to Mer (Murray Island) for incubation. From there, the hatchlings, which enjoyed much higher survival rates than they would in the wild, were dispersed to farms on various islands to be hand-fed in small pools. Juveniles were to be released to reinforce the wild population. Most hatchlings, however, were to be raised to adulthood as breeding stock for turtle restocking elsewhere (Bustard, 1972).

Sardines served as the primary food source for the large numbers of growing turtles. Sardines had always been a reliable and easily harvested food staple for Islanders. However, aggressive netting, an essential element in the maintenance of the turtle farm operation, resulted, according to Islander informants, in an unprecedented sardine population collapse at both Erub and Mer.[1] This in turn is said to have led to the retreat of formerly abundant species of large fish, particularly trevally (Carangidae), that normally pursue sardines onto the beaches of the home islands. Islander patience ran out when it was proposed that turtle farmers should turn to giant clams for turtle feed. Elders insisted that the project be terminated. According to Islanders, it took fifteen years for sardine populations to recover to former levels, but now trevally are again abundant along local beaches.

The search for evidence of an explicit conservation ethic among Torres Strait Islanders has produced limited returns. According to Johannes and MacFarlane, (1991), the 'issue of dwindling marine resources has arisen for the first time in Torres Strait only in the past century. ... Since such depletion is not part of the Islanders' past experience, it is not surprising that they do not possess a well-developed awareness of the vulnerability of their marine resources.' More productive avenues of inquiry, involving consideration of the likely motivations and practical outcomes of particular beliefs, values and practices, run counter to this interpretation and suggest

---

1. The high natural variability of sardines, combined with trawling in the Torres Strait and Gulf of Papua New Guinea, makes it difficult to ascertain how much of their decline can be attributed to turtle farming. The fact that local informants perceive this to have been the cause is however significant for the crisis-and-subsequent-learning model (Berkes and Folke, 1998) and the regard and respect Islanders hold for western science management models.

that Torres Strait Islanders developed a highly effective system for the sustainable management of marine resources (Nietschmann, 1989).

More than a hundred years of outsider intrusion, marked by a series of mostly profit-driven resource crises, most of which were created by non-Islanders, have stiffened local resolve to gain both ownership recognition and management jurisdiction of their home seas. For both ecological and social reasons, Eastern island fishers advocate limiting reef fisheries to locally controlled small-boat operations, in pursuit of diversified subsistence and commercial catches – principally tropical rock lobster, coral trout (*Plectropomus* spp.), Spanish mackerel (*Scomberomorus* spp.), red emperor (*Lethrinus miniatus*), sea cucumber (known locally as 'sandfish') and trochus shell. Rotational use of fishing spots, the distribution of fishing effort over multiple species, and seasonal shifts in wind and weather patterns that limit small-boat access to less than six months of the year are principal features in local management. These stand in marked contrast to the approach of larger, non-Islander commercial boats that target one or two species and can work intensively during all seasons in nearly any weather. There is also a major difference in economic imperative. Relentless accumulation is disparaged by Islanders; in the words of one informant:

'Them thing he happen on a needs basis, not on a craving for more and more. As soon as we satisfy, we stop and when the need come up again we go again.'

The rare individual who fishes hard at every possible opportunity is more likely to be the object of censure than praise.

STRATEGIES TO ENHANCE ISLANDERS' ROLE IN RESOURCE MANAGEMENT DECISION-MAKING

ISLANDER aspirations to assume primary control of resource and environmental management are being pursued along various avenues simultaneously. First, rights to use and to manage marine resources may be reshaped through Native Title recognition. Mer (Murray) Islanders gained High Court recognition of their ownership to land above the high water mark through the landmark Mabo decision in 1992. Through a series of Federal Court determinations on claims subsequently lodged with the Native Title Tribunal, most other Islander communities have gained similar recognition. The Torres Strait Regional Authority (TSRA) lodged a sea claim covering the entire Torres Strait region in late November 2001 on behalf of the Islanders (Mulrennan and Scott, 2001). This claim involved detailed documentation of the customary marine tenure system, which despite some assessments to the contrary (see Johannes and MacFarlane, 1991), continues to provide 'the foundational structures out of which [these] saltwater people's lives and cultural traditions are fashioned' (Sharp, 2002). Despite this, recognition of Islander title to reefs and seas below the high tide mark has already met with greater opposition than was the case for the land. Native Title rights to the Australian offshore, as recognized in the Croker Island Seas case in September 2001 (High Court of Australia, 2001), are weaker than terrestrial rights but Islanders hope that their own case will result in a more beneficial judgment, given the predominance of the sea for their cultural identity and economic prospects (TSRA, 2001).[2]

In the meantime, Islander concerns about the sustainability of certain fisheries and frustrations with the lack of economic benefits accruing from commercial fishing in their traditional waters have erupted in conflict with non-Islander commercial fishing interests and central government authorities. In the early 1990s, Eastern Islanders declared exclusive economic zones within their traditional waters, in line with demands for economic independence and the management of the seas in accordance with traditional law. Periodically, non-Islander commercial fishing boats have been evicted from this zone, although more recently a so-called 'gentlemen's agreement' has led to non-Islander boats generally avoiding waters within a radius of ten nautical miles from the home islands. The Islanders seek a thirty-mile radius, but there is nothing in official licensing or regulation to prevent entry even into the ten-mile zone, so incidents at sea have continued.

The declaration of exclusive economic zones also reflected Eastern Islanders' anxieties about potential fishing pressure from some of the larger islands in Western Torres Strait, where fishers use hookah gear to gain access to sandfish (*Holothuria* spp.) and tropical rock lobster (*Panulirus ornatus*) at greater depths.[3] These fishers are described as more 'cash-driven'. Eastern Islanders believe that these factors, together with insufficient regard for traditional marine territories, led to the 1997 collapse of the Warrior Reef sandfish (*Holothuria scabra*) population in the central strait, and subsequent closure of the fishery. For this reason they are adamant that their community territories must be respected, so that they may regulate access. They express some willingness to share with Western and Central Islanders, but on specific terms, including a ban on the use of hookah gear. For this reason, Eastern Islanders have made their participation in the blanket regional sea claim conditional on respect for community-level traditional territories.

A recent Cairns District Court decision dismissed armed robbery charges against an Islander who had used a crayfish spear to confront licensed commercial fishers operating in the traditional fishing territory of Mer. Ben Ali Nona's confiscation of AUS$ 600 worth of coral trout from the intruders was deemed not to be robbery,

2. The sea rights recognized in the Croker Island case are not exclusive but coexist with the public right to fish and the right to navigate (Sharp, 2002).
3. Hookah gear refers to the underwater breathing equipment used by professional fishers for harvesting lobster, sandfish and trochus shell.

on the grounds that he was acting on an 'honest right of claim' (Haigh, 1999). The acquittal is the outcome of a provision of the Queensland criminal code rather than recognition of Native Title sea rights. Nonetheless, it has fuelled grassroots support for a movement centred on the Torres Strait Fisheries Taskforce (TSFT), a body of young, energetic fishers determined to take control of fisheries management through the creation of a Torres Strait Regional Fisheries Council.

A Cultural Maritime Summit in March 2001,[4] in the wake of the Nona decision, was the venue for a regional statement of Islander demands. These included suspension of all fishing by non-indigenous commercial fishers throughout the Strait within a week (Anon, 2001a). The Commonwealth fisheries minister visited the Strait within days, warning against further interference with licensed fishing boats, but commencing political negotiations on important issues.

Islanders have particularly urgent concerns over the environmental impact of commercial prawn (*Penaeus* spp.) trawling, believed to be a major factor in the decline of tropical rock lobster, a resource that is vital to their own small-boat fishery. Over the years, large numbers of lobster on spawning migrations have been caught in prawning nets, and either sold illegally or returned to the water injured. Islanders for some time have been proposing government buy-back of prawning and rock lobster licences held by outsiders. On his visit to the Strait, the minister publicly rejected licence buy-backs, professing lack of government funds.[5] Behind closed doors, however, both state and Commonwealth governments have yielded important ground. They have afforded the chair of the Torres Strait Regional Authority (an Islander-elected regional self-governmental body) equal authority to themselves on the top-level fisheries decision-making committee: the Torres Strait Protected Zone Joint Authority (PZJA). In addition, an Islander TSFT representative has been granted observer status.

Discussions are also underway on the subject of prawn licence buy-backs, and other Islander proposals for dealing with the current crisis in the tropical rock lobster fishery. Recent stock assessments indicate that numbers of breeding stock and juveniles are among the lowest ever recorded, and that future recruitment may be too low to support the fishery (Torres Strait Rock Lobster Working Group, 2001). Many Islander fishers regard the total exclusion of prawn trawling vessels as their long-term objective. In the interim, however, they have agreed on a number of trial measures. First is a 50 per cent reduction of prawn trawling licences as a minimal condition for tolerating prawn trawling vessels in their waters. Although the Commonwealth has expressed support for a proposal to buy back thirty-nine of the seventy-nine licences in the region, there is much disagreement on how this buy-back arrangement should proceed. The Commonwealth has taken the position that the prawning industry itself

---

4. The Ngalpun Malu Kaimelan Gasaman Cultural Maritime Summit, 22–25 March 2001.
5. Each prawning licence is worth approximately AUS$ 800,000.

should purchase any buy-backs, but licence owners and the industry more generally are unhappy with this. For the moment the Commonwealth, industry and Islanders remain at loggerheads. One possible approach that has been taken elsewhere in Australia is for the Commonwealth to suspend the prawn fishery as a means of applying pressure on the industry to cooperate. In the interim, Islander fishers say they hold the option of escalated direct action in reserve.

Meanwhile, the TSFT has successfully lobbied the regional Islander leadership to rescind a promise of prawn licences to three private Islander enterprises, and restore them to the common benefit of Islanders. Some Islanders argue that prawn trawling on a reduced scale is environmentally sustainable, and acceptable if Islanders are afforded a stake in the industry.[6] One proposal is to establish an Islander trawling operation with one of the three licences, while renting the other two licences to provide financing, training and other support (B. Bedford, personal communication, 7 June 2001).

A second Islander demand has been for the exclusion of all prawn trawling from areas of lobster migration by imposing a seasonal rotation of the fished areas. Currently, trawlers sweep the whole of the prawning grounds from March to December (Prawn Working Group Meetings, December 2001). The Islander fishers' proposal would have all boats working only north of the 10° parallel in the first part of the year, and only to the south of the line in the second part. Each area would therefore be closed for a full seven months, closures timed to coincide with the clockwise migration of lobster through the eastern and central Strait. There has, as yet, been no official response to this demand, although the scientific merits of the proposal have received some consideration by the Australian Fisheries Management Authority (AFMA).

Islanders have stated that these demands are non-negotiable and have served notice of their readiness, if necessary, to close down the trawling grounds by laying barbed wire across the bottom, or by dumping old vehicles on the grounds to serve as 'rock lobster sanctuaries'. These would, of course, pose an inevitable risk of snagging and damage to trawling gear.

Islanders recognize that trawling is only one of several possible impacts on the lobster fishery, and are taking other measures as well. Of particular significance is Islander commitment to a total ban on the use of hookah gear. In the eastern islands, deeper waters inaccessible to free divers are regarded as sanctuaries. Eastern Islanders see a causal relationship between the use of hookah gear and the reduction of lobsters moving up onto shallower reef surfaces. Islanders feel that a ban on hookah gear would dissuade most non-Islander divers from participating in reef diving fisheries so that a reduction in total fishing effort would also result. It is

---

6. Research conducted by CSIRO marine scientists on the Great Barrier Reef suggests that the impacts of trawling depend on the intensity of trawling and on the type of trawl gear used (Pitcher, 1996).

interesting that Western Islanders, who do use hookah gear, have joined Eastern Islanders in supporting a total hookah ban throughout Torres Strait (at a Fisheries Meeting in Erub, 14 August 2001).

Similar concerns about the impact of 'technology creep' on lobster, coral trout and other stocks relate to the use of Geographical Positioning Systems (GPS) and depth sounders that allow the targeting of specific fishing locations. Restrictions on their use would tend to spread fishing effort, at the cost of increasing the fuel and time costs of looking for specific bottom features. Islanders recognize that the proposed restrictions would be a lesser hindrance to Islander than to non-Islander commercial fishers, as the latter are heavier users of these technologies and their local knowledge of productive sites is inferior to that of Islanders (Fisheries Meeting, Erub, 14 August 2001).

New arrangements in the rock lobster fishery, introduced in December 2001, have gone some distance towards addressing Islander concerns regarding rock lobster stocks. These include an increase in the minimum legal catch size for tropical rock lobster,[7] an extension of the existing two-month ban on the use of hookah gear by a further two months,[8] and a new two-month ban on all other forms of commercial fishing,[9] though 'traditional fishing' (i.e. fishing for subsistence or ceremonial purposes) by Islander fishers within the region is still permitted. While these measures only partly fulfil Eastern Islander aspirations, they herald the emergence of a more democratic approach to fisheries management in the region. Of particular significance was the fact that for the first time in its history, the PZJA meeting that endorsed this three-pronged approach to stock management was held as an open forum with invited stakeholders, including the TSFT, in attendance as observers (Anon, 2001b).

The endorsement of a new fisheries consultative structure in June 2002 was a development of particular significance for Islanders. In the words of TSRA Chairman Terry Waia: 'This proposed structure will mark a new chapter for the fisheries movement in the Torres Strait, because as I have said all along ... our aim is to have Torres Strait Islanders manage our fisheries for our benefit, and this model is the first step toward achieving that goal' (TSRA, 2002a). In this respect, the new consultative arrangements, which include Islanders at all levels of management, are a major departure from those established under the Torres Strait Treaty,[10] where Islander inputs were afforded a limited advisory role (Mulrennan and Scott, 2005).

---

7. The minimum legal catch size for tropical rock lobster is increased from 80 mm to 90 mm carapace length, or in the case of lobster tails, the minimum legal tail length is increased from 100 mm to 115 mm.
8. The ban on the use of hookah gear is extended from 1 October to 30 November for a further two months, to 31 January.
9. The new two-month ban is from 1 October to 30 November; this closure came into effect on 1 October 2002.
10. Specifically, the Torres Strait Fisheries Management Committee (TSFMC), the Torres Strait Fisheries Scientific Advisory Committee (TSFSAC), and the Environmental Management Committee (EMC).

The inclusion of active Islander hunter/fishers rather than a representative from the regional Islander leadership on the Torres Strait Fisheries Scientific Advisory Committee (TSFSAC) is viewed as another important concession. The integration of social science research expertise has also been recommended as a measure to enhance the cultural and socio-environmental aspects of natural resource management, particularly in relation to improved 'extension and transfer of research outcomes to stakeholders' (Sen, 2000). A major marine research programme with limited partnership arrangements with the TSRA and local island communities has also been established (TSRA, 2002b). While such changes are likely to result in the increased engagement of Islander knowledge and expertise in research and consultative processes, more substantial transformations in political structures of authority will be required to address the intransigent power asymmetries of former decades through the establishment of true partnerships in management decision making (Mulrennan and Scott, 2005).

CONCLUSIONS

As concerns for the condition of many of the world's fishery stocks increase, and the limitations of neoclassical management strategies rooted in economic efficiency and private property are exposed (Wingard, 2000), alternative approaches centred on community-based, small-scale fisheries have become the focus of much attention (Berkes, 2003). Until recently, most assessments of the potential contribution of these fisheries were concerned with the status of the customary marine tenure and traditional management systems that underpin these small-scale systems, and with questions of their contribution to conservation and contemporary management. The more guarded assessment of Johannes and MacFarlane (1991), similar to Polunin (1990), concludes that customary marine tenure has little potential in the contemporary context of transition from subsistence to commercial economies, while others see much value in their application (Wright, 1990; Hickey, this volume). Hyndman (1993, p. 100), for example, suggests that, 'sea-tenure estates are not broken down traditions but living customs which have always transformed and related to basic resource-management tasks.' Because they are 'diverse, flexible, dynamic and capable of regulating many kinds of subsistence and commercial activities associated with marine fishing, hunting, and gathering' (ibid.), they are expected to have a better chance of success in the management of local fisheries (Asafu-Adjaye, 2000).

An increasing number of researchers, less concerned with the cultural origins or traditional integrity of these fisheries, lend their support to small-scale fisheries approaches because they offer a much needed antidote to the limitations of conventional fishery management. As Berkes (2003) states:

'Instead of fishing-as-business, these alternative approaches focus on sustainable livelihoods; instead of top-down decision making, there is participatory management; instead of reductionism and positivism, there are complex system approaches; instead of sole reliance on expert-knows-best science, local and traditional knowledge are also used, instead of control-of-nature utilitarianism, there is emphasis on humans-in-ecosystem management.'

From an Islander perspective, local knowledge and management of marine resources has been continuous and evolving, and responsibility for their sea territories (even if inhibited by successive colonial regimes) has never been surrendered. Principles of resource conservation and management have a deep cultural history, and the application of these principles, together with specific knowledge, has evolved with changing conditions, including valuable lessons gained from resource crises, across a variety of fisheries. In contrast, conventional fisheries management approaches, based largely on single-species stock assessment, the achievement of optimal yields and command-and-control measures, have failed to address and respond to the social and ecological context of the Torres Strait region.

Recent achievements of Islanders in securing a greater role in fisheries management represent the outcome of a combination of legal actions, political negotiations, knowledge exchange, and – when progress along these avenues slowed – direct action at sea. Islander efforts to assert their role continue and, while it is still early days, these efforts appear to be resulting in positive changes to an otherwise unsustainable situation. Substantial political interests from within the Australian mainstream remain aligned against them; but with so much international attention now focused on the contribution of small-scale fisheries to the future of global fish stocks, it is to be hoped that Torres Strait Islanders, together with local fisheries communities elsewhere, can tap new alliances to help secure their cultural survival.

ACKNOWLEDGEMENTS

My thanks are due to the Erub Council, in particular Mr George Mye (past Chair) and Mr Elia Doolah (current Chair) for their support, and to the many other people of Erub for their knowledge, advice and friendship. Special thanks to Bluey Bedford, Kenny Bedford, James Bon, Kapua Gutchen, Douglas Jacob, Toshi Nakata and Peter Yorkston, who have allowed me to document their knowledge, attend their meetings, and record their concerns and positions.

I am grateful to Jeremy Beckett, Henry Garnier, Tony Kingston, Victor McGrath, Kenneth McKay, Jim Prescott and an anonymous reviewer for the valuable insights they have contributed to this paper. Thanks also to Nigel Haggan for expert editorial insights and to my graduate student and research assistant, Lauren Penney, for her tireless efforts in tracking down material and following up leads. Finally a special debt of gratitude goes to my husband and co-researcher, Colin Scott (anthropology, McGill University), for his contributions to this paper and to this research more broadly.

The research on which this paper is based is funded by a Social Sciences and Humanities Research Council of Canada (SSHRC) grant.

REFERENCES

ANON. 2001a. Ngalpun Malu Kaimelan Gasaman Cultural Maritime Summit. Unpublished Interim Report, March, 19 pp.

——. 2001b. Torres Strait cray industry to be protected. *Torres Strait News: The Voice of the Islands*, No. 467, 12–18 October, p. 5.

ASAFU-ADJAYE, J. 2000. Customary marine tenure systems and sustainable fisheries management in Papua New Guinea. *International Journal of Social Economics*, Vol. 27, No. 7, 8, 9, 10, pp. 917–24.

BECKETT, J. 1987. *Torres Strait Islanders: Custom and colonialism*. Cambridge, Cambridge University Press, 251 pp.

BERKES, F. 2003. Alternatives to conventional management: Lessons from small-scale fisheries. *Environments*, Vol. 31, No. 1, pp. 5–20.

BERKES F.; FOLKE, C. (eds.) 1998. *Linking social and ecological systems: Management practices and social mechanisms for building resilience*. Cambridge University Press, Cambridge, 459 pp.

BUSTARD, R. 1972. Turtle farmers of Torres Strait. *Hemisphere*, No. 16, pp. 24–8.

DYER, C.L.; MCGOODWIN, J.R. 1994. *Folk management in the world's fisheries*. Niwot, Colo., University Press of Colorado, 347 pp.

FAO. 1995. Code of conduct for responsible fisheries. Rome, Food and Agricultural Organization, 41 pp.

GANTER, R. 1994. *The pearl-shellers of Torres Strait: Resource use, development and decline 1860s–1960s*. Carlton, Victoria, Melbourne University Press, 299 pp.

HADDON, A.C. 1908. *Sociology, magic and religion of the eastern Islanders*. Vol VI of Reports of the Cambridge Anthropological Expedition to Torres Straits. Cambridge, Cambridge University Press, 315 pp.

HAIGH, D. 1999. Casenote: 'Fishing war' in the Torres Strait: The Queen v Bejamin Ali Nona and George Agnes Gesa. *Indigenous Law Bulletin*, Vol. 4, No. 22, pp. 20–21.

HIGH COURT OF AUSTRALIA. 2001. The Commonwealth v. Yarmirr; Yarmirr v. Northern Territory; *HCA* 56, 11 October.

HVIDING, E. 1996. *Guardians of Marovo Lagoon: Practice, place, and politics in maritime Melanesia*. Honolulu, Hawaii, University of Hawai'i Press, 473 pp.

HYNDMAN, D. 1993. Sea tenure and the management of living marine resources in Papua New Guinea. *Pacific Studies*, Vol. 16, No. 4, pp. 99–114.

JOHANNES, R.E. 1978. Traditional marine conservation methods in Oceania, and their demise. *Annual Review of Ecology and Systematics*, No. 9, pp. 349–64.

———. 1988. Research on traditional tropical fisheries: Some implications for Torres Strait and Australian Aboriginal fisheries. In: F. Gray and L. Zann (eds.), *Traditional knowledge of the marine environment in Northern Australia*. Canberra Great Barrier Reef Marine Park Authority, pp. 30–41.

JOHANNES, R.E.; MACFARLANE, J.W.M. 1991. *Traditional fishing in the Torres Strait Islands*. Hobart: CSIRO, 268 pp.

MABO AND OTHERS V. QUEENSLAND (NO. 2). 1992. 175 CLR 1 F.C. 92/014. Available online at http://austlii.edu.au/~graham/Slides/London/top20.html (last accessed 27 June 2005).

MATHEW, S. 2001. Small-scale fisheries perspectives on an ecosystem-based approach to fisheries management. Reykjavik Conference on Responsible Fisheries in the Marine Ecosystem, Reykjavik, Iceland, 1–4 October.

MULRENNAN, M.E.; SCOTT, C.H. 2001. Indigenous rights and control of the sea in Torres Strait. *Indigenous Law Bulletin*, Vol. 5, No. 5, pp. 11–15.

———. 2005. Co-management: An attainable partnership? Two cases from James Bay, Quebec and Torres Strait, Queensland. *Anthropologica*, No. 47, pp. 197–213.

NIETSCHMANN, B. 1984. Torres Strait Islander sea resource management and sea rights. In: K. Ruddle; R.E. Johannes (eds.), *The traditional knowledge and management of coastal systems in Asia and the Pacific*. Jakarta, UNESCO, pp. 126–54.

———. 1989. Traditional sea territories, resources and rights in Torres Strait. In: J. Cordell (ed.), *A sea of small boats*. Cambridge, Cultural Survival, pp. 60–93.

NOVACZEK, I.; HARKES, I.; SOPACUA, J.; TATUHEY, M. 2001. *An Institutional analysis of Sasi Laut in Maluku*. Malaysia, ICLARM, 327 pp.

PITCHER, R. 1996. Trawling nets $150M. CRC Reef Research Newsletter, December. Available online at: www.reef.crc.org.au/publications/news/news80.html (accessed on 26 May 2006).

POLUNIN, N.C.V. 1990. Do traditional marine 'reserves' conserve? A view of the Indonesian and New Guinea evidence. In: J. Ruddle and R.E. Johannes (eds.), *Traditional management of coastal systems in the Asia and the Pacific: A compendium*. Jakarta, UNESCO, pp. 191–212.

POMEROY, R.S. 1995. Community-based and co-management institutions for sustainable fisheries management in South-East Asia. *Ocean and Coastal Management*, Vol. 27, No. 3, pp. 143–62.

PRAWN WORKING GROUP MEETINGS. 2001. Islander prawn position on trawl/cray interaction. December. (Unpublished.)

RUDDLE, K.; HVIDING, E.; R.E. JOHANNES. 1992. Marine resources management in the context of customary marine tenure. *Marine Resource Economics*, No. 7, pp. 249–73.

SCOTT, C. 2004. 'Our feet are on the land, but our hands are in the sea': Knowing and caring for marine territory at Erub, Torres Strait. In: R. Davies (ed.), *Woven history, dancing lives: Identity, culture and history*. Canberra, Aboriginal Studies Press, Australian Institute of Aboriginal and Torres Strait Islander Studies (AIATSIS), pp. 259–70.

SEN, S. 2000. Improving the effectiveness of research in the Torres Strait: A review of the Torres Strait Scientific Programme. Report commissioned by the Australian Fisheries Management Authority (AFMA), 37 pp.

SHARP, N. 2002. *Saltwater people: The waves of memory*. Buffalo, Toronto, University of Toronto Press, 306 pp.

TORRES STRAIT REGIONAL AUTHORITY (TSRA). 2001. Torres Strait regional sea claim lodged. Press release, 23 November.

——. 2002a. TSRA Proposal to improve Islander input into Torres Strait fisheries management. Press release, 2 July.

——. 2002b. TSRA marine conservation receives a boost. Press release, 12 December.

TORRES STRAIT ROCK LOBSTER WORKING GROUP. 2001. Management proposals for the 2001/02 season. Meeting of the Torres Strait Rock Lobster Working Group. Agenda Item 5, Thursday Island, 1–2 August.

WINGARD, J.D. 2000. Finding alternatives to privatizing the resource: Community-focused management for marine fisheries. In: T.W. Collins and J.D. Wingard (eds.), *Communities and capital: Local struggles against corporate power and privatization*. Athens and London, University of Georgia Press, pp. 5–29.

WRIGHT, A. 1990. Marine resource use in Papua New Guinea: Can traditional concepts and contemporary development be integrated? In: K. Ruddle and R.E. Johannes (eds.), *Traditional marine resource management systems in the Pacific Basin: An anthology*. Study No. 2. Jakarta, UNESCO/ROSTEA, pp. 301–23.

CHAPTER 10 Sawen
Institution, local knowledge and myth in fisheries management in North Lombok, Indonesia
Arif Satria

ABSTRACT

SAWEN is a traditional resource management institution that originally integrated the management of forests, the sea and farmland using cognitive aspects (local knowledge and resource management principles), regulatory aspects (codes of conduct) and normative aspects (world views and belief systems). *Sawen* in North Lombok, Indonesia, was almost eradicated early in the Suharto regime, following an alleged coup attempt by the Indonesian Communist Party in the mid-1960s. Several concurrent and continuing factors contributed to the loss of *sawen*, including, but not limited to: first, Islamic orthodoxy aimed at implementing 'pure' Islamic rituals which opposesd *sawen* as practised by 'traditional' Moslems (who practised a syncretic mix of Islam and indigenous belief systems); and second, the replacement of many traditional management practices with national laws and regulations for natural resources including fisheries during the 1970s and 1980s. The failure to enforce these centralized fisheries regulations in many parts of the archipelago led to destructive fishing practices, conflicts among fishers, loss of property rights and loss of marine cultural identity.

The Indonesian reform movement of 1998 brought empowerment to many local communities in the archipelago and made it possible to attempt to revitalize *sawen* for fisheries management in North Lombok. The initiative came from the community and is compatible with the current government recent reform agenda in devolving power to local authority.

This chapter analyses the cause and effect of the cessation of *sawen* practices in marine resource management, and the recent attempt to revitalize it in Kayangan, a small coastal community in North Lombok. Preliminary findings suggest that revitalized *sawen* for marine resources was able to assist the local community in addressing issues of overexploitation, access rights and lack of enforcement of fishing regulations in their nearshore waters. More importantly, the revitalization of *sawen* has:

- restored the marine cultural identity of the community, which had ceased to exist over the three decades of the Suharto regime
- provided a 'protection institution' for small-scale fishers
- provided insights (local knowledge and wisdom) for implementation of local fisheries management
- created a legitimate institution of community-based fisheries management in the study area.

INTRODUCTION

THIS chapter discusses the revitalization of traditional ecological knowledge (TEK)[1] and traditional resource management systems in Indonesia, with specific reference to Kayangan, a coastal community in North Lombok, central Indonesia. The issue is timely, as depletion caused by anthropogenic factors is a real threat to the survival of resource-dependent communities. Charles (2001) stated that, 'it seems clear that one of the significant contributors to fishery collapse is the combination of (a) lack of knowledge in some cases, and (b) a failure to use all available sources of information and knowledge in other cases.' Opinions differ on how to deal with depletion, but one argument is that traditional management systems, based on local knowledge, offer an alternative approach for better management as a complement to the conventional management system. This argument is based on the fact that traditional management was successful in combating resource depletion and resource use conflicts in the past. Unfortunately, in Indonesia, many of these traditional systems ceased to exist due to political decisions during the New Order regime of former President Suharto (1966–98).

One such traditional management is *sawen*, an integrated traditional management system for forest, farmland and marine resources that was formerly practised in North Lombok. In the post-1998 Reform Era that brought decentralization of power, local communities began to pool their efforts and resources to revitalize *sawen* for the management of their marine resources.

The study area is in Kayangan Village, North Lombok, adjacent to Bali Island, in Nusa Tenggara Barat province, in central Indonesia (Figure 10.1). The village has a population of 4,952 people and is dominated by the *Sasak* ethnic group. Livelihoods include fisheries, agriculture, plantations and animal husbandry, with agriculture being the main employer and revenue generator. There are 112 people who still engage in traditional fishing.

1. 'The knowledge acquired through living in contact with the natural resources of a particular area over many generations' (Haggan and Brown, 2002).

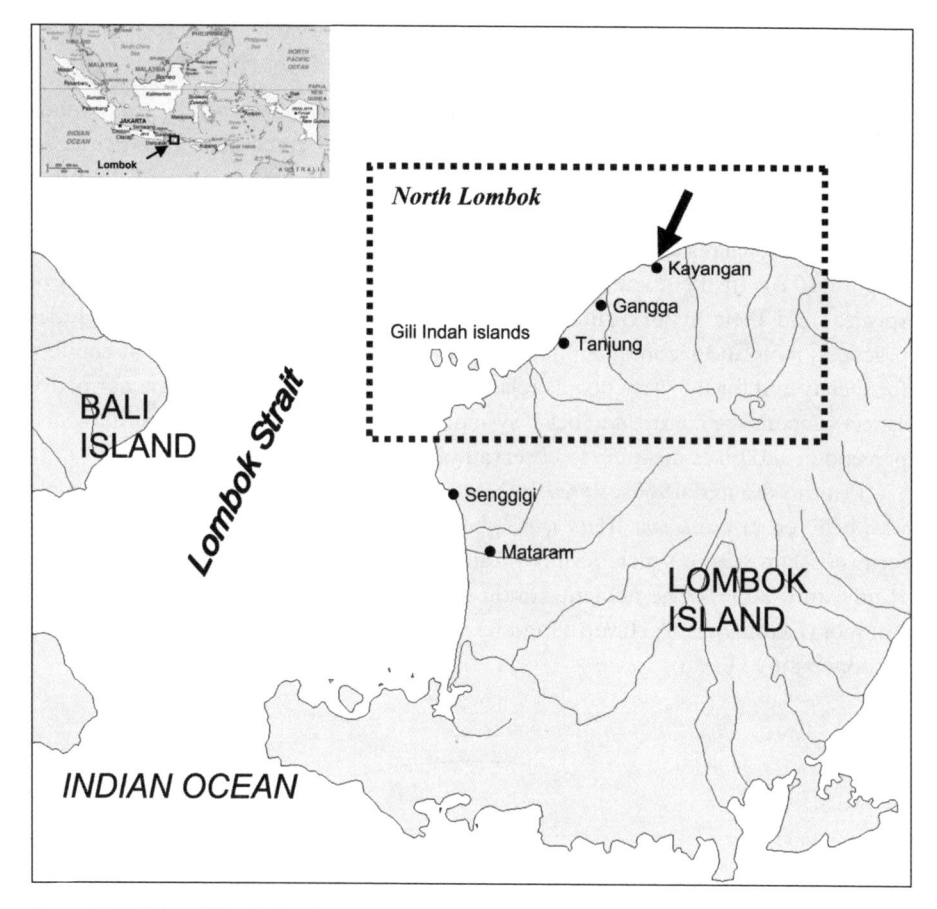

FIGURE 10.1 *Map of North Lombok, Nusa Tenggara Province, Indonesia.*

This chapter aims to analyse the past and present framework of *sawen*, and discern how and why revitalized *sawen* appears to work in Kayangan village, as opposed to other nearby examples, which are less successful. The chapter opens with a description of the historical background and practice of *sawen*. This is followed by an analysis of the factors that led to the cessation of *sawen* under Suharto, and how the 1998 Reform movement triggered revitalization of *sawen* for marine resources. It is also important to identify which elements of *sawen* are continuations of traditional practice, where it has had to adapt to present conditions, and the positive impacts of *sawen* on marine resources.

WHAT IS *SAWEN*?

SAWEN literally means 'boundary delineation'. Conceptually, *sawen* in North Lombok is a traditional resource management institution that embodies a *cognitive aspect* (local knowledge and resource management), a *regulative aspect* (rules of conduct), and a *normative aspect* (world view, belief systems, etc.) that coincide with Berkes' (1999) traditional resources management and Scott's (2001) institutional framework (Figure 10.2). In the cognitive aspect, local knowledge includes identification of species, and their history and behaviour, while resource management includes practices, tools and techniques.[2] The regulative aspect establishes codes of conduct for society and for the ways people relate to the natural world. Finally, the normative aspect contains the culture and belief system or worldview that shapes environmental perception and gives meaning to observations of the environment.

Prior to the mid-1960s, *sawen*, in North Lombok was practised by local people who believed in *wettu telu*. *Wettu telu*, a belief system unique to the study area, is a form of Islam with a syncretic mix of ancient Hinduism and indigenous animism (Budiwanti, 2000). The ancient Hindu influence came from the colonization of Lombok Island by the Hindu kingdom of Bali (the inhabitants of Bali are still predominantly Hindu).

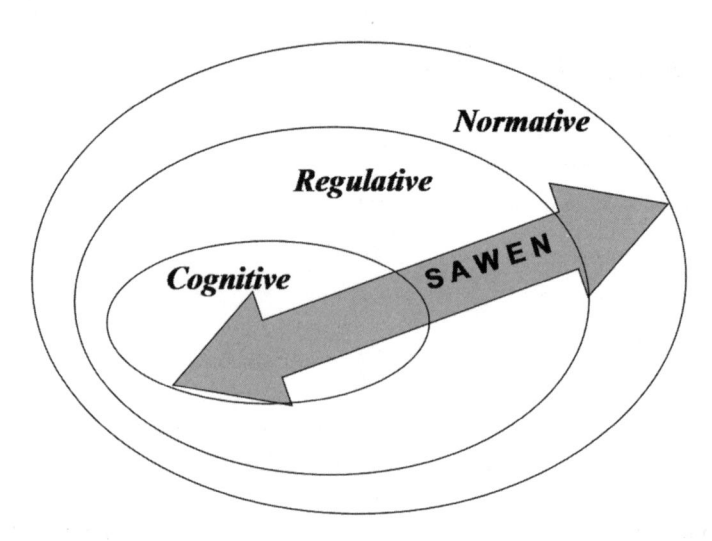

FIGURE 10.2 *Scope of* sawen.

---

2. There are common components of local knowledge. Kay and Alder (1999) and Ruddle (2000), for example, broadly divided components of traditional knowledge and practice into knowledge of the biophysical and biological resource characteristics.

In traditional management, as practised in the past, the community saw the world as two interrelated domains (Figure 10.3): the physical domain, which includes ecological characteristics, was embodied in the triad of forest, farmland and the sea; and the non-physical domain was encapsulated by a triad of authority figures held by political authority (*pemusungan*), religious authority (*penghulu*) and resource management authority (*mangku*). The existence of these two interrelated but not necessarily mutually inclusive domains are called the philosophy of *paer* (literally 'domain'); s*awen* is the underlying 'soul' within the ecological (or physical) domain and is connected to the resource management authority (*mangku*).

In the ecological domain, the forest is seen as 'the mother' (*buana alit*) of the triad. This perception stems from traditional knowledge that respects the forest as the source of water. If the forest is disturbed, then there will be an adverse cascading effect throughout the ecosystem that will affect farmland and the sea, harming the life of farming and fishing communities. The decline of forest resources would cause a decline of downstream resources in the farmland (for example by causing irrigation problems), and if farmland irrigation is disrupted, coastal resources will be threatened. This is indeed a basic philosophy of integrated resource management.

It must be emphasized that a resource management authority (*mangku*), as embodied in the traditional management worldview of the local people, is critical to the success of *sawen*. In the authority domain (Figure 10.3), we can see that the resource management authority (*mangku*), the political authority (*pemusung*) and the religious authority (*penghulu*) were autonomous, but no element was more important than the others. This *paer* philosophy was accepted in most villages in North Lombok.

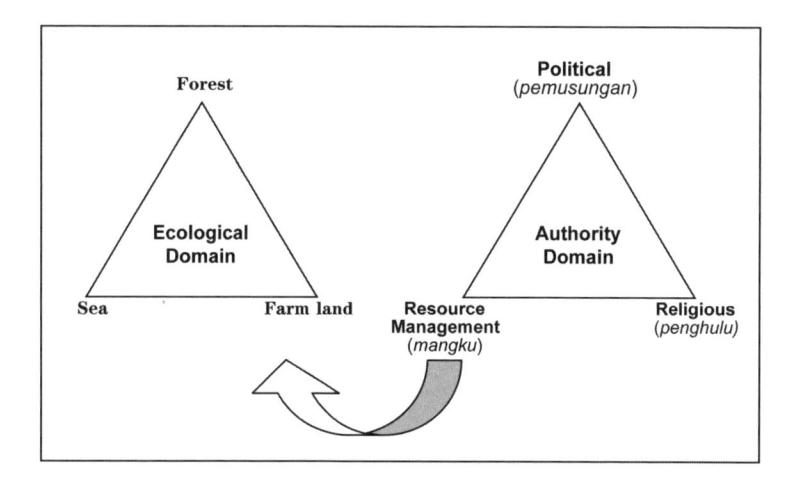

FIGURE 10.3 *Two domains of management in* paer *philosophy underlying sawen.*

Each ecological resource has its own management authority with distinct roles and responsibilities (Figure 10.4):

- *mangku alas*: forest resources authority
- *mangku bumi*: farmland resources authority
- *mangku laut*: marine resources authority.

Among these *mangkus* there was a strong commitment to the integrated management of resources; coordination and collaboration was a priority, resulting in functional interdependence (Figure 10.4).

Mangku was a hereditary resource management authority that could only be held by a descendant of a *mangku* family; in other words, it was an ascribed rather than an achieved status. This was due to the belief that *mangku* families had supernatural power as well as the knowledge to deal with resource management issues. The villagers' respect for the *mangku*'s power legitimized *mangku* and brought about voluntary compliance, as each *mangku*'s decision was perceived to contribute to a safe and peaceful life.

A *mangku* had two main roles. The first was to maintain the traditional value of social and human–nature relationships. This role served to achieve harmonious life in the community. The value of harmonious life is described by a popular North Lombok proverb: *aiq meneng tunjung tilah empaq bau* (water stays clear, the lotus can be kept intact, and fish can be caught). This means that each problem has to be tackled carefully and wisely without disturbing something else. The significance for sustainable resource management is that each attempt to extract something from the resources has to be done with consideration of the potential effects (possibly

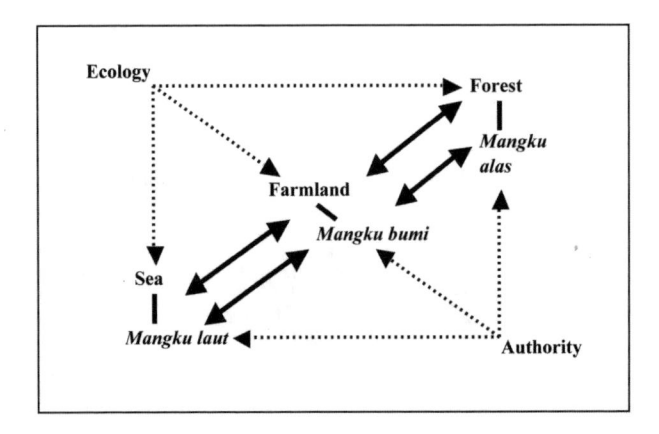

FIGURE 10.4 *Linkages of ecology and authority triad.*

unintended) that may result. This proverb mirrors the ecological triad concept. It has a strong parallel in Canadian aboriginal values of connection (Lucas, this volume).

The second role of the *mangku* includes three main resource management practices (Table 10.1):

- *menjango*: survey or observation
- *membangar*: visual mapping and boundary marking
- *membuka*: opening (Kamardi, 1999).

These practices were applied to forest, farmland and marine resource management, and were based on a combination of traditional knowledge and myth. Many of the religious ceremonies that preceded these practices show that myths were influential in resource management. These roles of the *mangku* were based on clear concepts, albeit resulting from a combination of traditional knowledge and myth, of resource management at the time.[3] These practices thus have significance for sustainable resources management.

TABLE 10.1 *Roles of all* mangkus *in traditional resources management practices.*

Roles	Resources		
	Forest (mangku alas)	Farmland (mangku bumi)	Marine (mangku laut)
Menjango (observation)	To assess the feasibility of forest resource utilization	To assess the feasibility of farming	To assess the marine zones that will be closed
Membangar (visual mapping and marking boundary)	To determine and delineate a particular area as a mark of exclusive access and use right	To determine and delineate farming area as a mark of exclusive access and farming right	To initiate closed season for particular areas and time (usually once a year)
Membuka (opening)	To initiate forest resource use	To initiate farming season in sixth month of traditional calendar	To initiate fishing season marked by religious ceremony

SAWEN IN MARINE FISHERIES

The practice of *sawen* (also known as *nyawen*) in the management of marine fisheries was identical to a closed season system (Table 10.2). After observing conditions of the sea, the *mangku laut* would decide whether the fishing season should be closed or not. To initiate a closure, the *mangku laut* would install two

---

3. 'A myth is a narrative account of the sacred which embodies collective experiences and represents the collective conscience' (Abercombie et al., 2000).

bamboo posts approximately 1.5 km from the shoreline, to mark the boundary of the closed area.[4] The closed season, usually lasting for around a month, was intended to lure the fish close inshore so that they could be easily caught during the open season (see also Hickey, this volume). The fishers depended on the nearshore area for their livelihood, as they lacked technology that would allow them to fish further afield, and in this case *sawen* can be seen as a way of dealing with scarcity of fish in the inshore area. Another function was to intercede with supernatural powers to protect the village from bad omens. The underlying myths were in fact based on scientific rationales, such as a closed season to enhance fish stocks. However, as the fishers had little formal education, the explanations were usually given in normative terms that prohibited or allowed particular activities, the so-called *pamali* or taboos. Following the issuance of a *sawen*, some rules were established, such as a prohibition on fishing in a particular area during the *sawen* period. Those who violated this rule suffered moral sanctions such as social ostracism. Such rules were easily enforced because fishers believed them to be sacred.

TABLE 10.2 *Some original* sawen *practices and their underlying scientific rationales.*

Sawen practices	Underlying myths	Scientific rationale
Nyawen (= installing bamboo posts in inshore waters to delineate boundary)	Luring the fish; Seeking supernatural protection for the village	Closed season to enhance fish stocks
Prohibition on felling trees on the beach	Respecting the supernatural power of the trees	Preventing erosion of the coastline
Fish have to be individually ungilled from the net prior to taking them home	Pamali (= taboo)	Unknown
Integration of forest, farmland, and coastal management by all *mangkus*	Respecting the forest as 'the mother of the resources'	Implementation of integrated resources management

When the closed season ended, the *mangku* officially opened the fishing season with a religious ceremony. The fish caught on the first day were offered for the ceremony, and nobody was allowed to sell fish or take them home on that day. This tradition enhanced social solidarity among fishers (see also Hickey, this volume). Other *sawen* ceremony-related rules or practices were designed to maintain a relationship either with the supernatural power or with nature (Table 10.2).

To enforce these rules and practices, the *mangku laut* appointed *lang-lang* (traditional coastal guards) to be responsible for the monitoring, control and surveillance (MCS) of each *sawen*. *Lang-lang* was a voluntary status and all fishers had an equal opportunity to become *lang-langs*, so most fishers had some experience

---

4. Another source said that the depth of bamboo installed in nearshore waters is around 20–30 m (Solihin, 2002).

in that role. In case of any violation, the *lang-lang* had a right to warn the violators and report the incident to the *mangku laut*, who would decide an appropriate sanction. Most sanctions were moral and designed to expose the violators to public embarrassment. However, fines were also levied for particular infractions.

The rules and practices of *sawen* were highly enforceable. In addition to the fact that the *mangku laut* and *lang-lang* had authority to enforce MCS, the rules themselves satisfy important criteria for legitimacy identified by Jentoft (1989). The content of *sawen* rules coincided with the fishers' worldview and their belief in the need to respect supernatural powers in order to ensure safe and peaceful lives. Moreover, *sawen* distributed resources in a way that was not biased to the interest of particular groups. In addition, even though the fishers had not been involved in the process of *sawen* establishment, they obeyed the rules because they had been handed down through generations. Finally, the fishers could be involved in the process of *sawen* implementation by becoming *lang-lang*. This sense of ownership of the rules strengthened compliance.

The above rules and practices demonstrate that local people had the wisdom and knowledge to maintain the integrity of their coastal environment and a harmonious social system. Rules were made on the basis of that wisdom and knowledge. The underlying myths that shaped such rules and practices were appropriate and contributed to sound resource management; this phenomenon parallels the 'theological society' of August Comte (1973).[5] The important point is that such wisdom and knowledge had been well maintained because they were handed down through generations. This intergenerational transfer of traditional knowledge was critical to resource sustainability throughout the history of North Lombok (see also Narcisse, this volume).

DECLINE OF *SAWEN* AFTER THE MID-1960S

UNFORTUNATELY *sawen* ceased to function as a traditional institution in promoting sustainability in the mid-1960s for two main reasons. The first was a theological conflict between adherents of *wettu telu*, a traditional local Islamic practice with a syncretic mix of ancient Hinduism and indigenous animism (Budiwanti, 2000), and the followers of *waktu lima*, an orthodox form of Islam that requires pure rituals in accordance with the Qur'an. As described above, *sawen* was practised by local people who believed in *wettu telu*. Despite its syncretic incorporation of ancient Hindu and indigenous animist elements and practices that differed from those prescribed

---

5. 'In this state of society, all theoretical conceptions, whether general or special, bear supernatural impress. The imagination completely predominates over the observing faculty, to which all right of inquiry is denied' (Comte, 1973).

by the Qur'an, people who adhered to *wettu telu* also considered themselves to be Moslems. However, the believers in *waktu lima* saw followers of *wettu telu* as heretics. The religious ceremonies of *sawen*, influenced by ancient Hinduism and indigenous animism values, were considered contrary to the doctrines of *waktu lima*. In some unfortunate instances, *waktu lima* believers were provoked to destroy the cultural and religious symbols of *wettu telu*, such as offering places for *sawen*, and to disrupt or suppress religious ceremonies of *sawen*, and any practices influenced by *wettu telu* doctrine.

The theological conflict between *wettu telu* and *waktu lima* was, in fact, a relic of the Dutch colonial policy of *divide et impera* (divide and conquer) that re-emerged in various forms in the mid-1960s, including attempts to destroy the cultural and religious symbols of *wettu telu*. *Waktu lima* adherents seized on the eradication of communism after the fall of the Indonesian Communist Party (PKI) in 1965 as a glorious opportunity to displace *wettu telu* adherents in both rural and resource governance. There was, in fact, no connection between *wettu telu* and the PKI, and no communist influence on the tenets of *wettu telu*.

The second factor was the massive process of modernization in Indonesia. The eradication of communism in 1966 marked the beginning of the New Order regime of President Suharto. During this period, waves of modernization in the economy, politics and culture engulfed the country, coinciding with modernization processes that were flooding the world at the time. As a result, many traditional management practices were rejected as inappropriate and old fashioned. Various national laws and regulations were enacted and imposed systematically downwards from the national to the local level. The *Undang-Undang No 5/1979* (The Rural Governance Law of 1979) is a good example of a national law that did not recognize traditional marine tenure and traditional fisheries management practices. It was intended to systematize the rural governance system, and had no place for local systems of resource management based on ancient custom and belief. Accordingly, *sawen* and its infrastructure in North Lombok was formally abandoned and replaced by 'modern' institutions.

FISHERIES MANAGEMENT WITHOUT *SAWEN* DURING THE NEW ORDER REGIME

Under the New Order, marine and fisheries matters were managed by the central government (Satria and Matsuda, 2004a). Fisheries Law No 9/1985 did not delegate any authority to the local government or local people. This centralization emulated industrialized nations, which offer no role to common-property regimes in fisheries (Gibbs and Bromley, 1989; Ruddle, 1996). Centralization brought in decision-making processes by central government departments, a shift from indigenous to 'scientific' knowledge systems, and nationalization of resources, which undermined and even dismantled local institutions (Berkes, 2002). Accordingly, there was no sense of responsibility, participation or stewardship to motivate local people to manage marine resources (Satria and Matsuda, 2004b). Moreover, the transaction costs of centralized monitoring and surveillance were high, and the available financial resources and personnel were inadequate. As a result, the regulations created by the central government were not enforceable. Thus the abandonment of *sawen*, coupled with the failure to enforce the new national laws that should have replaced it, led to a *de facto* open access situation, which in turn led to resource depletion and social conflict among fishers. This supports the idea that centralization fails to create effective and efficient fisheries management (Satria and Matsuda, 2004a).

In Lombok, many formal regulations made during the New Order regime were not enforceable. An example was Decree No. 607/1976 of the Minister of Agriculture, which set up a zoning system for capture fisheries. The decree was promulgated to overcome social conflicts, such as those of the later 1970s when modern trawlers were competing with traditional fishers in the archipelagic waters (Bailey, 1997; Satria, 2001). It appeared to be an ideal regulation to protect small-scale fishers, but it did not work properly; modern trawlers found ways to work round the decree. Social conflict and illegal fishing were the inevitable results.

A second example is the prohibition of destructive fishing practices, such as the use of poison and explosives. Regulations under the Fisheries Law of 1985 provide for court sanctions against those who use such destructive methods. However, these practices continued, and indeed seemed to increase (Satria and Matsuda, 2004a), as those responsible seemed to be 'untouched' by the law. In short, when *sawen* ceased to exist, the local fishers felt like strangers or aliens in their own home; they felt that they were 'helpless' and 'overpowered' in fisheries resource extraction practices as many outsiders came in to the area and exploited the resources in inappropriate ways.

TURNING THE TIDE: REVITALIZATION OF *SAWEN*

THE fall of President Suharto in 1998 marked the end of the New Order regime and the beginning of a Reform Era in Indonesia. This was a critical period, as political instability meant that the government lacked the accountability and authority to enforce formal rules in marine fisheries. The New Order's legacy of low enforcement was exacerbated by the turmoil of the early Reform movement, which created a phenomenon of 'stateless' areas throughout much of the archipelago. In this political vacuum, local people came to assume a new role as 'regulators'. The phenomenon of self-regulation in marine fisheries enabled local people to replace various formal rules by turning to their own traditional institutions. In the case of North Lombok, a revitalization of *sawen* resulted.

Several factors triggered the revitalization of *sawen*. First was the economic crisis that followed the 1998 Reform movement. This fuelled a 'gold-rush' attitude, leading to marine resource depletion. Destructive fishing practices were practised more extensively than ever before in North Lombok. Second, the concurrent political situation enabled the local people to become more autonomous in marine resources governance. The local fishers revitalized *sawen* to address issues of overexploitation, access rights and lack of enforcement of fishing regulations in their offshore and nearshore waters. A third factor was a growing awareness among the people that they needed to rediscover their marine local-cultural identity by reviving *sawen* practices that had been lost under the New Order regime. *Sawen* was recognized as a site-specific traditional resource governance system that comprises the local or traditional ecological knowledge, rules of conduct and worldview necessary for restoring local marine resources.

The process of *sawen* revitalization started in 1998 when the Reform era began, but was not formally enacted until August 2002 by the fishers' union in Kayangan with the village government's recognition. Revitalization of *sawen* was a truly bottom-up process, achieved purely by local people without any outside help or intervention.

Referring to Scott's (2001) analysis of the scope of institutions, the revitalized institution of *sawen* consists of three pillars: regulative, normative, and cognitive. The *regulative* pillar specifies what rules of conduct are established and how they are maintained. Behavioural and value standards are embedded in the second, *normative* pillar. Resources of knowledge for decision making reside in the third, *cognitive* pillar. Institutions fail when these pillars are weak, as noted by Jentoft (2004): 'the rules that regulate behaviour may be underdeveloped or poorly enforced; the normative standards may provide few incentives and little guidance; the knowledge that could inform decision-making may be inadequate or insufficient.'

The regulative pillar

The regulative pillar includes what Ruddle (1999) identified as territorial boundaries, rules, rights, authority, monitoring and sanctions. The territorial boundary of *sawen* extends 1.5 km out to sea and 4.5 km along the coast, following the jurisdictional borders of Kayangan village. A clear territorial boundary depicts clear ownership and is necessary to prevent overexploitation (Marten, 2001). It is also an indicator of an enduring institution (Ostrom, 1990; Dolsak and Ostrom, 2003).

When the revitalized *sawen* was formally enacted in August 2002, a set of rules was established called the *Awig-Awig Penyawen Kelompok Nelayan Pantura*.[6] The

---

6. The term awig-awig refers to local rules which are included in the regulative pillar. Accordingly in this paper awig-awig is part of sawen institutions.

rules contain two categories of orders and prohibitions, relating to fishing and to the environment:

a) It is forbidden to catch any marine organism using nets, cast-nets, traps, and catching juvenile fish during the closed season.
b) Fish have to be individually ungilled from the net prior to taking them home.
c) Fish may not be scooped from the water using a cloth and/or sarong.
d) Ornamental fish may not be caught.
e) The use of dynamite and poison fishing is forbidden.
f) Trawl-nets, gillnets, and drive-in (*muro ami*) nets may not be used in the area.
g) It is forbidden to mine sand, rocks and coral in the coastal area.

These constitute what Ostrom (1990) categorized as operational rules. Compared with the original *sawen* (Table 10.2), which mainly focused only on instituting closed seasons, these rules (Table 10.3) extend to other contemporary issues (such as coral mining) and are better suited to current conditions.

TABLE 10.3 *Goals of revitalized* sawen *rules and sanctions for violation.*

Rules	Social	Economy	Ecology	Myth	Sanctions
1. Prohibition on catching fish using nets, cast nets, traps, and catching juvenile fish during closed season			X	X	Public embarrassment
2. Fish must be individually ungilled from the net prior to taking them home				X	Public embarrassment
3. Prohibition on scooping out fish from the water using cloth and/or sarong				X	Public embarrassment
4. Prohibition on catching ornamental fish			X		Rp 500,000 (US$ 55)
5. Prohibition on practising dynamite and poison fishing			X		Rp 2,500,000 (US$ 277), and confiscation of boats and fishing gear
6. Prohibition on operating trawl-net, gillnet, and drive-in net in the area	X	X			Rp 15,000,000 (US$ 1,666) and confiscation of boats and fishing gear
7. Prohibition on harvesting sand, rocks and coral in the coastal area			X		The sanction will be determined once violation occurs

Source: Modified from Scott (2001), Nielsen (2003), Ruddle (1999) and Jentoft (2004).

Monetary penalties for those who violate the rules range from US$ 55 to US$ 1,666 (Table 10.3). These fines are especially targeted at violators from outside the community who do not respect moral sanctions. However, local fishers who violate, for example, rules (a), (b) and (c) will suffer moral-based penalties such as public embarrassment.

The rules are enforced by volunteers called *lang-lang laut* from the local fishing community. The *lang-lang*'s duties include monitoring, control and surveillance of the *sawen* institution. The *mangku laut* is the final authority for imposing sanctions, as well as the leader of religious ceremonies for marking closed and open fishing seasons. In revitalized *sawen*, both the *lang-lang* and the *mangku laut* are traditional authorities with similar characteristics to those in the original *sawen*. However, the new *mangku laut*, though still deriving authority through family lineage, lacks much of the supernatural power, wisdom and knowledge of earlier times. Nevertheless, he is seen as a legitimate authority, because the local fishers still believe that *mangku laut* descendants must have hidden capabilities and power, although such capabilities have not yet been empirically proved. Another difference is that, in the original *sawen*, the *mangku laut* was the sole authority, while in revitalized *sawen* the power is shared.

In the study area, the *mangku laut* is part of a more powerful fisheries management authority, the *Kelompok Nelayan Pantura Penyawen Teluk Sedayu* (Northern Coast Fishers Union) (Figure 10.5). This fishers' union has taken over some functions of the *mangku laut*, such as determining the length of the closed season. The role of the *mangku laut* is therefore limited to religious ceremony and moral sanctioning. In original *sawen*, knowledge was handed down through generations through an extremely effective 'learning by doing' mechanism. The thirty-year time lag between the termination and revitalization of *sawen* interrupted the intergenerational transfer and so weakened comprehension, although some knowledge transfer is facilitated by the fishers' association. The general regulative aspects of original and revitalized *sawen* are compared in Table 10.4.

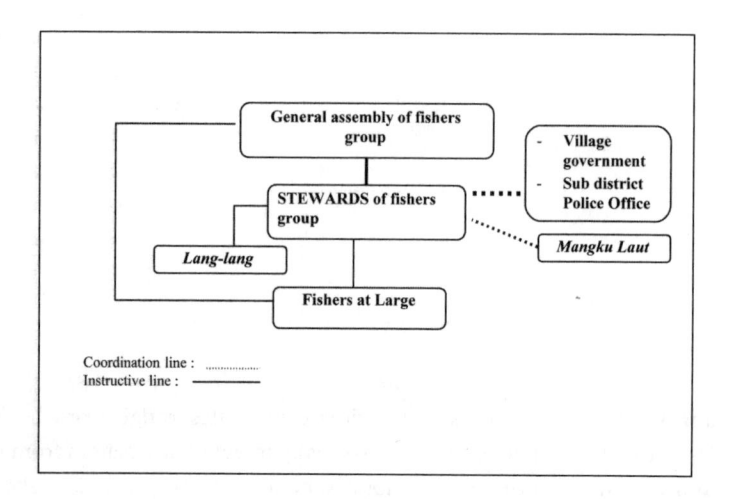

Figure 10.5 *Organization structure of* Kelompok Nelayan Pantura Penyawen Teluk Sedayu, Kayangan.

TABLE 10.4 *Comparison of original and revitalized* sawen.

Indicators	Original sawen	Revitalized sawen
1. Regulative		
Territorial boundary	Clear	Clear
Operational rules	Focus on closed system	Extended to cover destructive fishing practices and unjust fishing
Marine resources authority (*mangku laut*)	Ascribed status, wide scope of authority (fisheries management, ceremony of closed and open season, final decision of sanctions)	Ascribed status, but lacking spiritual and intellectual capability; limited functions in ceremony of closed and open season
Authority body	*Mangku laut* (personal)	Fishers' union (organizational)
Monitoring	By *lang-lang*, accountable	By *lang-lang*, accountable
Enforceability of rules	High	High
2. Normative		
Source of norms	*Sasak* culture	*Sasak* culture, market
Goals	Respect for supernatural power and ecological concern	Extended to create justice in fisheries, economic gains, and healthy coastal environment
Legitimacy	Due to rooted traditional norms	Due to consciousness of relevance of past traditional norms for current situation
Compliance	Avoiding moral sanctions	Based on legitimacy and economic gain (normative and instrumental)
Sanctioning	Moral	Moral, economic (fines, confiscation) and formal law
3. Cognitive		
Transfer of knowledge	Handed down through generations	Intergenerational transfer facilitated by fishers union
Comprehension of knowledge	Strong	Weak, due to the three-decade time lag
Application	Based on myths	Based on myths and common scientific rationales

Note: The indicators are modified from Scott (2001), Nielsen (2003), Ruddle (1999) and Jentoft (2004).

The perceived legitimacy of revitalized *sawen* makes the rules highly enforceable. Borrowing the criteria for good rules from Marten (2001), the rules of revitalized *sawen*: (a) are simple, so all local fishers know what is expected; (b) are fair, because no one will willingly to make sacrifices for the selfish gain of others; and (c) produce benefits that exceed the costs of operation. Legitimacy is further enhanced because most of the fishers are involved in setting the rules. The more directly involved the fishers are in installing and enforcing the regulations, the more the system will be accepted as legitimate (Nielsen, 2003).

The normative pillar

The normative pillar includes values (that is, conceptions of the preferred or the desirable) and norms (prescribing how things should be done and legitimate means to pursue valued ends). Both define goals or objectives, and are regarded as the basis of stable social order (Scott, 2001). The values and norms of revitalized *sawen* stem from the customs of *sasak*, a recent form of orthodox Islam that is more compatible with *wettu telu*. As a result, many ritual ceremonies of revitalized *sawen* are now influenced by the rituals of orthodox Islam, though some animistic practices still coexist.

The goals of the revitalized *sawen* are broadly defined as resource management that is ecologically sound, socially harmonious, economically just and culturally rooted. Table 10.3 shows that ecological goals predominate. This is similar to the original *sawen*, which contributed to ecological sustainability even though the ecological goals were formulated through mythological rationales that respected supernatural power. The dominance of ecological goals shows a revived sense of how humans should interact with nature. For example, ornamental fish are economically valuable, but fishing for them is prohibited because of the destructive methods involved. A sustainable way of catching ornamental fish in the North Lombok area has not yet been found.

In the normative sense, the revitalized *sawen* is legitimate because *sawen* values and norms have been modified to fit the present situation in which fishers' interests and market values operate. For example, the value of justice in fishing is embodied in the prohibition on trawls and drive-in nets in the *sawen* area. This creates an exclusive-access right for the local traditional fishers. Fines for those who violate such rules have now replaced the moral and spiritual aspects of *sawen* (Table 10.3). The original *sawen* was not based on monetary sanctions or fines, and social ostracism was the norm. The rise of monetary sanctions shows that revitalized *sawen* takes market value into account, making the rules effectively enforceable. A detailed comparison between the normative aspects of original and revitalized *sawen* is presented in Table 10.4.

The cognitive pillar

Table 10.3 also exemplifies the way that the rules of revitalized *sawen* are shaped by a different cognitive system, where the modern meaning of such rules replaces some myths (Table 10.4). Blast fishing, for example, is now not allowed because there is collective awareness that this method threatens resources, not for fear of supernatural sanctions. Contrast the original *sawen* where underlying myths dominated and collective actions were motivated by the dominant narrative account of the sacred.

The triad of forest–farmland–sea (Figure 10.3), central to the original *sawen*, is no longer applicable in the revitalized *sawen* of Kayangan, because of the decline of forest resources in North Lombok. The deterioration of the forest should have triggered the local people to manage the issue and incorporate forestry into the integrated resource management concept, as they did in the past. Yet the local people seem not to know how the forest should be managed in an integrated way with the marine resources. This may be because they were squeezed out of forestry under the New Order regime, when government took control of forest resource management and government-authorized companies took the lion's share of forest resources.

However, a new understanding of integrated resources management is emerging in the revitalized *sawen*. The triad of forest, farmland and sea is now replaced by a dual concept, where farmland and sea are understood as a unity, requiring integrated management. According to the new understanding, the condition of the sea affects the prospect for farmlands. If marine resources decline, farmlands will be in danger. The condition of farmlands becomes an indicator of the sea's condition. Two reasons underlie this concept. The first is a mythical belief that unseen or incorporeal beings which normally inhabit the sea will move to the land to take farming products if they cannot find enough fish and other living marine resources. Farm pests and diseases that attack the farms are seen as the effect of this mythical influence. The second belief, more rooted in the natural world, it is that a particular species of seabird (locally known as *burung cecerak)*, which usually lives off aquatic resources, will move to the land to take farm products if it cannot find its regular marine prey. This second reason seems quite convincing to the local people, although it needs further verification. Whatever the validity of these reasons, the important point is that they both motivate pastoral people to be concerned about the sustainability of marine resources, and fishers to take account of the needs of the land.

POSITIVE IMPACTS OF REVITALIZED *SAWEN*

IN what ways has the revitalized *sawen* benefited the area? It has had a number of positive impacts. The first is the restoration of a marine cultural identity for the local community through the revived values, norms, and cultural symbols (traditional ceremonies) of *sawen*; community pride in their way of life has clearly been renewed. This implies that fisheries are not considered only as a livelihood, but also as a way of life and a culture with a specific worldview.

The second benefit is protection for small-scale fishers. The prohibition of trawls, drive-in nets and other larger-scale fishing methods in Kayangan ensures exclusive-access rights for local traditional fishers. This also has indirect benefits such as a reduction in social conflict and a hypothetical improvement in the traditional fishers' income.

The third benefit is that the revitalized *sawen* provides insights (local knowledge and wisdom) for the implementation of local fisheries management. It also motivates the local people to recover traditional ecological knowledge as a complement to common or conventional scientific knowledge.

The revival of *sawen* has created a legitimate institution of community-based fisheries management (CBFM). Legitimacy is crucial to the ability of CBFMs to work effectively with government.

Finally, local fishers perceive local resources to be recovering, though this has yet to be scientifically proven. This ecological benefit is the result of prohibiting destructive fishing practices, coral and sand mining, and trawl-netting. The local perception that these rules have benefited the resources should be followed by empirical analysis in partnership with ecological scientists.

FACTORS AFFECTING THIS SUCCESS STORY OF REVITALIZED *SAWEN*

THE first and most crucial factor in the success of the revived *sawen* system is the prevalent bottom-up approach that has given it legitimacy. This is evidenced by the local people's initiative in revitalizing *sawen* without involving any external agencies such as local government or NGOs. The consensus was built through both territorial and functional representation.[7] Customary figures (*tokoh adat*), religious figures (*tokoh agama*), fisher figures (*tokoh nelayan*), village government, and representations of each sub-village (*dusun*) all helped to build this consensus.

The second factor is homogeneity, in terms of culture, livelihood and the economic scale of the fisheries. Kayangan is dominated by the *sasak* ethnic group. This homogenous cultural background facilitates consensus building, making it easier to revive values and norms. The local people realize that the past conflict between *waktu lima* and *wettu telu* need not recur or constrain present arrangements. There is mutual understanding among them. Their main and unifying concern is how to overcome present and future resource depletion. They agree that, as a traditional system, *sawen* is necessary and effective in the current resource situation. Homogeneity in the scale of the fisheries also helps to make *sawen* robust because it minimizes conflicts of interest among fishers.

The third factor is organizational networking. At the regional (North Lombok) level, the fishers' union (LMNLU)[8] was established in 2000 and its jurisdiction covers all villages including Kayangan (Satria and Matsuda, 2004b). Accordingly, the activities of the *Kelompok Nelayan Pantura Penyawen Teluk Sedayu* are coordinated by LMNLU, which represents fishers across all sub-districts (*Kecamatan*) in North

---

7. Territorial representation refers to a situation where the representatives are expected to speak for a particular geographical area, whereas functional representation refers to a situation where the representatives are expected to speak for particular activities (Jentoft, 2003).
8. Lembaga Musyawarah Nelayan Lombok Utara (LMNLU) or Representative Council of North Lombok Fishers, an organization established by the North Lombok fishers.

Lombok. One of the union's aims is to eliminate destructive fishing practices. This mission coincides with the rules devised in revitalized *sawen*, which is therefore fully supported by the LMNLU in Kayangan.

The fourth factor is local government recognition. This means that the rights of local people to devise their own institutions are not challenged by external governmental authorities. This is because *sawen* is, in fact, compatible with formal rules set by national and provincial governments. For example, the prohibition of destructive fishing practices is in line with the Fisheries Law (1985) and the Environmental Law (1992). Thus, supporting *sawen* assists the local government effort to enforce formal rules. Recently, the local government drafted a regulation *(perda)* recognizing and supporting institutions devised by local people. Such recognition is becoming more common in Indonesia now, as decentralization of marine resources management is required by the Local Autonomy Law No. 22/1999 (Satria and Matsuda, 2004a). Under this law, the local government has jurisdiction over marine resources management within 4 miles (6.4 km) of the coastline.

Meanwhile, revitalized *sawen* has some weaknesses: the three-decades that followed the demise of the original *sawen* diminished the capability of the newly appointed *mangku laut* and impoverished the community knowledge base, particularly relating to the integrated relationship embodied in the ecological triad.

On the credit side, s*awen* in Kayangan is more robust than, for example, the *awig-awig* community-based coral reef management system in Gili Indah.[9] This is partly because the process of revitalization of *sawen* has been more bottom-up, without intervention from external agencies such as government and NGOs. Also, the content of the rules coincides with the values, norms, and the interest of the fishers, and in terms of social structure, Kayangan fishers are more homogenous than Gili Indah fishers. Critically, however, one of the keys to success in Kayangan is the existence of traditional knowledge and historically rooted practices of resource management. Gili Indah does not have such a historical-traditional system.

CONCLUDING REMARKS

TRADITIONAL systems (local knowledge, method, rules and worldview) are the underlying cultural capital for current resource management. Knowledge, method, rules and worldview are interrelated. Local knowledge is an important cultural resource that guides the implementation of customary management systems (Ruddle, 2000). It is also the basis for the emergence of rules. However,

---

9. Based on presentation by author at 4th World Fisheries Congress, 6 May 2004 in Vancouver, entitled 'Awig-awig: Community Based Coral Reef Management in West Lombok, Indonesia'.

knowledge is shaped by culture, and knowledge, in turn, shapes culture (Ruddle, 2000). Knowledge and institutions depend on each other (Wilson, 2003). *Sawen* represents a traditional system that encompasses these interrelated components (local knowledge, method, rules and worldview). The success of *sawen* described in this chapter strengthens the evidence that traditional systems can be a viable alternative for future resource management and can complement the conventional 'scientific' approach. It is encouraging that collaboration between traditional ecological knowledge and science has been facilitated in recent years by the decentralization of fisheries management since the enactment of the Local Autonomy Law in 1999, and by increasing recognition of the benefits of parallel development of traditional and 'scientific' approaches under decentralization.

ACKNOWLEDGEMENTS

SOME of the ideas presented in this chapter were conceived during a visiting studentship in the summer of 2004 at the Fisheries Centre of the University of British Columbia (UBCFC) in Vancouver, Canada, and presented in a special seminar of the FISH 500 seminar series. I extend my gratitude to Professor Daniel Pauly and his colleagues, fellow students and staff at the Centre for receiving me with open arms and for the free exchange of ideas and fruitful collaborations. I thank Mr Nigel Haggan for inviting me to contribute this chapter, the co-editors for approval, and the referees for critical comments and fruitful inputs. My thanks also go to Eny Buchary for proofreading and criticizing the early draft of the manuscript and for helping me in conceptualizing the graphical representation of *Paer* philosophy as depicted in Figure 10.2. Also my thanks to Gaku Ishimura for his critical comments. Funding for my visiting studentship at UBCFC was mainly from the Monbugakusho (the government of Japan). My sincere thanks also go to Professor Dr Yoshiaki Matsuda and Asst. Prof. Dr Masaaki Sano for their supervision during my PhD at Kagoshima University Japan. I thank Her Excellency Madame Consul-General Binarti Fadjar Sumirat and her staff at the Consulate General of the Republic of Indonesia in Vancouver and His Excellency Ambassador Eki Syachrudin and his staff at the Embassy of the Republic of Indonesia in Ottawa for providing their generous support during my wonderful stay in Vancouver. More importantly, I would like to gratefully acknowledge the cooperation and support that I received from the fishing communities in Kayangan and other community members in North Lombok, Kamardi, Mahsum, Imtiha, Akhmad Solihin, Eva Angraini, Suhana and Fisheries Service officers of West Lombok. I acknowledge partial research funding from the Ministry of Marine Affairs and Fisheries-Indonesia, Bogor Agricultural University, Center for Coastal and Marine Resources Studies IPB, and Center for Agrarian Studies IPB.

REFERENCES

ABERCOMBIE, N.; HILL, S.; TURNER, B.S. 2000. *The Penguin dictionary of sociology.* London, Penguin, 480 pp.

BAILEY, C. 1997. Lesson from Indonesia's 1980 trawler ban. *Marine Policy,* Vol. 21, No. 3, pp. 225–35.

BERKES, F. 1999. *Sacred ecology.* Traditional ecological knowledge and resource management. Philadelphia, Taylor and Francis, 209 pp.

——. 2002. Cross-scale institutional linkages: Perspectives from the bottom up. In: E. Ostrom, T. Dietz, N. Dolsak, P.C. Stern, S. Stonich and E.U. Weber (eds.), *The drama of the commons.* Washington, DC, National Academy Press, pp. 293.

BUDIWANTI, E. 2000. *Islam* sasak: Wetu telu *versus* waktu lima. Yogyakarta, Indonesia, LKIS, 387 pp.

CHARLES, A. 2001. *Sustainable fishery systems.* Oxford, UK, Blackwell Science, 384 pp.

COMTE, A. 1973. The progress of civilization through three states. In: E. Etzioni-Halevy, Amitai Etzioni (eds.) *Social change: Sources, patterns and consequences.* New York, Basic Books.

DOLSAK, N.; OSTROM, E. (eds.). 2003. *The commons in the new millenium: Challenges and adaptation (politics, science, and the environment).* London, MIT Press, 369 pp.

GIBBS, C.J.N.; BROMLEY, D. 1989. Institutional arrangement for management of rural resources: Common-property regimes. In: F. Berkes (ed.), *Common property resources: Ecology and community-based sustainable development.* London, Belhaven, pp. 22–32.

HAGGAN, N.; BROWN, P. 2002. Aboriginal fisheries issues: The west coast of Canada as a case study. In: D. Pauly and M.L. Palomares (eds.), *Production system in fisheries management.* Vancouver, Fisheries Centre Research Reports, Vol. 11, No. 1, University of British Columbia, pp. 17–19.

JENTOFT, S. 1989. Fisheries co-management: Delegating government responsibility to fishermen's organization. *Marine Policy,* Vol. 13, No. 2, pp. 137–54.

——. 2003. Representation in fisheries co-management. In: D.C. Wilson, J.R. Nielsen and P. Degnbol (eds.), *The fisheries co-management experience: Accomplishment, challenge and prospect.* Dordrecht, Kluwer Academic, pp. 281–92.

——. 2004. Institutions in fisheries: What they are, what they do, and how they change. *Marine Policy,* Vol. 28, pp. 137–49.

KAMARDI. 1999. *Kearifan tradisional dan aspek ekologis* [Traditional wisdom and ecological aspects]. Paper presented at Seminar Pemberdayaan Masyarakat Adat Nusa Tenggara Barat. Mataram, 28 July 1999.

KAY, R.; ALDER, J. 1999. *Coastal planning and management.* New York, Routledge.

MARTEN, G. 2001. *Human ecology: Basic concepts for sustainable development.* London, Earthscan, 238 pp.

NIELSEN, J.R. 2003. An analytical framework for studying: Compliance and legitimacy in fisheries management. *Marine Policy,* Vol. 27, pp. 425–32.

OSTROM, E. 1990. *Governing the commons: The evolution of institutions for collective actions.* Cambridge, Cambridge University Press, 280 pp.

RUDDLE, K. 1996. Formulating policies for existing community based fisheries management systems. *The Journal of Policy Studies,* No. 1.

——. 1999. The role of local management and knowledge systems in small-scale fisheries. *The Journal of Policy Studies,* No. 7.

——. 2000. System of knowledge; dialogue, relationships and process. *Journal of Environment, Development, and Sustainability,* Vol. 2, pp. 277–304.

SATRIA, A. 2001. *Dinamika modernisasi perikanan: Formasi sosial dan mobilitas nelayan* [The dynamics of fisheries modernization: Social formation and fishers' mobility]. Bandung, Humaniora Press, 153 pp.

SATRIA, A.; MATSUDA, Y. 2004a. Decentralization of fisheries management in Indonesia. *Marine Policy,* Vol. 28, 361–450.

——. 2004b. Decentralization policy: An opportunity for strengthening fisheries management system? *Journal of Environment and Development,* Vol. 13, No. 2, pp. 179–96.

SCOTT, W.R. 2001. *Institutions and organizations* (2nd edition). Thousand Oaks, Sage.

SOLIHIN, A. 2002. *Analisis awig-awig dalam pengelolaan sumberdaya perikanan di kecamatan Gangga, Kabupaten Lombok Barat, Nusatenggara Barat* [Analysis of awig-awig of fisheries management in Gangga, West Lombok, West Nusatenggara]. Bogor, Jurusan Social Ekonomi Perikanan, Fakultas Perikanan dan Ilmu Kelautan IPB, undergraduate thesis, 94 pp.

WILSON, D.C. 2003. Fisheries co-management and the knowledge base for management decisions. In: D.C. Wilson; J.R. Nielsen; P. Degnbol (eds.), *The fisheries co-management experience: Accomplishment, challenge and prospect.* Dordrecht Kluwer Academic, pp. 265–79.

CHAPTER 11 Fishers' perceptions of the seahorse fishery in the central Philippines
Interactive approaches and an evaluation of results
Jessica Meeuwig, Melita Samoilys, Joel Erediano and Heather Koldewey

ABSTRACT

WE conducted a study in coastal communities in the central Philippines designed to involve seahorse fishers in research and conservation initiatives. The study comprised, first, an initial scoping survey to obtain data on the fishers and their fishery, including effort and habitat quality; and second, community meetings conducted as focus group discussions, in which results from the scoping study were fed back to the communities, questions were repeated, and information was gathered about fishers' knowledge and opinions with respect to the seahorse fishery, the state of their fishing grounds, and the condition of their livelihoods. Discussions on marine resource management were also held.

Participatory methods making much use of visual aids were designed to facilitate communication and discussion. The scoping survey collected information from 173 seahorse fishers in 19 communities on location and quality of fishing grounds, and fishing effort, while the community meetings collected information from 117 fishers in 10 focal communities. Average effort was reported in the scoping survey and community meetings as 111 and 192 trips (nights) per fisher per year, and 334 and 894 trips per fishing ground per year, respectively. Habitat quality of fishing grounds was generally assessed as good in the scoping survey and community meetings, but live coral was not commonly perceived as the dominant habitat type.

Responses differed markedly from independent ecological surveys of the same fishing grounds. A comparison of the answers provided by fishers in the scoping study and community meetings indicated that although absolute values differed, relative estimates of fishing effort per fishing ground and effort per fisher corresponded well across the two surveys. Fishers consistently described seahorse abundance, habitat quality and their livelihoods as in decline, and proposed a number of solutions. Through our participatory approach, seahorse fishers are playing a role in designing applied fisheries research, and in developing management plans for their fishery.

INTRODUCTION

STAKEHOLDER involvement in the planning and implementation of conservation initiatives is considered fundamental to the achievement of resource management objectives (Akimichi, 1978; Johannes, 1981, 1982; Polunin, 1983, 1984; Wright, 1985; Zann, 1985; Johannes, 1989; Bailey and Zerner, 1992; Ruddle et al., 1992; Ruddle, 1994; Jennings and Polunin, 1996; Walters et al., 1998; Neis et al., 1999; White and Vogt, 2000). Participatory approaches to resource management have a number of benefits:

- Stakeholders may have specialized knowledge relevant to resource management that is accessible only through collaborative approaches.
- The process transfers knowledge and builds stakeholder management capacity.
- Compliance with resource management decisions is more likely if stakeholders participate in their establishment.

There are a number of examples of stakeholder involvement in the management of tropical marine ecosystems. Local knowledge of fish behaviour has been harnessed in the management of South Pacific fisheries (Johannes, 1981, 1982; Jennings and Polunin, 1996; Cooke et al., 2000). Capacity building lies at the heart of community-based resource management initiatives in the Philippines (White, 1988; Vincent and Pajaro, 1997; Walters et al., 1998; Alcala, 1998, 1999; White and Vogt, 2000; Alcala, 2001). The integrity of community-based marine protected areas relies heavily on stakeholder compliance, which in turn increases with understanding and agreement based on involvement in the process of establishing these areas (Johannes, 1982; 1989; Gulayan et al., 2000; Pajaro et al., 2000; Alcala, 2001; Baelde, this vol.).

Interest in participatory approaches in resource management partly reflects the failure of top-down, centralized approaches to manage natural resources (Murdoch and Clark, 1994; Agrawal, 1995; Maguire et al., 1995; McClanahan et al., 1997; Sillitoe, 1998; White and Vogt, 2000). Bottom-up, community-based approaches (BOBP, 1990; Walters et al., 1998) that involve stakeholders may be more appropriate where resource exploitation is diffuse, as is typically the case with subsistence fisheries (Pauly, 1997), and where human and financial resources are limited (White and Vogt, 2000).

As part of a seahorse conservation programme (Project Seahorse, www.projectseahorse.org) we initiated a participatory research project in 1999 on the seahorse fishery of Danajon Bank, Bohol, central Philippines (Figure 11.1). There is so little official information on fishing in the region that basic information on the number of fishers, where, when and how much they fish was sought directly from the fishers. The process, particularly the initial scoping strategy, was relatively structured. By following the scoping interview with the community meetings, we were able to assess the consistency of the answers, and thus the degree to which we were asking questions in a way that made sense to the fishers. Filipino community organizers and biologists who have worked effectively with these communities over a long period provided invaluable assistance with the design and choice of questions.

Danajon Bank is a double barrier reef stretching approximately 145 km along the north-west coast of Bohol (Pichon, 1977; Figure 11.1). The reef system is shallow (approximately ≤10 m), silty, and composed of scattered and patchy coral reefs interspersed with *Sargassum* and seagrass (personal observation). Fishing is the primary source of income for communities located on islands in this system. Seahorse fishing began in the 1960s as part of a subsistence/cash-income fishery termed the lantern fishery. Fishers free-dive at night on shallow (1–5 m) fishing grounds, using a kerosene lantern strapped to the front of their small boats (4 m outrigger canoes called *bancas*) to illuminate prey items. They spear fish, catch crabs and hand-pick seahorses and holothurians (sea cucumbers) that they find. Lantern fishing is the main method used to collect seahorses in this region (Vincent and Pajaro, 1997), though not all lantern fishers collect seahorses. (Hookah divers also catch a limited number of seahorses.)

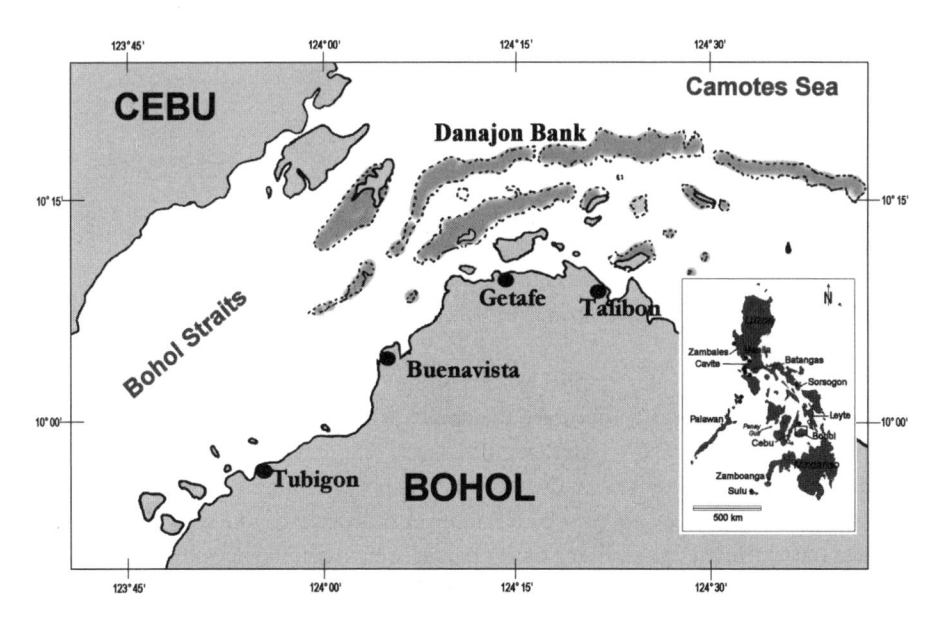

FIGURE 11.1 *Map of the Philippines showing the study area of Danajon Bank in northern Bohol, central Visayas.*

We developed a participatory approach that involved the exchange of information about marine resources on Danajon Bank between lantern fishers and researchers, and among fishers. Stakeholder inclusion was incorporated in the fisheries research programme to achieve three goals:

- obtain information about fishing effort and habitat quality of fishing grounds to aid in the design of the research component of the programme
- increase fisher awareness about marine conservation issues to build stakeholder resource management capacity
- develop an understanding of what fishers believe to be key marine conservation concerns and appropriate strategies for resolving these.

Our participatory approach was unusual in that it was also designed to allow evaluation of the accuracy and consistency of the information collected on fishing grounds. We did this by comparing two interview methods and by comparing fishers' perceptions of fishing ground habitat quality with ecological measures from underwater transects (Samoilys et al., 2001) conducted on a subset of the fishing grounds. This analysis evaluated the degree of correspondence between fishers' perceptions and ecological measures of habitat quality.

METHODS

The study consisted of two components:

- an initial scoping survey
- community meetings which involved
    - sessions in which the results of the scoping survey were fed back to the fishers and the survey was repeated
    - marine resource management discussions to collect information on fishers' knowledge, opinions and actions in relation to their fishery resources.

The scoping survey and community meeting methods were developed with input from community organizers (COs) with extensive experience in community-based marine resource management. COs are trained social workers who focus on community-level social issues as opposed to family or individual-level issues. They are an integral part of many community-based resource management programmes in the Philippines (Third World Center, 1990). The presence of a Filipino CO, who was fluent in the national language and supported by a local assistant fluent in the local language, was pivotal to communicating the research methods.

Scoping survey

The scoping survey, conducted from March to May, 1999, was designed to:

- determine the number of fishers involved in the seahorse lantern fishery on Danajon Bank and their distribution among villages
- identify the number of fishing grounds exploited in the seahorse lantern fishery
- quantify fishing effort per fisher and per ground
- assess habitat quality on the fishing grounds.

This information was subsequently used to identify twenty-eight coralline fishing grounds for the ecological research project (Samoilys et al., 2001).

The CO visited nineteen seahorse fishing communities in the municipalities of Getafe, Talibon, Bien Unido, Carlos P. Garcia, Ubay and Tubigon in northern Bohol, Central Philippines (Figure 11.1). In each fishing community, the CO first contacted village leaders to explain the project and ask permission to work in the community. Lantern fishers in the community were then identified, frequently by village leaders, and interviews requested. All fishers asked to participate agreed to do the interview, a total of 199 fishers – 9.1 ± 7.7 (sd) fisher per village (Table 11.1).

Each interview consisted of a brief verbal questionnaire. Limited information on the fisher (name, number of children) and gear (lantern or hookah, and paddled or powered boat) was collected (Figure 11.2a). Fishers were then asked to list all of the fishing grounds they visited, the number of hours spent fishing per trip, the number of trips per week, weeks per month, and months per year that they fished the ground. This information allowed the calculation of perceived annual total fishing effort (hours per year) for each fisher for each fishing ground. To indicate the total fishing pressure over time and current levels, fishers also indicated the year they began fishing each ground and the last year that they went there, if they no longer fished it. With respect to the habitat quality of these largely coralline fishing grounds, fishers were asked to indicate whether the site was 'good' (*ma'ayo*) or 'bad' (*guba*), identify the major habitat types (Figure 11.2b), and rank all of the sites they fished from best (= 1) to worst (= number of sites identified). For each site, we then calculated the following fishing ground indices:

- percentage good: the percentage of fishers that identified each fishing ground as 'good'
- percentage coral: the percentage of fishers that identified live coral as the dominant habitat component of a particular fishing ground
- fisher's relative rank (FRR): the average of the rank each fisher gives the fishing ground; each rank is relative to the total number of fishing grounds ranked by a fisher (e.g. 4th of 10 sites gives a relative rank of 0.4).

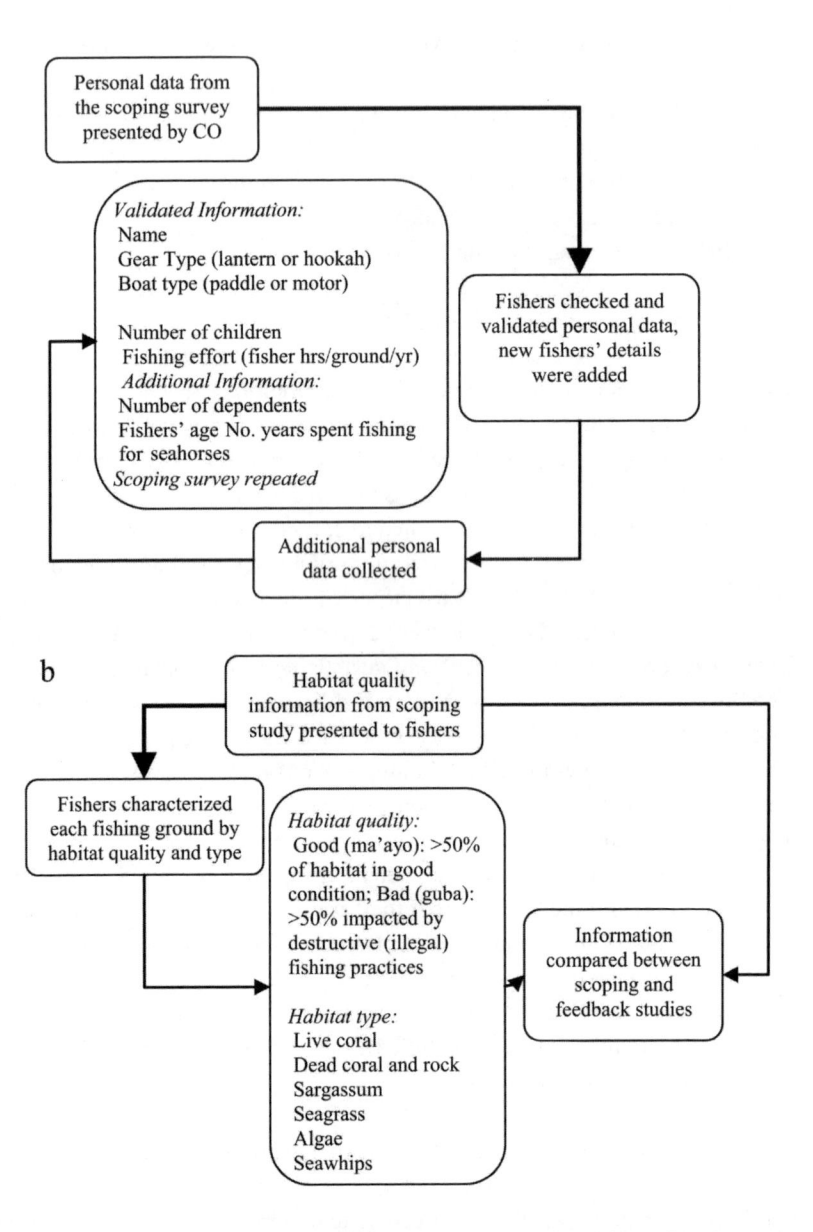

FIGURE 11.2 *Structure of (a) feedback sessions to validate personal and fishing effort data and repeat scoping survey for catch and effort data, (b) focus group discussions on fishing ground habitat type and quality.*

All three indices range from 0 to 1, where 1 indicates a good site (for example, all fishers think it is good, or all fishers identify live coral as the dominant habitat component, or it ranks at the top of their lists), and 0 indicates a poor site (say, no fishers think it is good, or no fishers identify live coral as the dominant habitat component, or it ranks at the bottom of their lists).

Community-based meetings: (a) feedback sessions

Community-based meetings were held from June to September 2000, except for one village (Alumar) which was visited in February 2001. Meetings were held with fishers in ten target villages for the feedback sessions (Table 11.1) and nine villages

TABLE 11.1 *List of villages participating in the scoping and community meetings.*

Village	Municipality	Equipment	No. of fishers interviewed	No. of lantern fishers	No. of fishing grounds/village	No. of lantern fishing grounds/village
Alumar	Getafe	Lantern and hookah	8	6	11	11
Banacon	Getafe	Lantern and hookah	6	5	7	7
Bansaan	Talibon	Lantern only	8	8	19	19
Batasan	Tubigon	Lantern and hookah	20	9	16	6
Calituban	Talibon	Lantern and hookah	4	3	3	3
Cataban	Talibon	Lantern only	15	15	7	7
Guindacpan	Talibon	Lantern only	13	13	9	9
Handay-Norte	Getafe	Lantern only	5	5	22	22
Handumon	Getafe	Lantern only	33	33	46	46
Jagoliao	Getafe	Lantern only	14	14	13	13
Nasingin	Getafe	Lantern and hookah	9	3	21	18
Nocnocan	Talibon	Hookah only	5	0	2	0
Paraiso	CPG	Lantern only	11	11	7	7
Pinamgo	Bien Unido	Lantern only	4	4	4	4
Sagasa	Bien Unido	Lantern only	3	3	2	2
Sagisi	CPG	Lantern only	4	4	5	5
Sinandingan	Ubay	Lantern only	20	20	22	22
Suba	Talibon	Lantern only	11	11	2	2
Lipata	CPG	Lantern only	6	6	6	6
			199	173	11.79	11.00

Note: Communities in bold participated in both components; others only in the scoping study.
CPG = Carlos P. Garcia municipality.

for the marine resource management discussions. These villages included those with the greatest number of lantern fishers (average of 12.6 fishers per village). The community meetings involved focus group discussions using highly visual but low-cost methods developed by one of the authors (JE) based on the Reflect method of community interviews. The effectiveness of this approach was demonstrated over the one or two-day duration of the community meetings by the relative ease of communication between researchers and fishers, who are highly knowledgeable but have low literacy levels.

The approach also allowed open-ended questions, a key characteristic for areas in which the researchers had little information. The community-based meetings also encouraged fishers and researchers to share ideas on marine conservation and fisheries management and explore the design of research programmes that would make the best use of both in the research process. The gathering of data used graphical symbols, such as cut-outs of seahorses and crabs of various sizes to indicate abundance. Fishers posted these symbols on large squared sheets with columns for each fisher (Figure 11.3). Throughout the meetings, fishers shared or validated information either through their individual worksheets or in group activities using graphic symbols and large squared sheets. In the group interactions, individual responses could still be tracked as graphic cards were uniquely numbered for each fisher.

The goals of the feedback sessions were to, first, share and validate the data collected in the scoping survey, and second, repeat the scoping survey, gather additional data, and add fishers who had been unable to participate in the scoping survey. The structure of the feedback sessions in each village is given in Figure 11.2.

FIGURE 11.3 *Focus group discussion methods using graphic symbols to solicit information from seahorse fishers.*

To repeat the questions in the scoping survey, a mixture of individual questionnaires and focus group discussions were used. The latter were used to solicit information on the lantern fishing grounds, in terms of habitat type (first identified in the scoping survey) and quality (Figure 11.2b).

Community-based meetings: (b) marine resource management discussions

The goals of the marine resource management discussions were to collect the fishers' views on:

- the relative importance of various marine resources
- the status of marine resources in the past, present and future
- the causes of resource degradation and their relative importance.

In this component of the meetings, fishers were asked to rank the six marine resources identified in the scoping survey in terms of their general economic importance both as a source of cash and food. These resources were grouped by fishers under widely differing taxonomic divisions, including order, family and genus: (i) crabs and other crustacea; (ii) fish; (iii) sea cucumbers; (iv) seahorses; (v) seaweed; and (vi) shells.

Fishers were also asked to provide information for the past (1990), present (2000) and future (2010), on three main topics: the status of their livelihood as fishers, the seahorse fishery, and the fishing grounds. Fishers were asked to assign their answers into categories. Fishing grounds were described as Good (>50 per cent of habitat is in good condition), Mixed (~50 per cent of habitat is in good condition), or Bad (>50 per cent of habitat has been damaged or destroyed). Seahorse populations were described as many, average, or few. Fishers' livelihood was described as Good (income from fishing is sufficient to support the family and provide food, education and recreation), Bad (income from fishing is barely enough to support basic necessities such as food), Very Bad (income is not sufficient to support the basic necessities). Collective discussions were then held to ask fishers for possible reasons for the trends and possible solutions, and to rank both reasons and solutions. The marine resource discussions also consisted of several sessions covering a range of topics, such as destructive fishing, particularly blast fishing, and how it affects their fishing grounds. Management options such as protected areas or sanctuaries were also discussed.

In most villages, the CO acted as facilitator for the entire group. However, for villages with more than twelve participants, fishers were subdivided into two or three groups with five or six members each, and groups were assigned different topics. A local facilitator was used for each sub-group, with the CO overseeing all groups. At the end, each sub-group reported and discussed its results with the whole group of fishers.

DATA ANALYSIS

The feedback sessions provided an opportunity to evaluate the accuracy and consistency of answers provided by fishers in the scoping survey. The two surveys differed both in terms of the fishers participating and the number of fishing grounds they considered. We analysed similarities between the two surveys for:

- All fishers and fishing grounds in the scoping survey (173 fishers and 67 fishing grounds, see fishing effort below), compared with 117 fishers and 25 fishing grounds in the feedback survey.
- Using only those fishers and fishing grounds common to both surveys, 71 fishers and 25 fishing grounds were common to both the scoping and feedback surveys.

The fishers' ranking of fishing grounds by habitat quality was compared to ecological survey data from underwater transects (Samoilys et al., 2001) conducted on a subset of these fishing grounds.

RESULTS

The ability to attract fishers was essential to the success of the community meetings. In all, 117 fishers, 68 per cent of all lantern fishers in ten villages, participated in the feedback sessions. In the marine resource management discussions, 114 lantern fishers in 9 villages participated. Feedback sessions were held in the morning and the resource management discussions in the afternoon, with 97 per cent attendance throughout the day's meeting. This high participation rate was attributed to the CO explaining the project and seeking permission in advance, the presence of a Filipino CO and the popular highly visual and graphic methods used by the CO.

Profile of Danajon Bank lantern fishers

Of the 199 fishers interviewed from nineteen villages across the Danajon Bank region, 87 per cent were exclusively lantern fishers (Table 11.1). In most villages, lantern gear was used exclusively, though hookah gear was also used in some. On average there were nine lantern fishers per village, accessing eleven lantern fishing grounds per village (Table 11.1). Fishing grounds were common to several villages.

Sixty percent of the lantern fishers in the scoping survey and 53 per cent of fishers participating in the feedback sessions still used non-motorised paddle-boats. The average number of children per fisher from the scoping survey was 4.1±2.4 (sd), and the average number of dependents from the feedback sessions was 5.2±3.0 (sd). On average, the number of children per fisher made up 80.5±35.4 per cent (sd, n = 70) of the total number of dependents. This relatively low number of children for the region probably reflects the relatively young age of the fishers: 33.6±10.8 (sd) years.

Fishers participating in the community meetings ranged from those who started fishing seahorses in 1961 to those who started in 2000. Nineteen of the fishers had stopped fishing seahorses between 1990 and 1999; the rest were still actively fishing.

Fishers gave names for 147 fishing grounds. However, reference to a map of the area indicated that these names represented 92 distinct fishing grounds, of which 73 per cent were predominantly used by lantern fishers (>95 per cent of the total effort per ground from lantern fishers), 16 per cent were used by both lantern and hookah fishers, and 11 per cent were exclusively used by hookah fishers. Nine fishing grounds were exploited in 1961, increasing to 67 in 1999, with the most rapid expansion occurring in the early 1970s (Figure 11.4). Only two grounds had been entirely abandoned (in 1999). On average, fishing grounds had been exploited for 14.5 years ± 5.7 (sd) (range 3–39).

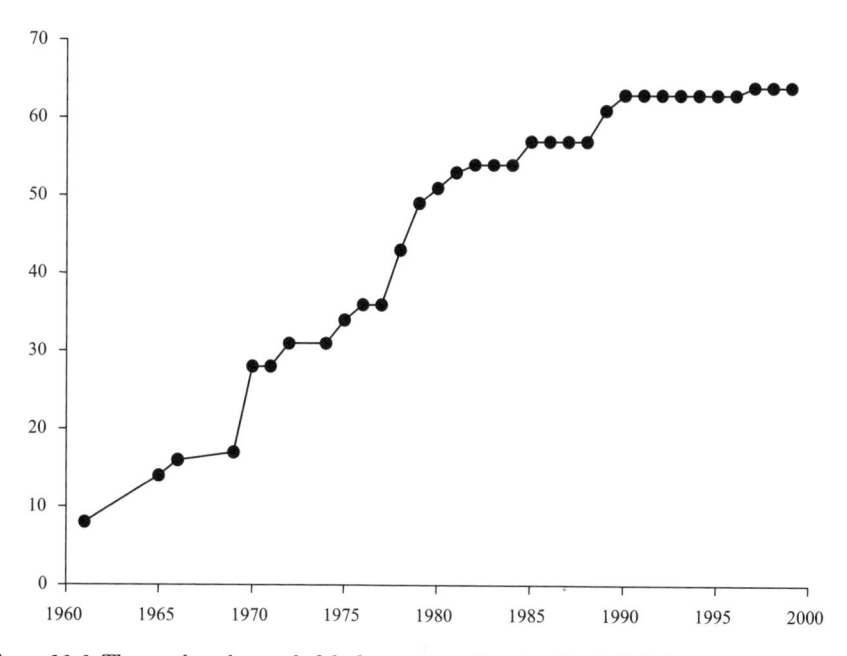

FIGURE 11.4 *The number of grounds fished per year on Danajon Bank, Bohol.*

Fishing effort

Reports of annual fishing effort per fisher and per fishing ground differed markedly between the scoping and feedback studies (Table 11.2). Considering the 67 grounds on which lantern fishing comprised at least 95 per cent of total annual effort, fishers in the scoping survey reported they were spending around 30 per cent of their nights fishing (111 fishing trips per year, Table 11.2). On average, each fishing ground was fished almost one trip per night for every night of the year (Table 11.2). In contrast, fishers in the feedback survey reported they were spending up to 50 per cent of their nights fishing on the 25 lantern fishing grounds considered (Table 11.2). Furthermore, these grounds were fished on average 2.5 trips per night for every night of the year.

Considering the subset of data for fishers and fishing grounds common to both studies, fishers in the feedback sessions reported total annual effort 2.6 times greater than that reported by the same fishers for the same grounds in the scoping study (45,665 hrs/yr vs. 17,513 hrs/yr, respectively). Annual effort per fisher within the overlapping group was significantly greater in the feedback group than in the scoping group (paired t-test, df = 70, p < 0.0005). Reported effort per fishing ground was also significantly greater in the feedback group than in the scoping group (paired t-test, df = 21, p = 0.027). Despite the absolute difference between the two groups, error estimates were relatively consistent, both by fisher (Figure 11.5a) and by ground (Figure 11.5b). Note that there was no correspondence between the estimates from fishers in Alumar and Bansaan villages, and these two outliers were therefore excluded from the analyses.

TABLE 11.2 *Annual lantern fishing effort on Danajon Bank as reported by fishers from the scoping and feedback surveys.*

	Fishing trip duration		Total fishing effort		Fishing effort per fisher		Fishing effort per ground	
	Hours	Hours	Trips	Hours	trips	Hours	Trips
Scoping survey (n = 173)	~4	76,562	19,141	444	111 (82)	1,334	334 (539)
Feedback sessions (n = 117)	3.5 (1.8)	75,114	21,653	671 (519)	192 (148)	3,129	894 (1,254)

Note: Figures in parentheses are standard deviations.
Fishing trip duration was not asked in the scoping survey: the value is an approximation.
n refers to the number of fishers interviewed.

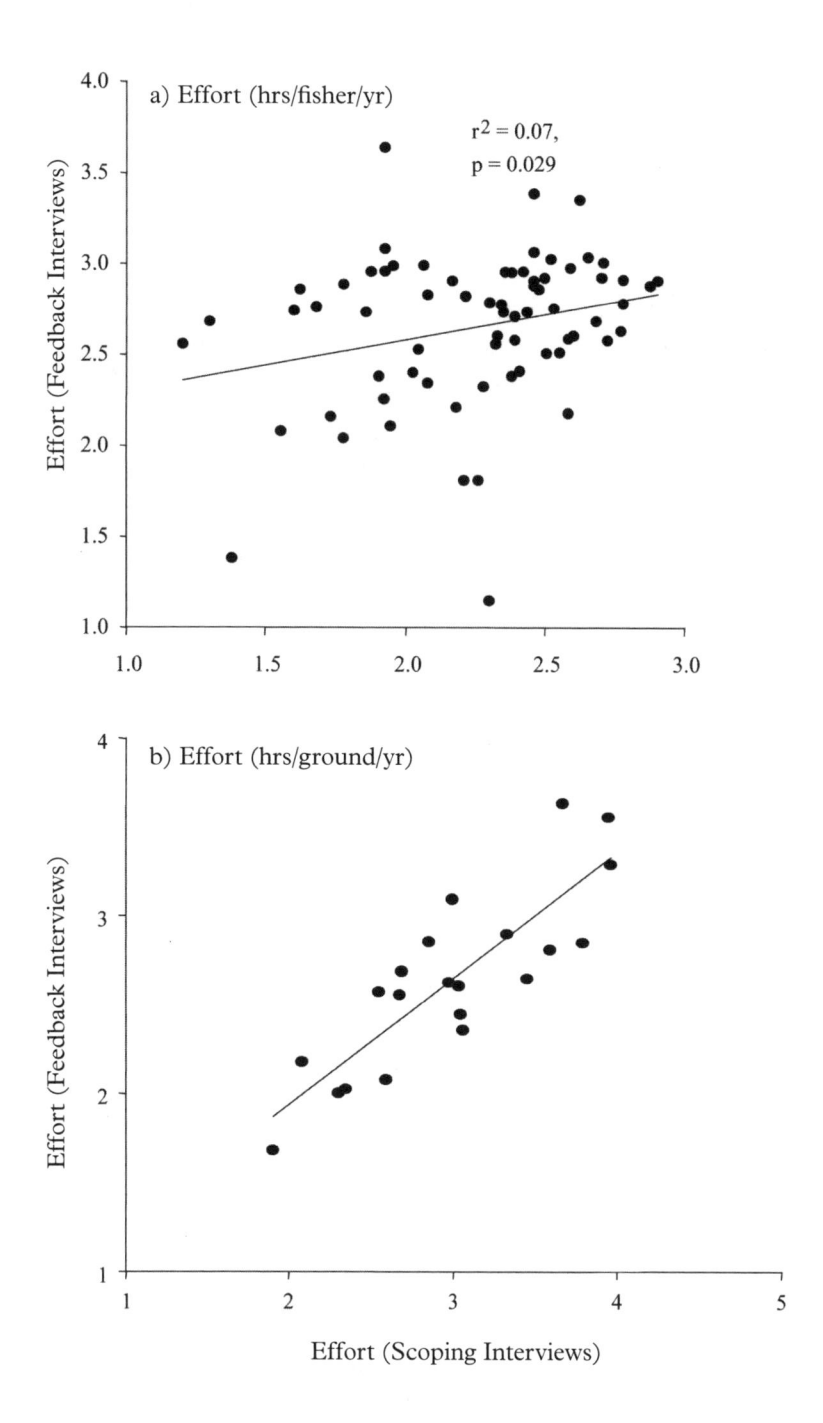

FIGURE 11.5 *Correlations of effort by (a) fisher and (b) fishing ground in the group of overlapping fishers (n = 71) and grounds (n = 25) for the Scoping (S) and Feedback (F) studies.*

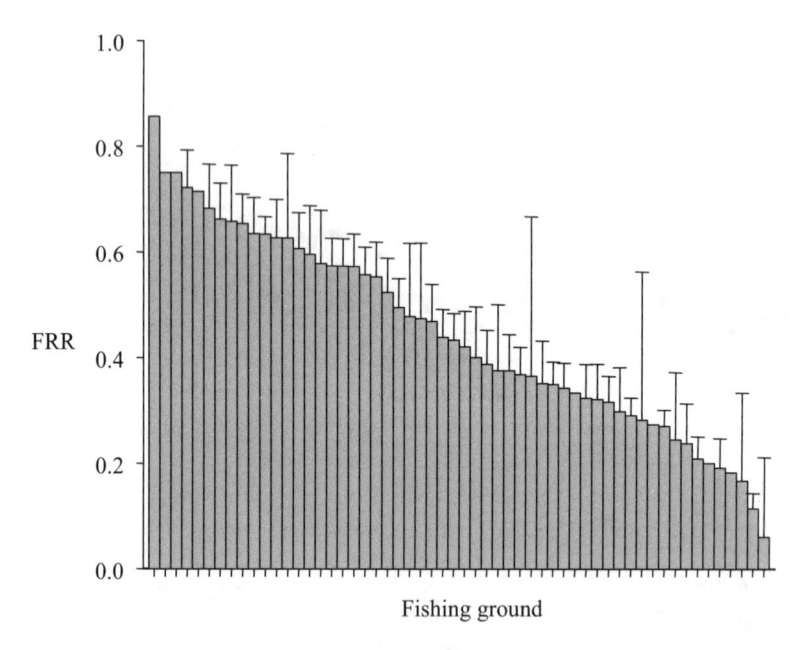

[Figure: bar chart showing FRR values from ~0.85 descending to ~0.05 across fishing grounds, with error bars]

FIGURE 11.6 *Mean fisher's relative ranking (FRR) of habitat quality by fishing ground with standard errors demonstrating general consistency of response among fishers for each ground.*

Fishing ground habitat quality

Habitat quality on the lantern fishing grounds was generally considered to be good by fishers in both the surveys. In the scoping survey, 78 per cent of fishers (± 28 per cent sd, range 0–100 per cent, n = 67 sites) said the fishing grounds were in good condition, while in the feedback sessions 75 per cent of fishers (± 35 per cent sd, range 0–100 per cent, n = 25 sites) said the fishing grounds were in good condition. If the group of fishers and grounds common to both studies are considered, 77.3 per cent±6.7 per cent and 81.4±7.4 per cent of the fishing grounds were described as 'good' by fishers in the scoping and feedback groups, respectively. No significant differences could be detected and indeed, when considering the responses of each fisher for each fishing ground (n = 128), 76 per cent of the answers were consistent between the two studies.

The Fishers Relative Ranking allowed sites to be ranked from high (FRR near 0) to low quality (FRR near 1). Although fishers' assessments varied both qualitatively and as a function of the number of fishing grounds fished, there was sufficient consistency to allow fishing grounds to be distinguished (Figure 11.6).

The assessment of habitat type was more problematic. In the scoping survey, on average, 45 per cent of fishers (± 31 per cent sd, range 0–100 per cent, n = 67 sites) said that the fishing grounds were dominated by live coral, as opposed to 26 per

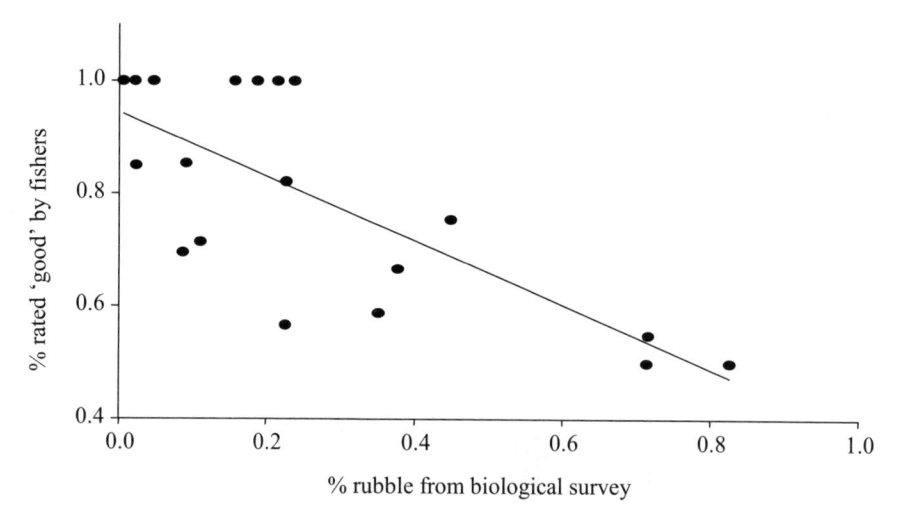

FIGURE 11.7 *Correlation between percentage of fishers indicating a site is 'good' and percentage of rubble cover measured on ecological surveys.*
Note: Points shown are from scoping survey.

cent of fishers (± 31 per cent sd, range 0-100 per cent, n = 25 sites) in the feedback survey. Using the same group of fishers and fishing grounds common to both studies, 49.2±6.5 per cent of fishers described fishing grounds as dominated by live coral in the scoping study, whereas only 22.1±6.8 per cent of fishers described the same fishing grounds as dominated by live coral in the feedback sessions. This difference was significant (paired t-test, n = 25, p = 0.007). When considering the responses of each fisher for each fishing ground (n = 128), only 20.9 per cent of responses were consistent between the two studies.

Fishers' assessments of habitat quality generally did not correlate with any formal measurements of habitat composition (percentage live coral, percentage *Sargassum*, percentage dead coral and the like) as measured by a biologist (Samoilys et al., 2001) using the line intercept method (English et al., 1997). The only significant relationship was that between the percentage of fishers indicating that a fishing ground was 'good' and the percentage of rubble cover (Figure 11.7). The fishers' assessment of habitat quality was significantly negatively correlated with the percentage of rubble cover for both surveys.

Resource management discussions

Food fish were ranked as the most economically important resource (mean rank = 1.61 (±0.11s.e.), followed by sea cucumbers (2.81±0.11), seahorses (3.04±0.16), crabs (3.60±0.11), seaweed (4.28±0.13) and shells (5.24±0.10). Notably, one seahorse genus

(*Hippocampus*), ranked third among orders and families of other organisms. The fishers' assessment of seahorse populations, fishing ground habitat quality and their livelihood indicates that these were largely healthy in the past (ten years ago), but conditions are felt to have deteriorated to the present with a poor outlook for the future (Figure 11.8).

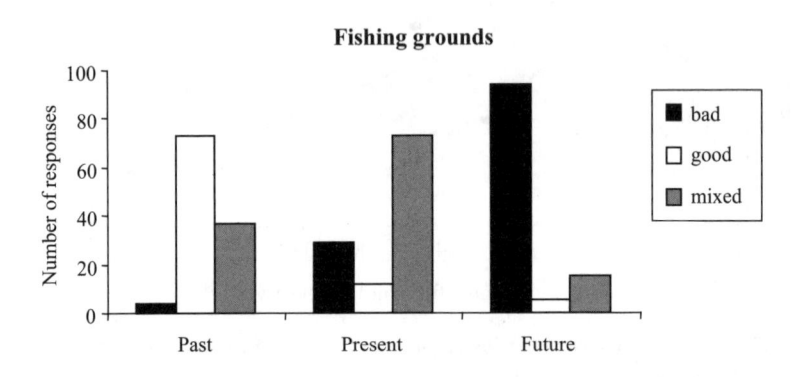

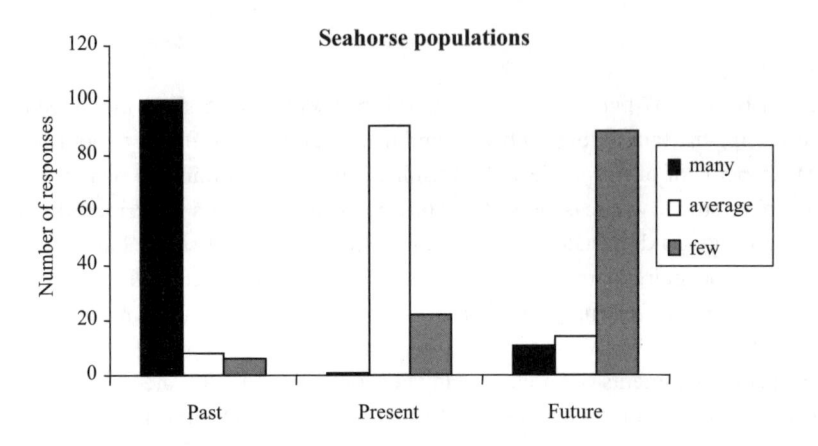

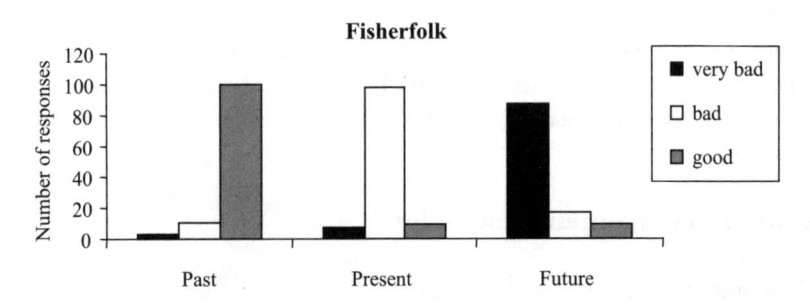

FIGURE 11.8 *Trends in status of (a) fishing ground condition, (b) seahorse populations and (c) fishers livelihood assessed by fishers from past (1990), present (2000) to future (2010).*

Reasons for the negative trends in fishing grounds, seahorse populations and fishers' quality of life were proposed and ranked, and suggestions for improvements were given (Tables 3–5). Fishers listed destructive (generally illegal) fishing as the most important reason for the poor condition of the fishing grounds. Dynamite ('blast' fishing), cyanide and *tubli*, a local plant poison, were the major illegal methods used (Table 11.3). Commercial fishing, primarily trawling and Danish seining (*liba liba*), was cited as the second most important reason for the degradation of fishing grounds. Both trawling and Danish seining are illegal within municipal waters. Fishers frequently used the terms commercial fishing and destructive fishing synonymously. Beach seining (*baling*), though legal in some municipal waters, was also cited as a destructive fishing method. Fishers stated emphatically that the fishing grounds were likely to deteriorate further, due primarily to continuing illegal and destructive fishing, and also to increasing numbers of fishers and a lack of concern regarding protection of the seas by fishers and government (Table 11.3). Some fishers stated that illegal fishing would continue because there was either no will on the part of government to enforce fishery laws, and/or because government officials were conniving with illegal fishers. Fishers in all villages listed the stopping of destructive and illegal fishing as the highest-ranking solution to the deterioration of their fishing grounds (Table 11.3). They suggested this should be done through strict and proper enforcement of fishery laws by local government units (at village and municipal level), through involvement of non-government organizations (NGOs) in fishery law enforcement, and through appointing more fish wardens.

Reasons for perceived declines in seahorse populations were more varied (Table 11.4). Fishers perceived the taking of pregnant seahorses and habitat destruction as primary reasons for the decline. Increased effort was also listed and was ascribed to an increase in the number of fishers, partly due to fishers switching from other fishery resources (such as finfish) that had declined. Fishers felt declines in seahorses are likely to continue due to insufficient numbers of adult seahorses, deteriorating habitat quality and a lack of juveniles (Table 11.4). To halt declines in seahorse populations, fishers most frequently suggested stopping destructive fishing and protecting pregnant seahorses (Table 11.4).

The major reasons for the poor condition of fishers' livelihoods were consistent among villages and included: less income derived from fishing, resulting in less disposable income for recreation; an increase in the costs of fishing; and an increase in costs of living. The remaining reasons were cited less consistently among villages. The reasons for the continuing decline in quality of life were rooted in the status of the fishing grounds, with destructive fishing cited as the main reason, followed by less catch and more fishing effort. Alternative livelihoods were perceived as the most important tool to improve the fishers' situation, with the need to stop destructive fishing as the second most important solution (Table 11.5).

TABLE 11.3 *Ranking of the responses from the marine resource discussions on the destruction of fishing grounds from most important to least important.*

Rank	Reasons for the destruction of fishing grounds	Reasons destruction will continue in the future	Solutions to arrest the destruction of fishing grounds
1	Destructive (illegal) fishing	Continuing destructive fishing	Stop destructive and commercial fishing
2	Commercial fishing	Increasing number of fishers	Establish more MPAs
3	Typhoons	Lack of concern in protecting the sea (fishers and/or government)	Stop buying destructively bought fish
4	Coral collecting	Increasing effort per fisher	Educate and inform fishers
5	Increasing # of fishers	Improved fishing methods	Maintain own MPA
6	Increasing # of outside fishers		Alternative livelihoods for fishers
7			Stop outside fishers
8			Organize fishers

Note: Ranking is based on the number of villages that identified a high, medium or low level of importance to the given reason or solution. Destructive fishing included both methods destructive to the habitat and illegal fishing such as trawling and seining in municipal waters.

TABLE 11.4 *Ranking of the responses from the marine resource discussions on declines in seahorse populations from most important to least important.*

Rank	Reasons for declines in seahorse populations	Reasons declines will continue in the future	Solutions to arrest declines in seahorse populations
1	Taking pregnant seahorses	Fewer adults for reproduction	Stop destructive fishing
2	Habitat destruction	Continuing habitat destruction	Stop catching of pregnant seahorses
3	Catching juveniles	Lack of good habitat (destroyed)	Caging of pregnant seahorses
4	Destructive fishing	Fewer juveniles	Stop fishing juveniles
5	Increased fishing effort	Increasing effort	Establish sanctuaries
6	Weather	Catching pregnant seahorses	Moratorium on seahorse fishing
7	Indiscriminate catching		Regulation of trade and catch
8	Catch during spawning season		Marine protected area management
9	Pollution		Protect habitat
10			Seasonal closures
11			Fishers to cooperate with local government units and nongovernmental organisations

Note: Ranking is based on the number of villages that identified a high, medium or low level of importance to the given reason or solution.

TABLE 11.5 *Ranking of the responses from marine resource discussions on the status of fishers' livelihoods.*

Rank	Reasons for deterioration of fishers' livelihoods	Reasons livelihood deterioration will continue	Solutions to arrest the deterioration of fishers' livelihoods
1	Less income	Destructive fishing	Alternative livelihood
2	Difficulty meeting basic food needs	Less catch	Stop destructive fishing
3	Increased operating costs	Increased no. of fishers	Alternative income
4	Increased price of commodities	Increased operating costs	Fishers' cooperative
5	No alternative livelihoods	Travel further to fish	Improve technology
6	Inability to improve gear technology	Destroyed fishing grounds	
7	Difficulty funding children's schooling	No alternative livelihoods	
8	Bad weather		
9	Travel further to fishing grounds		

Note: Ranking is based on the number of villages that identified a high, medium or low level of importance to the given reason or solution.

DISCUSSION

THE participatory approaches of the focus group discussions generated a lot of interest among the lantern fishers of Danajon Bank. The highly visual, graphical methods of conveying data were very effective in engaging the fishers and soliciting responses. The method is particularly well suited to fishers who are highly knowledgeable but semi-literate. For example, only 11 per cent complete elementary school in Handumon village (D.Y. Buhat, personal communication). High participation rates indicated this element of the programme was successful.

One issue in the focus group discussion approach is the validity of the responses obtained from the group. Bias towards answers provided by dominanting fishers, with other fishers copying, is likely. In the present study we were able to examine this by comparing reported fishing effort data obtained from the conventional questionnaire-based approach (the scoping survey) with the focus group discussions of the feedback survey. Although there were differences in the absolute values obtained, trends in fishing effort among fishing grounds were significantly correlated between the two surveys. Similarly there were no significant differences between the two methods in the description of the overall quality of the fishing grounds.

Most of the fishing communities of Danajon Bank that we visited had not been involved in our conservation programme, and this study served to integrate the CO into the communities and to engage the fishers in our research and management initiatives. One objective of the study was to generate discussions on resource management, and though at times dominated by key members in the fisher communities, group discussions served as opportunities for sharing ideas, particularly between the CO and the communities. This step of educating, informing and mobilizing fishers (called 'conscientization' in Filipino CO terminology) is vital in the community organizing process (Third World Studies Center, 1990). It is also fundamental to stakeholder involvement in conservation and management initiatives (Ruddle, 1994; Walters et al., 1998; Alcala, 1999; Cooke et al., 2000; White and Vogt, 2000).

A much higher estimate of fishing effort was obtained from the feedback survey than from the scoping survey. This may reflect bias in the group discussions or the difference in sample size. Sixty-seven fishing grounds were included in the scoping survey and only 25 in the feedback survey. However, with a change in CO during the feedback survey, we found that not all fishers had responded to the questions about fishing effort during the scoping survey, and that estimates were in fact based on only around two fishers per village. Therefore it is likely that the feedback survey, which collected effort estimates from each fisher in each village (mean = 9 fishers per village), provides a more accurate estimate of fishing effort. An average of 2.5 fishing trips per night per lantern fishing ground throughout the year was recorded, which is high considering the fishing grounds were less than 1 km^2 in size (Samoilys et al., 2001) and fishing trips lasted for 3.5 hours.

Estimates of fishing effort from interviews with fishers are renowned for their inaccuracy in terms of absolute value (Rawlinson et al., 1994; Die, 1997). However they provide useful relative estimates, and can be used to plot trends over time. This is well demonstrated in the present study. Highly consistent relative estimates of fishing effort per fishing ground were obtained between the two surveys. Effort per fisher was less consistent, and therefore presumably less reliable, but still significantly correlated between the two surveys.

We suggest that long-term blast fishing and other destructive fishing methods in this region mean that fishers' perceptions of a healthy fishing ground have changed and now differ markedly from ours. Fishers described their fishing grounds as being in good condition in the scoping and feedback surveys. In contrast, independent transect surveys revealed average percentages of live coral cover of 15 per cent and of rubble/dead coral cover (an indication of blast fishing damage) of 37 per cent for the same fishing grounds (Samoilys et al., 2001), suggesting the fishing grounds are in poor condition. This discrepancy indicates that fishers and ecologists are using different standards to assess fishing ground habitat quality. There is a difference in baseline (Pauly, 1995, 1996) in terms of healthy habitat quality, with the fishers' baseline being substantially lower. Fishers may use the extent of rubble cover as an indication of habitat quality, since the relationship between fishers' perceptions of good habitat was significantly negatively correlated with per cent rubble cover from independent surveys. A fishing ground was not considered to be in bad condition by fishers until rubble cover exceeded 50 per cent, a value that would be considered very high by ecologists (Gomez et al., 1994; Chou, 2000).

Our results also indicate potential difficulties in composing suitable questions when interviewing fishers. Fishers may interpret questions quite differently from the way they were intended by the interviewer, and results can be easily misinterpreted if care is not taken to cross-reference them against other answers and to consider the answers in light of the fishers' experiences. This is a common problem when conducting interviews and focus group discussions with subsistence fishers. In our study, habitat 'quality' was poorly defined, and was open to many interpretations. This may explain why the fishers described their fishing grounds as being in poor condition during the marine resource status discussions, because they were linking habitat quality directly to fishery condition. Such questions need to be defined as specifically as possible and considered in light of a range of answers if fishers' knowledge is to be accurately interpreted.

The marine resource discussions revealed that twenty-year trends (1990–2010) in the status of the fishing grounds, seahorse populations and the lantern fishers' livelihoods were all negative. In many cases there was strong consensus across villages about the reasons and the solutions for these trends. For example, illegal fishing (primarily blast fishing) was cited as the primary cause of the poor state of the fishing grounds, with its corollary of stopping illegal fishing as the primary solution. In other cases there was less consensus amongst fishers. For example fishers assessed their livelihood as being bad for a number of different reasons, though most of these did relate to an increasing need for cash which their livelihoods could not provide. In all cases it was clear that fishers recognized their problems and had informed ideas on how to alleviate them, though perceiving themselves as largely powerless to effect change.

It was overwhelmingly clear that stopping illegal fishing, especially blast fishing, and finding alternative livelihoods for the fishers were key solutions to the problems in the Danajon Bank lantern fishery. These results provide us with useful backing when directing our conservation efforts, though neither result is surprising. The prevalence and problem of blast fishing in the Philippines is well recognized (Alcala and Gomez, 1987; Yap and Gomez, 1988; Bryant et al., 1998; Chou, 2000). Furthermore, the lantern fishers of Danajon Bank are marginalized, comprising a relatively small proportion (nine fishers per village) of the total village population, with the lowest average income in the region, living well below the national poverty level (D.Y. Buhat, personal communication). Considering the fact that they fish for up to 50 per cent of their nights in arduous conditions, using paddle canoes and spending on average 3.5 hours in the water per night with no protection, it is not surprising that they would gladly welcome a supplementary livelihood.

The fishers' views are guiding us in our fishery management planning with various stakeholders (Martin-Smith et al., 2004). The fishers demonstrated a good understanding that gravid seahorses are important for population sustainability, citing the taking of pregnant seahorses as the primary cause of population depletion, and that the ensuing lack of adults and juveniles will contribute to further decline. It was not clear whether they knew that the pregnant seahorses were males (Vincent, 1994), but the option of protecting pregnant seahorses through fishery regulations is clearly understood (Martin-Smith et al., 2004). Fishers also linked population decline directly to habitat destruction. Fishers from the village of Handumon, where Project Seahorse has been active since 1995 (Vincent and Pajaro, 1997), provided the same range of reasons and solutions to their problems as other villages. One village, Guindacpan, consistently provided more answers and appeared more informed. The reasons for some of the differences between villages require further study.

Fishers' knowledge can guide conservation initiatives. We are acting on their knowledge and formalizing it. These results have also been central to establishing appropriate research programmes to determine sustainable catch levels, identify priority areas for MPA establishment, and support general management efforts. Fishers involved in the surveys have continued to participate in resource management through completion of catch calendars that track daily seahorse catch, effort and fishing location. Such monitoring programmes are essential for an adaptive management approach that allows the efficacy of management initiatives to be determined and modified as necessary. This involvement was greatly facilitated by the capacity building that occurred through this participatory programme.

The lantern fishers demonstrated that they are aware of conservation and management issues, are concerned about their marine resources and their livelihoods, recognize the negative trends, and know the reasons for their demise. However, they feel powerless to do anything about it, and see the government as being responsible but ineffective. These results have been instrumental in our initiatives to introduce supplementary livelihoods, and to facilitate the formation of a fishers' alliance across Danajon Bank to provide seahorse fishers with their own institution with which they can effect change.

ACKNOWLEDGEMENTS

FUNDING for this research was provided by the Community Fund, UK (formerly the National Lottery Charities Board). We are grateful to our partner Haribon Foundation for Natural Resources for its commitment and support. We would like to thank Julie Noble, Milo Socias, Roldan Cosicol, Paulette Apurado and Hari Balasubramanian for their valuable assistance in this work. The manuscript benefited from comments from Amanda Vincent. We thank the fishers and communities for their participation and support.

REFERENCES

AGRAWAL, A. 1995. Dismantling the divide between indigenous and scientific knowledge. *Development and Change*, Vol. 26, pp. 413–39.

AKIMICHI, T. 1978. The ecological aspect of Lau (Solomon Islands) ethno ichthyology. *Journal of the Polynesian Society*, Vol. 87, pp. 310–26.

ALCALA, A.C. 1998. Community-based coastal resource management in the Philippines: A case study. *Ocean and Coastal Management*, Vol. 38, pp. 179–99.

——. 1999. The role of community in environmental management. In: T.E. Chua, T.E.; N. Bermas (eds.), *Challenges and opportunities in managing pollution in the East Asian seas*. MPP-EAS Conference Proceedings of the 12th PEMSEA Conference, Vol. 1, pp. 529–37.

——. 2001. *Marine reserves in the Philippines: Historical development, effects and influence on marine conservation policy*. Makati City, Bookmark, 115 pp.

ALCALA, A.C.; GOMEZ, E.D. 1987. Dynamiting coral reefs for fish: A resource-destructive fishing method. In: B. Salvat (ed.), *Human impacts on coral reefs: Facts and recommendations*. French Polynesia, Antenne Museum E.P.H.E., pp. 51–60.

BAILEY, C.; ZERNER, C. 1992. Community-based fisheries management institutions in Indonesia. *Maritime Anthropological Studies*, Vol. 5, pp. 1–17.

BAY OF BENGAL PROGRAMME FOR FISHERIES DEVELOPMENT. 1990. *Helping fisher folk to help themselves: A study of people's participation in fisheries development*. Madras, Affiliated East-West Press (P).

BRYANT, D.; BURKE, L.; MCMANUS, J.; SPALDING, M. 1998. *Reefs at risk: a map based indicator of threats to the world's coral reefs*. Washington, DC, World Resources Institute, 56 pp.

CHOU, L.M. 2000. Southeast Asian reefs – status update: Cambodia, Indonesia, Malaysia, Philippines, Singapore, Thailand and Vietnam. In: C. Wilkinson (ed.), *Status of coral reefs of the world*. Townsville, Australian Institute of Marine Science, pp. 117–29.

COOKE, A.J.; POLUNIN, N.V.C.; MOCE, K. 2000. Comparative assessment of stakeholder management in traditional Fijian fishing-grounds. *Environmental Conservation*, Vol. 27, pp. 291–3.

ENGLISH, S.; WILKINSON, C.; BAKER, V. 1997. *Survey manual for tropical marine resources*, 2nd edition. Townsville, Australian Institute of Marine Science, 390 pp.

DIE, B. 1997. Fishery (CPUE) surveys. In: M.A. Samoilys (ed.), *Manual for assessing fish stocks on Pacific coral reefs*. Brisbane, ACIAR Fisheries Programme, pp. 30–37.

GOMEZ, E.D.; ALINO, P.M.; YAP, H.T.; LICUANAN, W.R.Y. 1994. A review of the status of Philippines reefs. *Marine Pollution Bulletin*, Vol. 29, pp. 62–68.

GULAYAN, S.J.; ANCOG, I.; PAJARO, M.G.; BRUNIO, E.O. 2000. Management of marine sanctuaries in Bohol, central Philippines. Abstract. 9th International Coral Reef Symposium, Bali, Indonesia, October 2000. Lawrence, Allen.

JENNINGS, S.; POLUNIN, N.V.C. 1996. Fishing, fishery development and socio-economic in traditionally managed Fijian fishing grounds. *Fisheries Management and Ecology*, Vol. 3, pp. 101–13.

JOHANNES, R.E. 1981. Working with fishermen to improve coastal tropical fisheries and resource management. *Bulletin of Marine Science*, Vol. 31, pp. 673–80.

——. 1982. Traditional conservation methods and protected marine areas in Oceania. *Ambio*, Vol. 11, No. 5, pp. 258–61.

——. (ed.). 1989. *Traditional Ecological Knowledge: A Collection of Eessays*. Gland, Switzerland and Cambridge, UK, IUCN, 77 pp.

MAGUIRE, J.J.; NEIS, B.; SINCLAIR, P.R. 1995. What are we managing anyway? The need for an interdisciplinary approach to managing fisheries ecosystems. *Dalhousie Law Journal*, Vol. 18, No. 1, pp. 141–53.

MARTIN-SMITH, K.M.; SAMOILYS, M.A.; MEEUWIG, J.J.; VINCENT, A.C.J. 2004. Collaborative development of management options for an artisanal fishery for seahorses in the central Philippines. *Ocean and Coastal Management*, Vol. 47, No. 3–4, pp. 65–93.

MCCLANAHAN, T.R.; GLAESEL, H.; RUBENS, J.; KIAMBO, R. 1997. The effects of traditional fisheries management on fisheries yields and the coral-reef ecosystems of southern Kenya. *Environmental Conservation*, Vol. 24, No. 2, pp. 105–20.

MURDOCK, J.; CLARK, J. 1994. Sustainable knowledge. *Geoforum*, Vol. 25, No. 2, pp. 115–32.

NEIS, B.; SCHNEIDER, D.C.; FELT, L.; HAEDRICH, R.L.; FISCHER, J.; HUTCHINGS, J.A. 1999. Fisheries assessment: What can be learned from interviewing resource users? *Canadian Journal of Fisheries and Aquatic Science*, Vol. 56, pp. 1949–63.

PAJARO, M.G.; NOZAWA, C.M.; LAVIDES, M.; GUTIERREZ, S. 2000. Status of marine protected areas in the Philippines: Better management of coral reefs and the coastal areas in the tropics. Abstract. *9th International Coral Reef Symposium*, Bali, Indonesia, October (2000). Lawrence, Allen.

PAULY, D. 1995. Anecdotes and the shifting baseline syndrome of fisheries. *Trends in Ecology and Evolution*, Vol. 10, No. 10, p. 430.

——. 1996. Biodiversity and the retrospective analysis of demersal trawl surveys: A programmatic approach. In: D. Pauly and P. Martosubroto (eds.), *Baseline studies of biodiversity: The fish resources of western Indonesia*. Manila, ICLARM studies and reviews, no. 23, p. 16.

——. 1997. Small-scale fisheries in the tropics: Marginality, marginalization and some implications for fisheries management. In: E.K. Pikitch, D.D. Huppert and M.P. Sissenwine (eds.), *Global trends: Fisheries management*. American Fisheries Society Symposium. Bethseda, Md., American Fisheries Society, pp. 20, 40–49.

PICHON, M. 1977. Physiography, morphology and ecology of the double barrier reef of north Bohol (Philippines). *Proceedings of the 3rd International Coral Reef Symposium*, Miami, 1976, pp. 261–67.

POLUNIN, N.V.C. 1983. The marine resources of Indonesia. *Oceanography and Marine Biology Annual Review,* Vol. 211, pp. 455–531.

——. 1984. Do traditional marine 'reserves' conserve? A view of the Indonesian and New Guinean evidence. In: K. Ruddle and T. Akimichi (eds.), *Maritime institutions in the Pacific.* Senri Ethnological Studies, Vol. 17. Osaka, Japan: National Museum of Ethnology, pp. 267–84.

RAWLINSON, N.J.F.; MILTON, D.A.; BLABER, S.J.M.; SESEWA, A.; SHARMA, S.P. 1994. A survey of the subsistence and artisanal fisheries in rural areas of Viti Levu, Fiji. Canberra, Australia Centre for International Agricultural Research, Monograph No. 35, 13 pp.

RUDDLE, K. 1994. *Traditional community-based marine resources management systems in the Asia-Pacific region: Status and potential.* Manila, Philippines: ICLARM, 330 pp.

RUDDLE, K.; HVIDING, E.; JOHANNES, R.E. 1992. Marine resource management in the context of customary marine tenure. *Marine Resource Economics,* Vol. 7, pp. 249–73.

SAMOILYS, M.A.; MEEUWIG, J.J.; VILLONGCO, Z.A.D.; HALL, H. 2001. Seahorse fishing grounds of Danajon Bank, central Philippines: Habitat quality and seahorse densities. Indo-Pacific Fish Conference, Durban South Africa, May 2001, Program Abstracts. Durban, Oceanographic Research Institute, p. 51.

SILLITOE, P. 1998. The development of indigenous knowledge: A new applied anthropology. *Current Anthropology,* Vol. 39, No. 2, pp. 223–52.

THIRD WORLD STUDIES CENTER. 1990. *The language of organising: A guidebook for Filipino organizers.* Quezon City, University of the Philippines, 147 pp.

VINCENT, A.C.J. 1994. Seahorses exhibit conventional sex roles in mating competition, despite male pregnancy. *Behaviour,* Vol. 128, No. 12, pp. 135–51.

VINCENT, A.C.J.; PAJARO, M.G. 1997. *Community-based management for a seahorse fishery.* In: D.A. Hancock, D.C. Smith, A. Grant and J.P. Beumer (eds.), Proc. World Fisheries Congress, Brisbane, 1996. Brisbane, CSIRO, pp. 761–66.

WALTERS, J.S.; MARAGOS, J.; SIAR, S.; WHITE, A.T. 1998. *Participatory coastal resource assessment: A handbook for community workers and coastal resource managers.* Cebu City, Philippines, Coastal Resource Management Project and Silliman University, 113 pp.

WHITE, A.T. 1988. The effect of community managed marine reserves in the Philippines on their associated coral reef fish populations. *Asian Fisheries Science,* Vol. 2, pp. 27–41.

WHITE, A.T.; VOGT, H.P. 2000. Philippine coral reefs under threat: Lessons learned after 25 years of community-based reef conservation. *Marine Pollution Bulletin,* Vol. 40, No. 6, pp. 537–50.

WRIGHT, A. 1985. Marine resource use in Papua New Guinea: Can traditional concepts and contemporary development be integrated? In: K. Ruddle and R.E.

Johannes (eds.), *The traditional knowledge and management of coastal systems in Asia and the Pacific.* Jakarta: UNESCO, Regional Office for Science and Technology in Southeast Asia, pp. 79–100.

YAP, H.T.; GOMEZ, E.D. 1988. Aspects of recruitment on a northern Philippine reef. Proceedings of the 6th International Coral Reef Congress. *Australia,* Vol. 2, pp. 279–83.

ZANN, L.P. 1985. Traditional management and conservation of fisheries in Kiribati. In: K. Ruddle and R.E. Johannes (eds.), *The traditional knowledge and management of coastal systems in Asia and the Pacific* Jakarta, UNESCO, Regional Office for Science and Technology in Southeast Asia, pp. 7–38.

CHAPTER 12 Local ecological knowledge
and small-scale freshwater fisheries
management in the Mekong River
in Southern Laos
Ian G. Baird

ABSTRACT

SMALL-SCALE fishers possess a vast amount of local ecological knowledge (LEK) about the fishes and fisheries of the Mekong River and tributaries in southern Laos. Between 1993 and 1999, a community-based fisheries co-management programme was implemented for the sustainable management and conservation of living aquatic resources in the Siphandone Wetlands area of the Mekong River in Khong District, Champasak Province. Sixty-three villages established varied regulations related to living aquatic resources, and local government officially recognizes these as constituting 'village law'. Independent evaluations indicate that local people believe that aquatic resources and fishers have benefited from the regulations adopted, and it seems likely that the system will function into the future, although the effectiveness of local management measures vary from village to village. The management measures in Khong are based largely on LEK, which has played an important role in establishing regulations and monitoring activities, and in adapting management practices to meet local conditions. This chapter describes how LEK has been practically applied, disseminated and strengthened to improve freshwater fisheries management in rural communities.

INTRODUCTION

SMALL-SCALE fishers in many parts of the world have a vast amount of local ecological knowledge (LEK) about fishes and fisheries, which they depend upon for their livelihoods (Johannes, 1981; IIRR, 1996; Poulsen and Valbo-Joergensen, 2000; Johannes, 2001). While it is now generally recognized within the scientific community that fishers have such LEK, there has been only limited research into the ways local people use this for fisheries management purposes, or how it can be adapted in order to improve the management of wild capture fisheries (Johannes, 2001).

This chapter describes how fishers have applied their LEK to improve the management of wild capture fisheries in the Mekong River and tributaries in the Siphandone Wetlands area in Khong District, Champasak Province, in the Lao People's Democratic Republic (Lao PDR or Laos) (see Figure 12.1).

BACKGROUND

At 4,400 km, the Mekong is the tenth longest river in the world, and the fourteenth largest in terms of total annual discharge. However, it ranks third (after the Amazon and the Brahmaputra) when it comes to maximal flow (Baran et al., 2001). The diverse habitats of the Mekong River Basin support approximately 1,200 fish species, one of the richest faunas in the world (Rainboth, 1996; Van Zalinge et al., 2000), although many species have not yet been taxonomically described. A large number of species seasonally migrate to Laos and Thailand from as far away as the Great Lake in Cambodia and the South China Sea in Vietnam (Lieng et al., 1995; Roberts and Baird, 1995; Baird et al., 1999a; Warren et al., 1998; Van Zalinge et al., 2000; Baird et al., 2001a; Baird et al., 2003). Others are relatively sedentary or only locally migratory (Baird et al., 1999a; Baird, 2001; Baird et al., 2001b).

Fish provides a significant proportion of both protein and cash income for the bulk of rural lowland people in Laos (Baird et al., 1998; Garaway, 1999; Hubbel, 1999; Sjorslev, 2000), and there are a large variety of fisheries, with differing habitats and seasons, harvesting methods and fisher livelihoods. Fishing methods vary with the species of fishes being targeted, and fishers' knowledge of the biology and behaviour of the fishes (Claridge et al., 1997; Baird et al., 1998). Certainly, the LEK of fishers contributes greatly to their ability to feed themselves and their families, and to generate income (Baird et al., 1999a, 1999b). In recent years, population growth, improvements in fishing gear, better access to markets, and infrastructure projects of various types, especially the construction of large dams, have had a negative impact on fisheries in the Mekong (Roberts, 1993b, 1993c; Roberts and Baird, 1995; Baird 1999a, 1999b; IRN, 1999). Although there are few official data available, individual fishers widely report that they are experiencing significant declines in catches (Roberts, 1993a; Roberts and Warren, 1994; Roberts and Baird, 1995; Lieng et al., 1995; Hogan, 1997; Baran et al., 2001; Baird et al., 2001a, 2001b).

The Siphandone (4,000 islands) Wetland area, in the southernmost part of Laos (Figure 12.1), is one of the most complex ecosystems in the mainstream Mekong River. It is made up of large and small inhabited and uninhabited islands, channels, seasonally inundated forests, deep-water pools, rapids and waterfalls (Claridge, 1996; Daconto, 2001). Largely situated in Khong District, adjacent to Cambodia,

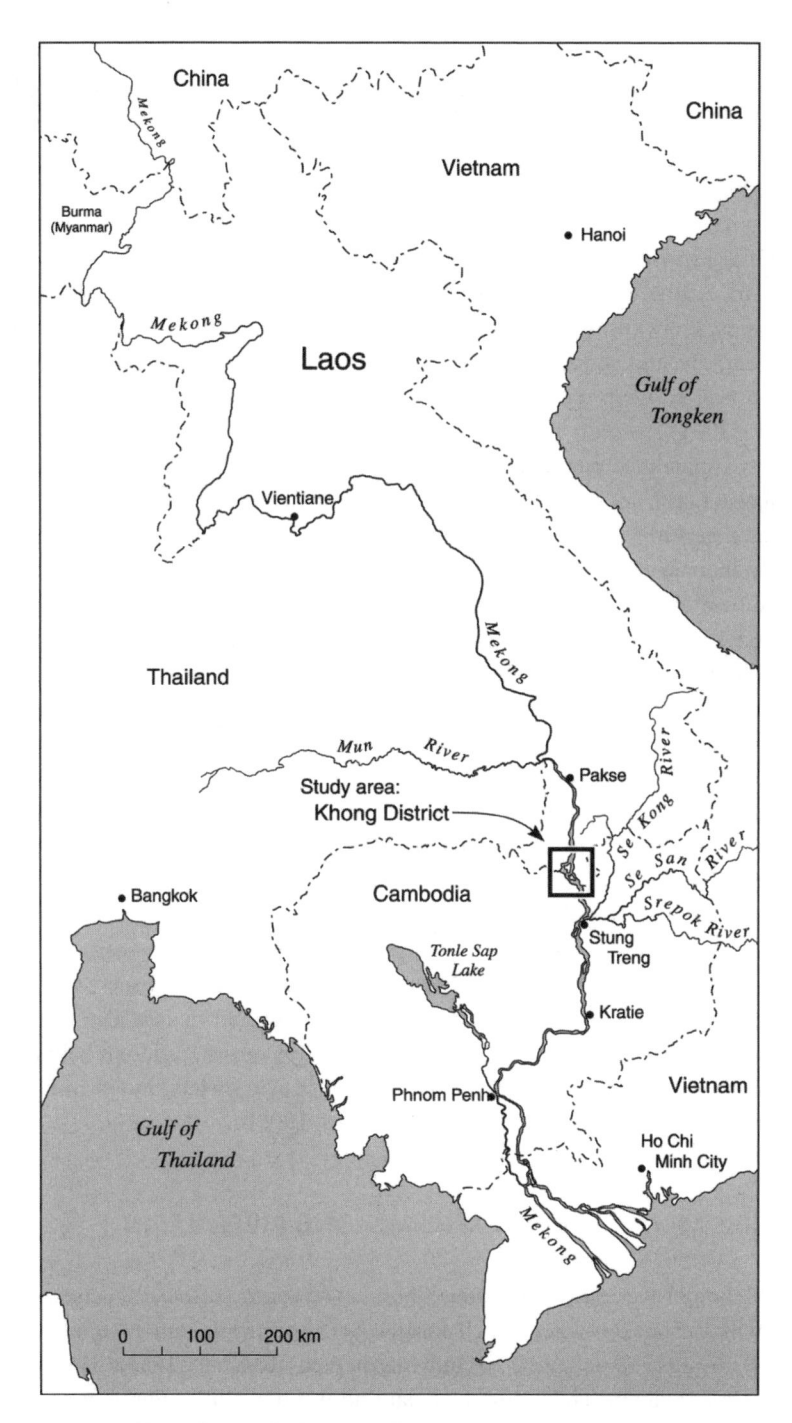

LOCAL ECOLOGICAL KNOWLEDGE AND SMALL-SCALE FRESHWATER FISHERIES MANAGEMENT IN THE MEKONG RIVER

FIGURE 12.1 *Map of Laos showing Siphandone Wetlands area in Khong District.*

the area is characterized by high biodiversity and productivity. So far, 201 fish species have been recorded from fish catches from the mainstream Mekong River at one location just below the Khone Falls in Khong, of which about 165 can be considered economically significant to local fishers (Baird, 2001).

In 1995 there were 65,212 people living in Khong, the vast majority being ethnic Lao, rural subsistence-oriented peoples with a long history of inhabiting the area. Of the 136 villages in Khong, 86 are situated on islands, and most of the others are located on the eastern bank of the Mekong River. There is little immigration into the area, and considerable emigration, but the human population is nonetheless growing rapidly, and agricultural land on the islands is now quite scarce. In the long term, this will certainly be important for natural resource management in Khong (Daconto, 2001).

Approximately 94 per cent of families in Khong participate in artisanal fisheries at a subsistence level or as a way of generating income, and in 1996/97 it was estimated that 4 million kg of wild fish were caught in Khong District, and that over US$ 1 million worth of wild fish and fish products originating from Khong were exported from the district to outside markets. This is a huge amount of money for an area where the average family income is the equivalent of just a few hundred dollars a year. The average person caught 62 kg of fish per year. Aquaculture is virtually non-existent in the area. This indicates that wild capture fisheries in Khong may be more important to local people than in any other district in Laos (Baird et al., 1998).

LEK related to fish and other living aquatic resources in Khong is widespread, and most of the people in the district spend a considerable amount of time on or near the water. Most people have lived all their lives near the Mekong, and their livelihoods are greatly influenced by the changing hydrological conditions of the river. Moreover, fishing traditions are strong. Fishing begins very early in life for many children, and especially young boys, and most are seasoned fishers by the time they are teenagers (See Nsiku, this volume). However, old people also fish, as do women and girls, and different age and gender groups rely on different habitats and fishing methods, thus adding to the complexity of LEK in society, and its importance in understanding how ecosystems function (Baird, 1999b).

FISHERIES MANAGEMENT IN THE MEKONG RIVER BASIN

In the Mekong River Basin, including Khong, wild capture fisheries management is faced with various obstacles and challenges. Scientifically documented information about the resource is very limited and fragmented (Roberts, 1993a; Roberts and Warren, 1994; Hill and Hill, 1994; Kottelat and Whitten, 1996; Baran et al., 2001). Furthermore, the Mekong system is characterized by having a large number of fisheries, some large but most small, each operating differently and adding to the complexity of management (Hill and Hill, 1994; Claridge et al., 1997; Ahmed et al., 1998; Baird et al., 1998). Many of these fisheries are located in relatively remote areas, making direct government management extremely difficult, potentially costly and generally unrealistic (Cunningham, 1998). The large number of highly migratory fish species in the Mekong basin that move between two or more countries also makes it difficult to manage many species at only a local level (Roberts, 1993c; Roberts and Baird, 1995; Warren et al., 1998; Hirsch, 2000; Baird et al., 2001a).

COMMUNITY-BASED FISHERIES CO-MANAGEMENT

CENTRALLY imposed natural resource management systems typically increase the monitoring and regulatory responsibilities of governments. Unfortunately, the fisheries departments in non-industrialized nations are typically understaffed and underfunded (Cowx, 1991; Kottelat and Whitten, 1996; Johannes, 1998; Cunningham, 1998). Given the pressing need for improved natural resource management, alternative decentralized management models, including 'co-management' (CM) and 'community-based natural resource management' (CBNRM), are being increasingly proposed in South-East Asia and other parts of the world (Berkes and Kislalioglu, 1993; Christy, 1993; Kuperan and Abdullah, 1994; Clay and McGoodwin, 1995; McCay and Jentoft, 1996; Ali, 1996; Pomeroy and Carlos, 1997; Hogan, 1997; Johannes, 1998; Johnson, 1998; Pomeroy, 1998; Hirsch and Noraseng, 1999; Masae et al., 1999), although not everyone has been enthusiastic about the prospects of applying CM to solve management problems (Brechin et al., 2003). In any case, CM is very significant in terms of putting fishers' knowledge to work, as it provides fishers with a more formalized role in management, and more opportunities to use their LEK in ways that benefit them and the resources.

Natural resource co-management has been defined as, 'the collaborative and participatory process of regulatory decision-making among representatives of user-groups, government agencies and research institutes' (Jentoft et al., 1998). The term co-management (CM) is useful for demonstrating that fisheries management is often a joint effort between resource users and governments. However, some CM programmes remain strongly government dominated, with little real decision-making power being given over to resource users (Glaesel and Simonitsch, 2003). Because of the uncertainty of who controls management decisions when it comes to CM, some prefer to use the term community-based natural resource management (CBNRM), as it emphasises that the communities are the centres of management structures. However, the term CBNRM is limited because it does not imply the involvement or recognition of governments. Nor does it specify whether there are any partnerships or agreements between governments and users. In reality, most fishing communities require and desire some level of government support in order to be able to effectively defend community resource areas covered under local management regulations. Therefore, it seems preferable to use the term 'community-based fisheries co-management' (CBFCM) to convey the message that management systems and decision-making structures are centred in communities, with users having considerable management powers. The government is nevertheless participating in the process, and recognizes the validity of the community-based management systems, and user tenure over resources. Essentially, the systems in Khong are good examples of CBFCM, and this type of management regime holds considerable promise in the Mekong Basin for applying LEK to improving fisheries management.

PUTTING LEK TO WORK IN FISHERIES MANAGEMENT

Local people in Khong have a highly developed folk taxonomy for fishes, as they come into regular contact with a large number of species. All medium and large-sized species have specific local names, even when there are only small differences in outward appearances. These names are widely known within the general population, as discussions in communities are often centred on fishing activities. The average fisher is familiar with well over 100 local names, which are used to describe the approximately 165 species of fish that are economically significant to local people (Baird, 2001).

Photographs of fish found in Khong shown to children as young as 5 or 6 years old elicit many local names, indicating that many children of that age already have a vocabulary of fifty or more local names for fish. However, at such young ages, children are not easily able to match local names with fish photographs, compared with teenagers or adults, although they do recognize some.

As a testament to the accuracy of their folk taxonomy, when a foreign ichthyologist visited one village in Khong in 1993, he heard of three local names for fish in the genus *Micronema*. At the time, he believed that these names indicated an over-differentiation of local names for describing the two species that he believed actually occurred there (Roberts, 1993a). However, it has since been confirmed that the villagers were correct, and that there are actually three species of *Micronema* in Khong, each corresponding with a single local name *(pa nang khao, pa nang ngeun* and *pa sa-ngoua* in Lao) (Baird et al., 1999a).

Local people in Khong possess a considerable amount of LEK about fish behaviour, including migration and feeding patterns (see, for example, Baird et al., 1999a, 1999b; Baird and Phylavanh, 1999). However, in the past most scientists interested in LEK have been more concerned with documenting it than in strengthening and disseminating it to make it more practically useful for local fisheries management. This is because LEK has often been seen as useful for biologists rather than being particularly important for informing fishers in determining management measures. This understanding relates to important issues of epistemology, which have resulted in many natural scientists seeing LEK as being only really believable once it has been verified through scientific testing (Johannes, 1981).

COMMUNITY-BASED FISHERIES CO-MANAGEMENT IN KHONG DISTRICT

BETWEEN 1993 and 1999, sixty-three villages in Khong District established regulations to manage and conserve inland living aquatic resources, including fish, in the Mekong River, streams, backwater wetlands and rice paddy fields (see Figure 12.2). The CBFCM systems in Khong have been supported by two projects supported by non-governmental organizations (NGOs), the Lao Community Fisheries and Dolphin Protection Project (LCFDPP), which was implemented between 1993 and 1997, and the follow-up Environmental Protection and Community Development in Siphandone Wetland Project (EPCDSWP), between 1997 and 1999 (Baird, 1999b).

Local government has endorsed the process, so that villages can incorporate their LEK in the design, implementation and enforcement of regulations. These regulations are consensus based and can be altered in response to changing circumstances. Recognized as 'village law' (*kot labeeap ban* in Lao), the regulations established in each of the villages are different. Nevertheless, many communities have adopted similar regulations, with slight variations (Baird, 1999b).

Both personal observations on the water and information provided by close relatives, friends, elders and other community members are the basis for the LEK of fishers in Khong District. Without their detailed LEK, the regulations developed through CBFCM would certainly be much weaker and less appropriate and effective. The most commonly adopted regulations relate to:

1. The establishment of permanent or seasonal 'no-take' fish conservation zones (FCZs) in deep parts of the Mekong River. The largest FCZ is 18 ha, the smallest is 0.25 ha, and the mean size is 3.52 ha. The deepest FCZ is approximately 50 m in the dry season, the shallowest is approximately 2.5 m. The average depth is 19.5 m. These are especially important as low-water fish refuges for protecting large brood stock in the dry season. As of 1999, a total of sixty-nine FCZs had been established in Khong. Sometimes two or three villages jointly manage single FCZs, while in other cases individual villages manage up to three FCZs on their own (see also Sultana and Thompson, this volume).

2. Banning the blocking of streams with fish traps at the beginning of the rainy season to prevent the harvesting of fish that are making short spawning migrations into inundated rice fields and other wetlands, and/or to allow them to spawn before being caught.
3. The banning of 'water banging' fishing, where a long wooden pole with a metal piece at the end of it is used to bang the surface of the water in order to chase small cyprinid fishes like *Henicorhynchus* spp. *(pa soi* in Lao*)* and *Paralaubuca typus (pa tep* in Lao*)* into small-meshed gillnets. This ban has been implemented because it is believed that the method results in fish leaving local areas, leading to lower catches for those fishers who set stationary gillnets without chasing fish into them.
4. The banning of spear fishing with lights at night. This ban has been implemented because it is seen to be too effective a fishing method, catching large quantities of brood fish. It is also unpopular because people who use this method sometimes steal fish from nets and traps at night and have also been known to steal chickens and other household items from villagers.
5. The banning of catching juvenile snakeheads (*Channa striata*) *(pa kho* in Lao*)*, especially when they are less than about two-weeks old and are still travelling in schools. These juveniles are very vulnerable to scoop-net fishing, but the amount of fish harvested is very small due to their small size. Villagers supporting the ban believe that it makes more sense to allow the fish to grow before harvesting them, thus increasing total production.
6. The banning of frog (*Rana* spp.) (*kop* in Lao) catching at the beginning of rainy season, when they spawn, and in some cases, at other times of year. The spawning season ban is especially important, because frogs croak loudly at that time, making them very easy to locate and catch. Moreover, if frogs are harvested before they can spawn, recruitment may be reduced, leading to population declines. The banning of certain harvesting methods such as frog traps, frog hooks and lights at night has also been advocated by dozens of village communities due to the belief that frog harvesting for commercial sale is too intense, leading to population declines. Local farmers also see frogs as important for controlling insect attacks on their rice crops.
7. The banning of tadpole (*Rana* spp.) catching at the beginning of the rainy season after spawning takes place. The principle of protecting these small juveniles is the same as for protecting juvenile snakeheads in point 5 (above).
8. The protection of inundated forest habitat by encouraging villagers not to cut down wetland trees and bushes in the mainstream Mekong River.

LEK has been of critical importance in the development of regulations in Khong. In fact, it has been the most important (although not the only) factor associated with the establishment of specific regulations. For example, the establishment of

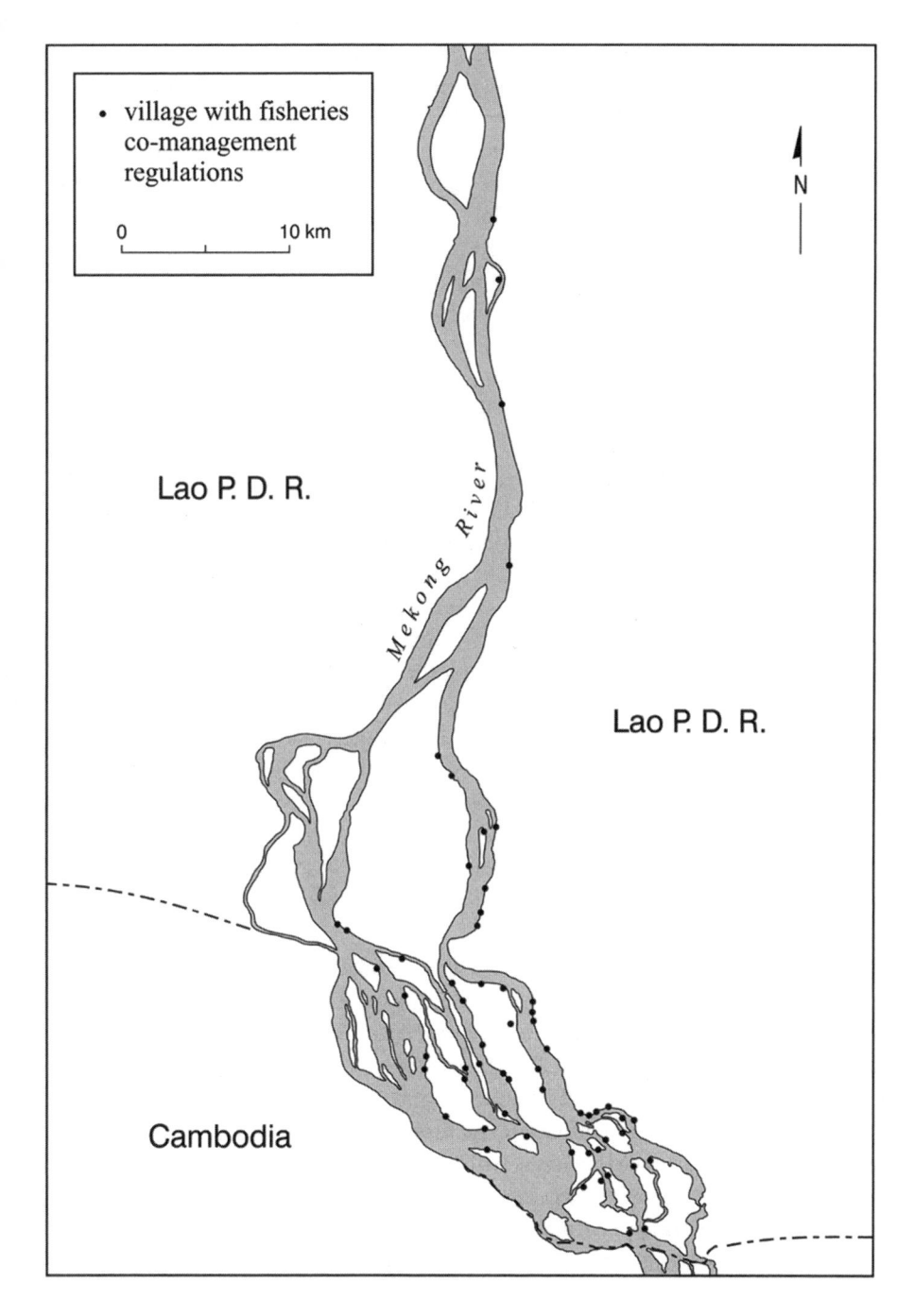

FIGURE 12.2 *Khong district showing sixty-three villages with regulations to manage and conserve fish and other aquatic resources in the Mekong River, streams, wetlands and rice paddies.*

FCZs was an idea that fishers came up with by themselves, based on observations that during the dry season many fish species, especially large ones, congregate in deep-water areas, where they are potentially vulnerable to gillnetting. While fishers in Khong began protecting deep-water FCZs in Khong in 1993, the validity of this strategy was only confirmed by scientists years later, although what fishers knew long ago is now widely recognized (Baird et al., 1999b; Chomchanta et al., 2000; Poulsen, 2001; Kolding, 2002; Baird and Flaherty, 2005). Prior to the scientific verification of the importance of deep-water pools in the Mekong River, many fisheries scientists (especially non-Laotians) were very critical and sceptical about establishing FCZs in deep-water areas, believing that there was no 'scientific evidence' to support the LEK of fishers. Some even appealed, albeit unsuccessfully, to senior provincial officials to prevent fishers establishing FCZs, which they described as being 'unscientific'. But fishers were firm in their convictions, and confident in their LEK, and critically, local government strongly supported the views of the fishers.

The banning of stream blocking at the beginning of the rainy season is another idea that fishers came up with based on their LEK, which indicated to them that spawning fish moving up to their rice fields were being caught by this method, and that fishers would probably harvest more fish in the long term if they allowed snakeheads (*Channa* spp) (*pa kho* in Lao) and walking catfish (*Clarias batrachus*) (*pa douk* in Lao) to spawn before fishing for them.

The decision of many villages to ban 'water banging' fishing was also based on fisher LEK. Fishers observed that certain fishes, especially small cyprinids, tend to move out of areas where this method is used, because of the noise. This resulted in reduced catches for other fishers in the area.

Many villages have also banned spear fishing at night due to detailed observations that showed the drastic decline of certain species of large fish, such as *Channa micropeltes* (*pa meng phou* in Lao), that tend to stay close to shore and in relatively shallow water during the night.

LEK is also the key for why juvenile *Channa striata* snakeheads are often protected when they are still grouping. Again, fishers observed that fishing during this period leads to a severe decline of larger snakeheads later on, and thus have applied their LEK to improve their chances of getting better catches in the long term.

The various measures developed to protect frogs (*Rana* spp.) in individual villages also indicate how regulations based on LEK are specifically adapted to local ecological and socio-economic conditions.

Fishers have also applied their LEK in protecting certain wetland forests that they recognize as being ecologically important for maintaining the populations of certain fish species.

Although generational transfer is a vitally important component of its development, it is critical to recognize that LEK is not a stagnant state of knowledge.

It is situated knowledge developed through the actual experiences of individual fishers, and is therefore not uniform within groups of fishers. In general, those who spend more time on the water know more than others, but much also depends on individual powers of observation and dispositions for learning, which may differ considerably among individuals. Differences in LEK are also based on the particular habitats, species and fishing methods used by different fishers. It is often based on what people need to know in order to be successful at fishing. For example, line fishers may have a considerable amount of LEK about those fish species caught on hooks, but know much less about those caught with other gear. However, most fishers in Khong use a wide variety of fishing gear and methods, based mainly on seasonal appropriateness and habitat diversity, and therefore have a considerable breadth of LEK about a broad range of species and habitats.

Still, there are certainly limitations to LEK, probably the most important being related to scale. That is, LEK is often locally specific, and may be less applicable outside the particular locations where it has developed. Therefore, those with localized LEK may lack a broader geographical perspective.

Even the most knowledgeable fishers are generally very eager to learn more, and are quite receptive to integrating new information into their LEK, provided that the source is credible and the information makes sense in light of what they already know. Although they rarely meet outsiders, especially fisheries scientists, when they do, the fishers with the most LEK are often the ones most interested in learning from them. That is how they got to know so much: by being inquisitive, and ready to integrate new information into their LEK. This willingness to learn makes it possible for fishers, scientists, government officials and NGO workers to collaborate in the design of CBFCM programmes and regulations that maximize the benefits of LEK, so that fishers responsible for managing fisheries can make more informed decisions. Local government and supporting NGO projects have played important roles in strengthening CBFCM by providing additional scientific information to local fishers to augment their LEK, as well as through helping to facilitate the transfer of LEK from community to community (See Meeuwig et al., this volume). For example, scientific knowledge about long-distance fish migrations provided to fishers has been useful in helping fishers determine management strategies for those species; the willingness of fishers to integrate this information has helped to legitimise their strong role in management with local government.

This is not to say that fishers were not already making good decisions before outside support was provided to them, but LEK-based management is particularly vulnerable to changing environmental, political and social conditions, including increasing pressure on migratory species from outside local areas, and in the Mekong, from other countries. Since LEK is often very locally relevant, while lacking a broad and regional perspective, it may be useful for outsiders to inform local fishers of external factors, as a way of helping to improve local fisheries management decisions, and to build support for the cessation or at least mitigation of developments with serious environmental impact potential for critical parts of the system. In this way, the LCFDPP and later the EPCDSWP have helped to support the CBFCM programme in Khong. However, it is important to remember that effective transfer of knowledge is based on respect. If outside experts want to influence the LEK of fishers, they need to be willing to learn what the fishers know, how they use their knowledge, and how they communicate LEK among themselves, as this will help the outsiders know what new information can be communicated in context with already accepted LEK. However, this often takes time – more than most researchers have.

Facilitating the exchange of LEK between communities is another important way in which the LCFDPP and the EPCDSWP have helped to strengthen local fisheries management in Khong. This type of activity has proved very useful, as it is usually very easy for fishers to accept information provided by other fishers who are in similar socio-economic and cultural situations, and speak the same language. In Khong, many of the regulations chosen by communities were adopted after other communities first implemented them. Indicative of this, villages that entered the CBFCM programme in Khong early on generally have far fewer regulations than those that got involved later. However, some early entrants have also added regulations after hearing that later villages had adopted similar ones. The relative homogeneity of communities in Khong makes this process of information exchange easier. Thus, the programme has evolved through the dissemination of LEK, as well as through adding new information to it (Baird, 1999b; Baird et al., 1999b).

'Peer review' is of critical importance to ensuring the relevance of LEK-based CBFCM. As almost all villagers are highly experienced fishers with a good grasp of LEK, regulations that do not make ecological sense will not be adopted, because people in the community are likely to quickly realize the deficiencies of such regulations and object. However, to allow for proper peer review, it is important to separate men from women in order to provide space for women to participate. Also, it is common for particular informal leaders of particular social, age or gender-based groups to represent the LEK of a broader group of people in meetings. While peer review in Khong has its own particular way of functioning, it is clear that it is not just for academics, and that the peer review process in Khong has helped to ensure high-quality regulations.

It is significant that most of the government officials responsible for fisheries management in Khong are of the same ethnic group as the fishers themselves, and most originate from rural villages in Khong. The officials and the fishers thus have similar backgrounds, and similar LEK about fisheries. This is important, as it is generally easy for the fishers and local officials to understand each other, and officials can easily relate to the regulations that communities adopt. For example, both groups realize that while many fish species are highly migratory, it is still possible for locally applied regulations to benefit many species.

One of the important reasons why CBFCM has been successful in Khong is that villagers have a strong sense of belonging to their communities, and a firm belief that their children and grandchildren will be living in the same villages in the future. This has helped to encourage a conservation ethic, and to ensure that many locals manage resources for the long term. A long-term sense of belonging often leads to good community-based resource management (Ostrom, 1990; Pomeroy, 1998).

ADAPTIVE MANAGEMENT AND LEK

ADAPTIVE management is critical for successful natural resource management, especially over the long term (Walters, 1986). When fishers are involved in making management decisions, as they are in Khong, strengthening LEK is a critical part of supporting the adaptive management process. Adaptive management requires making management decisions, implementing them, monitoring and analysing the results of implementation, and then altering management decisions on the basis of the results, gradually improving and adjusting them over time. This is commonly done by locals involved in the management of all kinds of natural resources, and is common in relation to fisheries management in Khong.

Fishers monitor the success of FCZs in various ways, some of which are based on specific observations of natural processes. While observations of changes in fish species and quantities of fish caught are certainly very important, other tools for understanding FCZ success are more difficult for outsiders to understand. For example, fishers monitor the populations of some algae-grazing fish species like *Mekongina erythrospila (pa sa-i* in Lao), *Morulius* spp. *(pa phia* in Lao) and *Labeo erythopterus (pa va souang* in Lao) by observing shallow rocky areas adjacent to FCZs. If the rocks are covered with algae, this indicates that there are few algae-eating fish in the adjacent FCZ. On the other hand, when fish graze on the algae on the rocks, fishers can see what species have fed there, since the width of the grazing lines differ according to the species involved and the sizes of individual fish. This method of observing fish presence and absence is little known within the scientific community, at least in the context of the Mekong River basin.

Another innovative and little known method of monitoring fish in FCZs used in Khong relates to fish rising to the surface of the water for oxygen or other purposes. This is especially common during the height of the hot season, when water levels are at their lowest, and fish tend to concentrate in deep-water areas. Villagers have a considerable amount of LEK about what fish species rise to the surface at what times of the day, and where, although this specific aspect of the ecology of particular fish species is not understood by scientists. Villagers also exhibit a considerable amount of skill in their abilities to recognize those fish that rise to the surface, even though the non-experienced eye and ear are unlikely to be able to identify them.

Villagers also assess the size of populations of particular migratory species like *Henicorhynchus lobatus* (*pa soi houa lem* in Lao) by looking at the size of individual fishes. If they are small, the population size may be large, as competition for food is greater and growth is slower. However, if the fish are big, the population may be small, providing ample opportunities for increased feeding and growth.

Local people also monitor populations of the smallscale croaker *Boesemania microlepis* (*pa kouang* in Lao), which are important beneficiaries of certain deepwater FCZs in Khong. During their spawning season in the dry season, these large fish make loud croaking sounds that are audible even out of the water. Local people gauge the amount of croaking that occurs each year, and in that way they have a good sense whether populations are increasing or declining. This skill has helped in their management of particular FCZs where *B. microlepis* spawn (Baird et al., 2001b).

DATA COLLECTION IN FISHERIES CONSERVATION ZONES

In Khong, the EPCDSWP has helped to develop a more formalized data collection programme to monitor the results of management decisions related to the establishment of FCZs. This was done to help communities improve their management strategies, and also to provide government agencies with quantitative data useful for assessing the value of FCZs (Baird et al., 1999b).

Initially, eight villages in Khong participated in the programme. In each village, locals developed hypotheses about what fish species had already benefited from FCZs, based on past observations. As specific FCZs protect different microhabitats of importance to different species, locals first hypothesized what species would benefit, and then determined what fisheries to monitor to test whether those species had really benefited. Between five and twenty fishers were selected by villagers in each community to record daily fish catch data for the selected fisheries. After months of data collection, the data from different individuals were pooled and statistically analysed. Although not all data were correctly recorded, most were useable in the analysis. The data were then returned to the villagers to be reviewed and verified. During this verification process the villagers were able to add a considerable amount of context and depth, and the data were often altered as a result. The data verification process acted as an important tool for helping the fishers to understand how effective management strategies have been for specific fish species, although there is still much more to learn (Baird et al., 1999b).

The data were also used to test the knowledge of the fishers regarding their understanding of the catch structure of particular fisheries. It was found that in Khong most fishers are able to rank the top ten species of fish caught in fisheries based on total weight quite reliably, thus showing their deep understanding of the fisheries (Baird et al., 1999b).

The process of adaptive management in Khong has also been strengthened through various other activities at the community level, the most important being periodic village meetings to review regulations informally among community members and discuss ways to improve regulations and their implementation. For example, observations made by fishers have been important in informing committees how they might benefit by changing the boundaries of FCZs, an example of adaptive management.

CONCLUSIONS

While not all the villages in Khong have been equally successful in their aquatic resource management efforts, for biological, geographical and social reasons, most villagers have widely reported increased stocks of certain aquatic animals, as well as increased fish catches, since the adoption of CBFCM regulations. Some rare and endangered species of fish have apparently also made comebacks at least partially due to the regulations, thus benefiting biodiversity (Cunningham, 1998; Chomchanta et al., 2001; Ait Aqua Outreach, 1997; Baird and Flaherty, 2005). Improved solidarity and coordination within and between rural fishing and farming villages and the government has also been observed. The transaction costs to government of managing fisheries are minimal, since local people do most of the work. Therefore, the local government advocates the system, and hopes to expand the work to other villages in the future, although not all villages have opted to get involved, and some do not have appropriate habitat near their village for establishing FCZs. In any case, the initiative has been quite successful, since both local people and the environment are benefiting.

This chapter has also indicated that in cases where fishers are given a high level of authority over management decisions; as is the case in Khong, it is important to make maximum use of LEK to improve fisheries management, and in this context it is often useful to disseminate and strengthen LEK in various ways.

However, the situation may not always be as straightforward as it appears to be in Khong, especially when one is dealing with less ethnically and socially homogenous communities, or with strong central governments unfamiliar with local issues. But, even when less homogenous communities are the focus, CBFCM may still be the most viable option for improving management, especially when one considers small-scale fisheries in remote rural areas where few scientific data are available. The critical importance of LEK should be recognized, as it has clearly been demonstrated, at least in the context of Khong, to have considerable potential for strengthening the local management of living aquatic resources.

ACKNOWLEDGEMENTS

THANKS to Mr Phongsavath Kisouvannalath, Mr Vixay Inthaphaisy, Mr Bounpheng Phylavanh, Mr Bounhong Mounsouphom, Mr Khamsouk Xaiyamanivong and Mr Bounthong Senesouk for their important contributions to developing and monitoring the aquatic resource community-based fisheries co-management programme in Khong District. Thanks also to the government in Khong District, which has strongly supported local people and LEK. Mr Ole Heggen from the Geography Department, University of Victoria, Canada, prepared the maps. Thanks to the three anonymous referees who took the time to review and help improve this chapter. Special thanks go to all the villagers from Khong District. They have shared, over the years, their immense LEK about fish and fisheries, and this chapter is dedicated to them.

REFERENCES

AIT AQUA OUTREACH. 1997. *Participatory evaluation workshop of the LCFDPP*. Savannakhet, Lao PDR, AIT Aqua Outreach, 7 pp.

ALI, A.B. 1996. Community-based management in inland fisheries: Case studies from two Malaysian fishing communities. In: I.G. Cowx (ed.), *Stock assessment in inland fisheries*. London, Fishing News Books, pp. 482–93.

AHMED, M.; NAVY, H.; VUTHY, L.; TIONGCO, M. 1998. *Socioeconomic assessment of freshwater capture fisheries in Cambodia: Report on a household survey*. Phnom Penh, Mekong River Commission, 186 pp.

BAIRD, I.G. 1999a. Fishing for sustainability in the Mekong Basin. *Watershed*, Vol. 4, No. 3, pp. 54–56.

——. 1999b. The co-management of Mekong River inland aquatic resources in Southern Lao PDR. *Proceedings of the International Workshop on Co-Management*, 23–28 August 1999, Penang, 43 pp.

——. 2001. Aquatic biodiversity in the Siphandone Wetlands. In: G. Daconto (ed.), *Siphandone wetlands*. Bergamo, Italy, CESVI, pp. 61–74.

BAIRD, I.G.; INTHAPHAISY, V.; PHYLAIVANH, B.; KISOUVANNALATH, P. 1998. *A rapid fisheries survey in Khong District, Champasak Province, Southern Lao PDR*, 31 pp. Environmental Protection and Community Development in Siphandone Wetland Project. Pakse, Lao PDR, *CESVI*.

BAIRD, I.G.; PHYLAVANH, B. 1999. Fishes and forests: Fish foods and the importance of seasonally flooded riverine habitats for Mekong River fish. Environmental Protection and Community Development in Siphandone *Wetland Project*. Pakse, Lao PDR, *CESVI*, 46 pp.

BAIRD, I.G.; INTHAPHAISY, V.; KISOUVANNALATH, P.; PHYLAVANH, B.; MOUNSOUPHOM, B. 1999a. *The fishes of southern Lao*. Lao Community Fisheries and Dolphin Protection Project. Pakse, Lao PDR, Ministry of Agriculture and Forestry, 162 pp. (In Lao.)

BAIRD, I.G.; INTHAPHAISY, V.; KISOUVANNALATH, P.; VONGSENESOUK, B.; MOUNSOUPHOM, B. 1999b. *The setting up and the initial results of a villager based system for monitoring fish conservation zones in the Mekong River, Khong District, Champasak Province, Southern Lao PDR*. Environmental Protection and Community Development in Siphandone Wetland Project. Pakse, Lao PDR, *CESVI*, 44 pp.

BAIRD, I.G.; HOGAN Z.; PHYLAVANH, B.; MOYLE, P. 2001a. A communal fishery for migratory catfish *Pangasius macronema* in the Mekong River. *Asian Fisheries Science*, Vol. 14, pp. 25–41.

BAIRD, I.G.; PHYLAVANH, B.; VONGSENESOUK, B.; XAIYAMANIVONG, K. 2001b. The ecology and conservation of the smallscale croaker *Boesemania microlepis* (Bleeker 1858–59) in the mainstream Mekong River, Southern Laos. *Natural History Bulletin of the Siam Society*, Vol. 49, pp. 161–76.

BAIRD, I.G.; FLAHERTY, M.S.; PHYLAVANH, B. 2003. Rhythms of the river: Lunar phases and migrations of small carps (Cyprinidae) in the Mekong River. *Natural History Bulletin of the Siam Society*, Vol. 51, No. 1, pp. 5–36.

BAIRD, I.G.; FLAHERTY, M.S. 2005. Mekong River fish conservation zones in Southern Laos: Assessing effectiveness using local ecological knowledge. *Environmental Management*, Vol. 36, No. 3, pp. 439–54.

BARAN, E.; VAN ZALINGE, N.; NGOR PENG BUN. 2001. Floods, floodplains and fish production in the Mekong basin: Present and past trends. Contribution to the Asian Wetlands Symposium 2001, 27–30 August 2001, Penang.

BERKES, F.; KISLALIOGLU, M. 1993. Community-based management and sustainable development. In: *FAO/Japan expert consultation on the development of community-based coastal fishery management systems for Asia and the Pacific*. Rome, Food and Agriculture Organization of the United Nations (FAO), pp. 619–24.

BRECHIN, S.R.; WILSHUSEN, P.R.; FORTWANGLER, C.L.; WEST, P.C. 2003. *Contested nature: Promoting international biodiversity with social justice in the 21st century.*, Albany, State University of New York Press, 321 pp.

CHOMCHANTA, P.; VONGPHASOUK, P.; SOUKHASEUM, V.; SOULIGNAVONG, C.; SAADSY, B. 2000. *A preliminary assessment of Mekong fishery conservation zones in the Siphandone area of southern Lao PDR, and recommendations for further evaluation and monitoring*. Vientiane, The Living Aquatic Resources and Research Centre (LARReC), 13 pp.

CHRISTY, F. 1993. A re-evaluation of approaches to fisheries development: The special characteristics of fisheries and the need for management. In: *FAO/Japan expert consultation on the development of community-based coastal fishery management systems for Asia and the Pacific*. Rome, FAO, pp. 597–608.

CLARIDGE, G. (COMPILER). 1996. *An inventory of wetlands of the Lao PDR*. Vientiane, Wetlands Programme, IUCN – the World Conservation Union, 287 pp.

CLARIDGE, G.F.; SORANGKHOUN, T.; BAIRD, I.G. 1997. *Community fisheries in Lao PDR: A survey of techniques and issues*. Vientiane, IUCN – The World Conservation Union, 70 pp.

CLAY, P.; MCGOODWIN, J. 1995. Utilizing social sciences in fisheries management. *Aquatic Living Resources*, Vol. 8, No. 3, pp. 203–7.

COWX, I. 1991. Catch effort sampling strategies: Conclusions and recommendations for management. In: I.G. Cowx (ed.), *Catch effort sampling strategies: Their application in freshwater fisheries management*. London, Fishing New Books, pp. 404–13.

CUNNINGHAM, P. 1998. Extending a co-management network to save the Mekong's giants. *Mekong Fish Catch and Culture*, Vol. 3, No. 3, pp. 6–7.

DACONTO, G. (ed.) 2001. *Siphandone wetlands*. Bergamo, Italy, CESVI, 192 pp.

GARAWAY, C.J. 1999. *Small waterbody fisheries and the potential for community-led enhancement: Case studies in Lao PDR*. London, T.H. Huxley School for the Environment, Earth Sciences and Engineering, Imperial college of Science Technology and Medicine, University of London, 414 pp.

GLAESEL, H.; SIMONITSCH, M. 2003. The discourse of participatory democracy. In: N. Haggan; C. Brignall; L. Wood (eds.), *Putting fishers' knowledge to work*. Fisheries Centre Research Reports, Vol. 11, No. 1. Vancouver, University of British Columbia, pp. 157–67.

HILL, M.T.; HILL, S.A. 1994. *Fisheries ecology and hydropower in the Mekong River: An evaluation of run-of-the-river projects*. Bangkok, Mekong Secretariat, 106 pp.

HIRSCH, P. 2000. Managing the Mekong commons: Local, national and regional issues. In: M. Ahmed; P. Hirsch (eds.), *Common property in the Mekong: Issues of sustainability and subsistence*. ICLARM Stud. Rev. 26. Penang and Sydney, ICLARM – The World Fish Centre, and Department of Geography, University of Sydney, Sydney, pp. 19–25.

HIRSCH, P.; NORASENG, P. 1999. Co-management of fisheries in diverse aqua-ecosystems of southern Laos. *Proceedings if the International Workshop of Fisheries Co-Management*, Penang, Malaysia, 23–28 August 1999, Pakse, Lao PDR.

HOGAN, Z. 1997. Aquatic conservation zones: Community management of rivers and fisheries. *Watershed*, Vol. 3, No. 2, pp. 29–33.

HUBBEL, D. 1999. Food for the people: natural fisheries of the Mekong River. *Watershed*, Vol. 4, No. 3, pp. 22–36.

INTERNATIONAL RIVERS NETWORK. 1999. *Power struggle: The impacts of hydro-development in Laos*. Berkeley, International Rivers Network, 68 pp.

IIRR. 1996. *Recording and using indigenous knowledge: A manual*. Silang, Cavite, Philippines, International Institute of Rural Reconstruction, 211 pp.

JENTOFT, S.; MCCAY, B.J.; WILSON, D.C. 1998. Social theory and fisheries co-management, *Marine Policy*, Vol. 22, No. 4–5, pp. 423–36.

JOHANNES, R.E. 1981. *Words of the lagoon: Fishing and marine lore in the Palau district of Micronesia*. Calif., University of California Press, 245 pp.

——. 1998. Dataless management. *Trends in Ecology and Evolution*, Vol. 13, No. 6, pp. 243–6.

——. 2001. *The need for a centre for the study of indigenous fishers knowledge*. Contribution to Wise Coastal Practices for Sustainable Human Development Forum. Available online at: http://www.csiwisepractices.org/?read=388 (last accessed 26 June 2005).

JOHNSON, C. 1998. Beyond community rights: Small-scale fisheries and community-based management in southern Thailand. *TDRI Quarterly Review*, Vol. 13, No. 2, pp. 25–31.

KOLDING, J. 2002. *The use of hydro-acoustic surveys for the monitoring of fish abundance in deep water pools and fish conservation zones in the Mekong River, Siphandone area, Champassak Province, Lao PDR*. Vientiane, LARReC, 32 pp.

KOTTELAT, M.; WHITTEN, T. 1996. Freshwater biodiversity in Asia with special reference to fish. *World Bank Technical Paper*, No. 343. Washington DC, The World Bank, 59 pp.

KUPERAN, K.; ABDULLAH, N.M.R. 1994. Small-scale coastal fisheries co-management. *Marine Policy*, Vol. 18, No. 4, pp. 306–13.

LIENG, S.; YIM, C.; VAN ZALINGE, N.P. 1995. Fisheries of the Tonlesap River Cambodia, I: the bagnet (Dai) Fishery. *Asian Fisheries Science*, Vol. 8, pp. 258–65.

MASAE, A.; NISSAPA, A.; BOROMTHANARAT, S. 1999. An analysis of co-management arrangements: A case of fishing community in southern Thailand. *Proceedings of the International Workshop on Fisheries Co-Management*, Penang, 23–28 August 1999. Songkhla, Thailand, CORIN and ICLARM.

MCCAY, B.J.; JENTOFT, S. 1996. From the bottom up: Participatory issues in fisheries management. *Society and Natural Resources*, Vol. 9, pp. 237–50

OSTROM, E. 1990. *Governing the commons: The evolution of institutions for collective action*. Cambridge, U.K., Cambridge University Press, 280 pp.

POMEROY, R.S. 1998. Fisheries co-management and community-based management: Lessons drawn from Asian experiences. *Proceedings of the Crossing Boundaries, Seventh Conference of the International Association for the Study of Common Property*, 10–14 June 1998, Burnaby, Canada, 13 pp.

POMEROY, R.S.; CARLOS, M.B. 1997. Community-based coastal resource management in the Philippines: A review and evaluation of programs and projects, 1984–1994. *Marine Policy*, Vol. 21, No. 5, pp. 445–64.

POULSEN, A.F. 2001. *Deep pools as dry season fish habitats in the Mekong River basin*. Report of the Assessment of Mekong Fisheries Component (AMFC). Vientiane. Mekong River Commission. CD-Rom

POULSEN, A.F.; VALBO-JOERGENSEN, J. (eds.). 2000. *Fish migrations and spawning habits in the Mekong mainstream: A survey using local knowledge.* AMFC Technical Report. Vientiane, Mekong River Commission, 132 pp.

RAINBOTH, W.J. 1996. *Field guide to the fishes of the Cambodian Mekong.* Rome, Food and Agriculture Organization of the United Nations, 265 pp.

ROBERTS, T.R. 1993a. Artisanal fisheries and fish ecology below the great waterfalls of the Mekong River in southern Laos. *Natural History Bulletin of the Siam Society,* Vol. 41, No. 1, pp. 31–62.

——. 1993b. Environmental impact assessment: the EIA on EIA. *Asian Soc. Environmental Protection Newsletter,* Vol. 9, No. 3, pp. 1–3, 10.

——. 1993c. Just another dammed river? Negative impacts of Pak Mun dam on fishes of the Mekong basin. *Natural History Bulletin of the Siam Society,* Vol. 41, pp. 105–33.

ROBERTS, T.R.; BAIRD, I.G. 1995. Traditional fisheries and fish ecology on the Mekong River at Khone waterfalls in southern Laos. *Natural History Bulletin of the Siam Society,* Vol. 43, pp. 219–62.

ROBERTS, T.; WARREN, T. 1994. Observations on fishes and fisheries in southern Laos and northeastern Cambodia, October 1993–February 1994. *Natural History Bulletin of the Siam Society,* Vol. 42, pp. 87–115.

SJORSLEV, J.G. (ED.). 2000. Luangprabang fisheries survey. Vientiane, AMFC/MRC and LARReC/NAFRI, 45 pp.

VAN ZALINGE, N.P.; NAO THUOK; TOUCH SEANG TANA; DEAP LOEUNG. 2000. Where there is water, there is fish? Cambodian fisheries issues in a Mekong River Basin perspective. In: M. Ahmed and P. Hirsch (eds.), *Common property in the Mekong: issues of sustainability and subsistence.* ICLARM Stud. Rev. 26. Penand and Sydney, ICLARM — The World Fish Centre, and Department of Geography, University of Sydney, pp. 37–48.

WALTERS, C. 1986. *Adaptive management of renewable resources.* New York, Macmillan, 374 pp.

WARREN, T.J.; CHAPMAN, G.C.; SINGHANOUVONG, D. 1998. The upstream dry-season migrations of some important fish species in the lower Mekong River in Laos. *Asian Fisheries Science,* Vol. 11, pp. 239–51.

CHAPTER 13 Use of fishers' knowledge in community management of fisheries in Bangladesh

Parvin Sultana and Paul Thompson

ABSTRACT

COMMUNITY-BASED fisheries management has been promoted in part as a way of combining local fishers' knowledge with expert or scientific advice. A key issue is how to mobilize fisher knowledge to improve fisheries management. This chapter discusses experience in Bangladesh of participatory planning and community measures to restore and manage fisheries at three sites.

In Ashurar Beel the fisher community assessed their problems and the performance of management measures taken by the government in the past. In 1997, with NGO assistance, they established a management committee, a fish sanctuary and other conservation measures. Monitoring indicates this has enhanced catches of higher-valued species and strengthened user roles in decision-making.

In Shuluar Beel a structured participatory planning process was followed. This highlighted the fact that fish and other aquatic resources had increasing significance for people's livelihoods despite declining catches, and resulted in a detailed analysis and consensus among fishers, farmers and the landless of problems and possible solutions. This has been the starting point for initiating local management institutions and collective action for sustainable use.

In Dhampara area, a network of local fish sanctuaries was established. The project shared knowledge by means of exchange visits by community representatives to other sites with similar community management.

The case studies show that local knowledge has an important role in better management of Bangladesh's inland fisheries, as does sharing of experience among communities. Communities can agree on and implement actions to improve fishery management, but in the long term this needs a framework of government support that maintains local fisher community initiatives when they are threatened by local elites.

INTRODUCTION

COMMUNITY-BASED management has become a common strategy for improving management of natural resources and empowering local communities in the past two decades. This is based on co-management, using local and traditional ecological knowledge, recognizing local institutions, and establishing common-property regimes (Pomeroy and Berkes, 1997; Berkes and Folke, 1998; Berkes et al., 1998; Ostrom, 1990). Although design principles for community management institutions (Ostrom, 1994) and factors linked with sustainable common-property regimes and institutions (Agrawal, 2001) have been studied, there is still a question of how best to initiate such regimes and the role of local knowledge. The examples presented here are based on participatory planning, NGO initiatives to build and modify institutions, and existing local knowledge.

BANGLADESH CONTEXT

FISH are a vital part of the diet and culture of Bangladesh: in the words of a traditional saying – *machee bhatee bangali* – fish and rice make a Bangladeshi. Over half of the country comprises floodplains, and these seasonal and permanent wetlands form a major capture fishery and source of livelihoods. About 40 per cent of all fish consumed come from the floodplains (Department of Fisheries, 2000). Over 70 per cent of households in the floodplains catch fish for income or food (Minkin et al., 1997; Thompson et al., 1999).

These inland fisheries are declining due to conversion of wetlands for agriculture, blocking of fish migration routes by flood control embankments, and overfishing. In response, since the mid-1980s policy changes and projects have been designed to improve inland fisheries management. However, these have tended to be top-down administrative decisions over leasing fishing rights, or technical measures to enhance fisheries such as stocking carp (Islam, 1999). There was little participation in planning by the fishing communities or role for local fishers' knowledge.

Interest in indigenous knowledge of natural resources in Bangladesh has grown recently (Sillitoe, 2000), with a focus on local technical knowledge within a scientific framework. Understanding of local floodplain resources and problems, for example in Dixon et al. (2000), has focused on stakeholder differences and their perspectives of the resource base. Methods of local planning that focus on common shared knowledge and solutions to problems, while recognizing differences in interests, have been developed (Sultana and Thompson, 2003), and practical measures by communities to improve fisheries management have been strengthened.

CASE STUDY SITES

Lessons from local planning and incorporation of fishers' knowledge in resource management are considered from three locations in Bangladesh (Figure 13.1).

Ashurar Beel

A *beel* is a floodplain depression that may hold water all year or just in the monsoon (wet season). Ashurar Beel is a perennial waterbody in north-west Bangladesh with a maximum, wet season, area of about 400 ha and dry season area of over 100 ha. During the dry season water is left in seven depressions locally known as *daha*. The name Ashurar is derived from the Bangla word *ashi* meaning eighty (tradition has it that there are eighty inlets and outlets). Fourteen fishing and non-fishing villages surround the *beel*, inhabited by 2,466 households.

The *beel* came under the first phase of the Community-Based Fisheries Management Project (CBFM-1) in 1996 when a national NGO (CARITAS) organized the professional fishing households into groups. Fishers were scattered and did not cooperate before the project; many had migrated into the area because of river erosion elsewhere. They had just experienced a project that imposed fishery enhancement through stocking of carp in the *beel*. Local fishers had had to pay part of the costs although they were not involved in planning and said that most of the fish escaped or were caught soon after stocking. This *beel* is a *jalmohal* (a water estate owned by the government, where the lease of fishing rights is usually auctioned to the highest bidder), but little revenue has been collected in recent years.

Dhampara project

This is a Bangladesh Water Development Board (BWDB) project located in Netrakona District in north-east Bangladesh, where a flood-control embankment and water control structures were built in 1998–2001. Following a decade of developing guidelines for participation in the water development sector (Ministry of Water Resources, 2001), the project aimed from the outset to involve the residents in its planning and to mitigate any adverse effects. Fisheries are widely regarded as having been adversely affected by past flood-control projects (Ali, 1997) and so were an obvious target for mitigation. The project covers 15,000 ha, including rivers, canals, *beels* and floodplains. The NGOs who supported the fisheries component focused on conservation measures, promoting a network of forty small fish sanctuaries within one year (BWDB, undated).

INDIGENOUS AND ARTISANAL FISHERIES

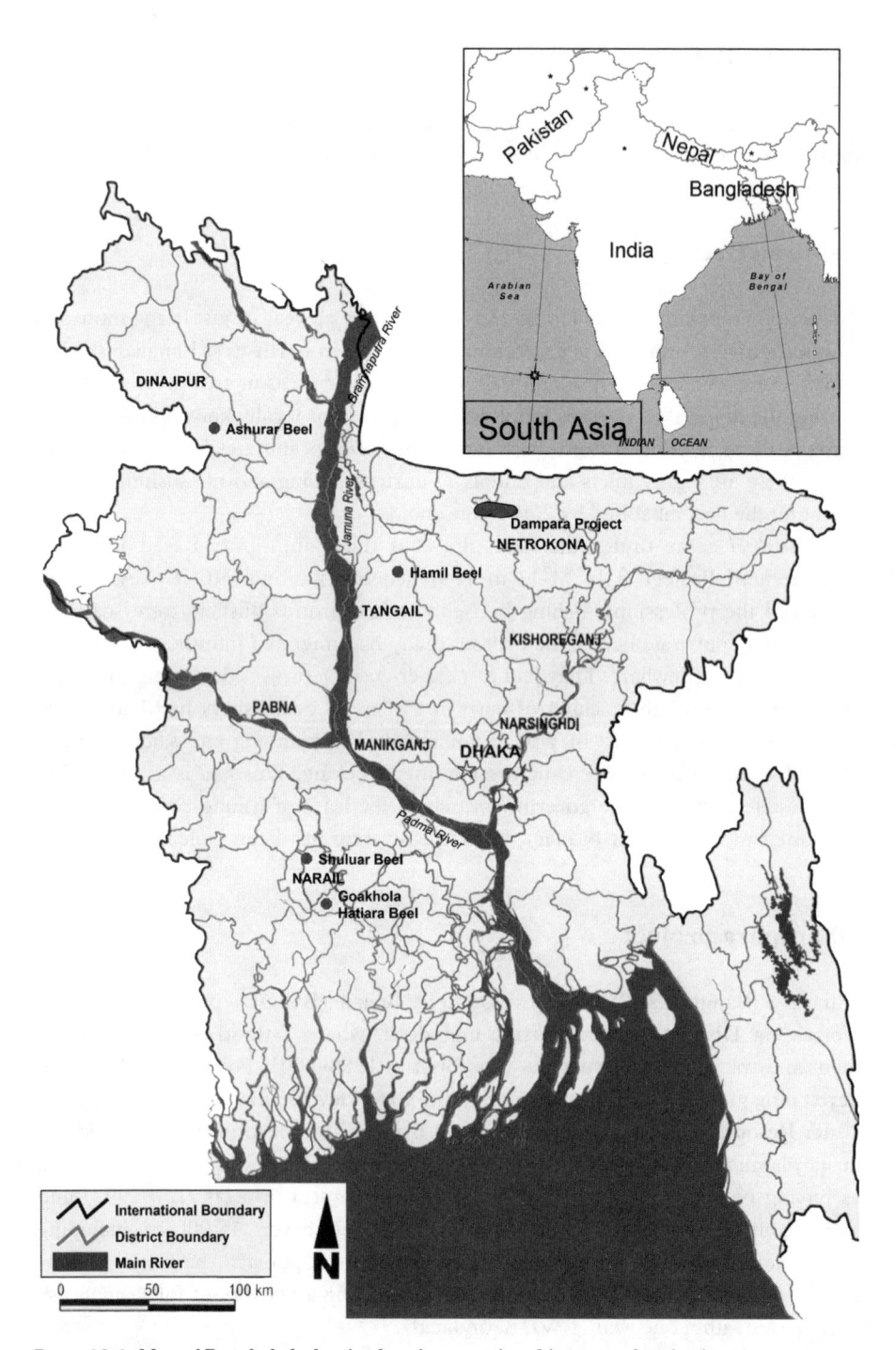

FIGURE 13.1 *Map of Bangladesh showing locations mentioned in text and main river system.*

Shuluar Beel

This seasonal *beel* in south-west Bangladesh covers at its maximum 1,000 ha. The *beel* is connected by a canal to a secondary river, but rainfall is its main source of water. All the land in the *beel* is private and is mainly cultivated with rice. The government has no direct role in management of aquatic resources here, as there is no *jalmohal*. On the river side is a flood control embankment. About 1,800 households live in five villages around the *beel* and have open access to fishing during the monsoon; almost all the households catch fish at some time in the year. Half of the households are very poor and depend on fish and other aquatic flora and fauna for income; the other half fish for their own consumption. The Community-Based Fisheries Management project Phase 2 (CBFM-2) started to work here in late 2001 through a regional NGO (Banchte Sheka), and in mid-2002 a participatory planning process was initiated.

USE OF INDIGENOUS FISHERY AND FLOODPLAIN KNOWLEDGE FOR MANAGEMENT

HERE we compare the outcomes from a recent participatory planning process, which made explicit use of the knowledge and analyses of fishers and other floodplain users, with outcomes at sites where fishers were involved in deciding on management improvements but the process of using local knowledge was less structured. The local institutional arrangements that use fishers' knowledge in decision making are also summarized.

Ashurar Beel: locally set fishery rules

A Beel Management Committee (BMC) was formed in September 1997, comprising the leaders of each fisher group organized by the NGO. In 1999, the BMC was dissolved and reformed to represent additional groups, and the four executive officers were elected by the general members. The election was overseen by government staff, the NGO and local elites. This committee is accountable to the members of the groups. The committee formulates the strategy for fishing in the waterbody, making and attempting to enforce their own rules.

In 1997, the fishers agreed to stop building private *katas* (brushpiles made by placing tree branches in the water) in the *beel*. Traditionally, better-off people or fishers claim an area by making a brushpile where they then catch fish two or three times a year. *Katas* are built because local people know that they shelter and attract larger, more valuable fish species. However, the older fishers observed that the *katas* were increasingly being completely harvested, leaving no brood fish, whereas previously some fish had been left in them. Also, *katas* restricted the access of most fishers (who were now better organized), and it was said that *katas* were causing faster siltation. The fishers proposed to demolish individual brushpiles and end the practice. They collectively constructed a fish sanctuary by moving all the tree branches from the existing *katas* into the deepest 8 ha of the *beel* (known as Bhurir Doha) to make a year-round fish shelter and sanctuary (see the discussion of fish conservation zones, Baird, this volume). In 1999, the BMC decided to replace the sanctuary materials as the old ones were rotten. Through general meetings, they decided that each group would provide a tree branch and that each participant would provide a bamboo pole.

It is common knowledge among the fishers that the *beel*-resident fish breed in the early monsoon when the water level rises, and that other fish species move in from the river to breed. The fishers therefore thought that if they stopped fishing, there would be more fish later in the monsoon and the fry would grow to a larger and more valuable size. From 1997 onwards the BMC made a rule on behalf of all fishers that there would be a closed season for about three months in the pre- and early monsoon (although some groups started seine-net fishing before the end of this season). All types of net were banned from about mid-April, except for traditional gillnets with mesh sizes of 10 cm (4") or more (which do not target juvenile fish).

In May 1998 CARITAS organized a two-day workshop at the *beel* where the fishers formalized their new rules. The workshop was opened by the Minister for Fisheries and Livestock, and the Director General of the Department of Fisheries. Through the workshop, the main strategies (sanctuary and closed season) were discussed with officials and the appropriateness of the fishers' plans was recognized. Their plans combined the general principles of government polices (on protection of fish in the breeding season) with the specific knowledge of the Ashurar Beel fishers and the novel idea of a large, voluntary brushpile sanctuary. Some other rules adopted by the community in the workshop were:

- Any decisions about the fishery will be taken on a participatory basis.
- Anyone not following the BMC's rules would be punished. For fishing in the sanctuary the sanctions would be to warn the offender first. A second offence would incur a fine, and a third offence would be taken to court.
- Outsiders (people who are not from the participant villages) cannot fish for income in the *beel*, but local people who are not group members are allowed to fish with small devices (such as push-nets) for food so long as they do not break the other rules.
- Gillnets will not be set at right angles to the flow of water. The fishers observed that fish normally swim with or against the current and were easily targeted by nets set across the water flow. They wanted to avoid catching all fish migrating through the *beel* and to balance the catch between users of different methods.

Shuluar Beel: assessment of trends in fisheries and natural resources

A participatory action plan development (PAPD) workshop was held with six stakeholder groups. The participants were drawn at random from a census of households in villages using this *beel*, stratified by stakeholder groups. The workshop followed the process described in Sultana and Thompson (2003). This facilitated structured process works through a combination of individual stakeholder sessions and plenary sessions. It enables the specific interests and concerns of each stakeholder group to be articulated, and also finds where there is common ground among the community on problems and solutions. The types of household (referred to as stakeholder groups above) identified through reconnaissance surveys and consultations with key local informants were:

- people who fish for an income (mostly part-time fishers who are also labourers)
- other landless and marginal landowner households with landholdings of up to 0.4 ha
- farming households (owning over 0.4 ha of land)
- *kua* owners (landowners who also get an income in the dry season from the fish trapped in their *kuas* – ditches deliberately dug in the floodplain as fish-aggregating devices).

In addition the research team invited a group of women from the landless households to the PAPD to reflect differences in interests according to gender and because these poor women (unlike women from better-off households) gather aquatic resources. The participants in each group made a problem census and an analysis of potential solutions, and assessed natural resource status and trends.

The participants explained that almost all natural resources in the *beel* are becoming scarcer day by day, and that the amounts they could catch or gather are now less than fifteen years earlier (Table 13.1). They said that increasing human population, overexploitation, habitat loss and conversion of wetland to agriculture have adversely affected natural resources. Fish of course are the most important resource for the fishers, who claimed that fish were abundant fifteen years back when, apart from traditional (Hindu) fishers, other people (Muslims) only fished for food. At that time fish were less important than now for people's livelihoods (particularly the Muslims) as most fish were caught for home consumption, fish were more abundant and anybody could fish anywhere. Now, however, fish are scarcer (and their value is higher), people fish for income, and landowners do not allow people to fish in their water (land) area during the late monsoon. During high flood periods fishing is difficult even though access is unrestricted, but when water starts to recede and fish get trapped in the *kuas*, landowners guard their land. Fishers can then only fish in other areas, such as public land.

TABLE 13.1 *Changes in natural resource status, use and access in Shuluar Beel.*

	Use	Status (%) 2002	Status (%) 15 years earlier	Importance to livelihood (rank) 2002	Importance to livelihood (rank) 15 years earlier	Access 2002	Access 15 years earlier
Fish	Consumption, Sale for income	33	100	9	3	Landowners do not allow fishing in their private areas	No restriction
Water lily	Consumption (vegetable), Sale & income	40	60	7	3	Free access	Free access
Snail	Fish feed, Duck feed, Sale & income	30	65	8	2	Free access	Free access
Grass/Sechi/Kolmi	Cattle feed, Human food	25	75	2	7	No restriction on collecting, but livestock decreased	Free access, lots of livestock
Jhinuk (mussel)	Fish feed, Duck feed, Lime production	7	93	2	10	Use decreased, replaced by artificial products	Free access
Shaluk (aquatic plant fruit)	Sale (Tk.8–10/kg), Consumption	40	60	10	4	Land owners do not allow harvest	Free access
Trees/forest	Furniture, firewood, fruits, give oxygen, cattle feed, fish feed	60	30	10	4	Good income source; can cut trees from own land. Very few common-property trees	Free access to forest on public land
Land (soil and agriculture)	Crop cultivation, House building, Pottery	40	60	10	6	Very little common property, access to public land is controlled by local influential people	Poor could use public land

Assessments averaged/consolidated over all five groups consulted.
Status is an indicator from group discussions (mainly with fishers and landless) compared with the most abundant situation they could ever recall (= 100%).
Rank = score out of 10 for importance to livelihood where 10 is the maximum.

Edible water plants, snails and mussels are also important. Access to these resources is open for all and free. People collect them for sale or for their own consumption. Snails are now a valuable resource. With the increase in shrimp farming in nearby areas in the past decade, demand for snails as shrimp feed has increased. Previously children collected snails to feed domestic ducks. Now commercial snail collection is common, and a gatherer can earn Tk.4,000–5,000 (about US$ 80) per month during the peak season (June to August). People reported that the snail population has decreased compared with fifteen years ago because there are no restrictions on collection.

Water lilies are also valuable. The flower stems and the fruits of the *shaluk* (giant lily) are used as human food. In the past people only collected these for their own consumption, but nowadays poor people collect them to sell for Tk.8–10 per kg, and the plants have become scarcer due to overexploitation. However, better-off households do not collect these resources. Earlier there were small areas of forest in the villages, which are now fully exploited. Common public land is either lost due to encroachment by powerful people, or the local authority has given the usage rights not to poor people (as it is authorized to do) but illegally to others. The only trees left are privately owned, and since the poor have little or no land they have virtually no trees.

Table 13.2 shows the fishers' ranking of species according to their catch and importance. Small fish dominate in the floodplain environment now, whereas fifteen

TABLE 13.2 *Changes in fisher's ranking of fish species (importance in catch) in Shuluar Beel.*

	2002			15 years earlier (about 1987)	
Rank	Local name	Scientific name	Rank	Local name	Scientific name
1	Puti	*Puntius* spp.	1	**Galda Icha***	*Machrobrachium rosenbergii*
2	Gura Icha	*Machrobrachium styliferus*	2	**Magur***	*Clarius batrachus*
3	Taki	*Channa punctata*	3	**Rui***	*Labeo rohita*
4	Shol	*Channa striata*	4	**Shing**	*Heteropneustes fossilis*
5	Koi	*Anabas testudineus*	5	Shol	*Channa striata*
6	Roina	*Nandus nandus*	6	Bele	*Glossogobius giuris*
7	**Shing**	*Heteropneustes fossilis*	7	Tengra*	*Mystus vittatus*
8	Tepa	*Tetradon cutcutia*	8	Sarputi*	*Puntius sarana*
9	Tak Chanda	*Leignathus equulus*	9	Falui	*Notopterus notopterus*
10	Bajari Tengra	*Mystus tengara*	10	**Gazar**	*Channa marulius*
11	Guchi Baim	*Macrognathus pancalus*	11	Tara Baim*	*Macrognathus aculeatus*
12	Khalisa	*Colisa labiosus* (?)	12	Roina	*Nandus nandus*
13	Pabda	*Ompok pabda*	13	**Boal***	*Wallago attu*
14	Falui	*Notopterus notopterus*	14	Bacha Mas*	*Eutropiichthys vacha*
15	Bara Baim	*Mastacembelus armatus*	15	Chelenda*	*Silonia silondia*
16	Bele	*Glossogobius giuris*	16	Tepa	*Tetradon cutcutia*
17	**Gazar**	*Channa marulius*	17	**Chital***	*Chitala chitala*
18	Kakila	*Xenentodon cancila*	18	Puti	*Puntous* spp.
19	Chela	*Salmostoma phulo*	19	Koi	*Anabas testudineus*
20	Chusro	*Colisa fasciatus* (?)	20	Batasi*	*Pseudeutropius atherinoides*

Larger/higher value species are in bold.

* = species that dropped out of the top 20.

Rank 1 = most important.

years ago there was more diversity, and bigger and higher-valued species were more dominant. At that time fish could move freely from the river along the canal to the *beel*. Bigger fishes, which move to the floodplain to breed, cannot now enter the area because of the embankment and because rich and influential people trap fish in fences across the canal entrance. Due to stagnant water, the fishers say that fish disease has become severe in some years. They hope that natural populations of major carp can be replenished if there is proper management of the size of sluice gates. Furthermore, they believe that conserving fish in the ditches for a few years could rehabilitate some species.

Representatives of the five different stakeholder groups discussed problems (particularly those related to natural resources and their trends) and potential solutions (see also Meeuwig et al., this volume). Table 13.3 shows the prioritization of problems according to the number of votes given by participants in separate sessions (each person had five votes to allocate as he/she wished among problems identified by the group). Declining fish catches and natural resources, including problems of water management, clearly dominated for both poor and non-poor people. Local knowledge was the basis of planning. The PAPD participants identified the causes of trends in natural resources and the solutions that they thought to be appropriate and feasible. Table 13.4 shows the problem–cause–solution analysis of the main natural resource problems. In the plenary sessions, local elected representatives and fisheries officers attended and supported the analysis and recommendations. Implementation will depend on the farmers, fishers and labourers who use the *beel*. The PAPD resulted in agreement on a local committee to represent all user villages and stakeholders. This committee has since protected some *kuas* as dry-season fish refuges, and plans to change the operation of the sluice gates and fishing in the canal.

EXCHANGE OF KNOWLEDGE AND KNOWLEDGE TO PRACTICE

EXCHANGES of knowledge and ideas among fisher communities and with facilitators and external advisors are important aspects of experience in Bangladesh.

Dhampara project

The Dhampara project fisheries mitigation component had a limited time to improve fisheries management and compensate for any adverse impacts of the embankment and water-control structures. For this reason, the project focused on conservation of natural fish. Most of the area is seasonal floodplain. Fish are trapped after the monsoon in the landowners' *kuas*, which are then pumped out to catch all the fish, leaving little or no water for fish to survive in the dry season.

One of the project consultants visited Goakhola-Hatiara Beel (under the CBFM-1 project near to Shuluar Beel) and was impressed by activities for fish conservation: from 1997 onwards, the community had leased *kuas* and protected them as dry-season fish sanctuaries. He identified people from seasonal *beels* in the Dhampara area who were personally interested in the fishery, owned more than one *kua* and were socially respected and philanthropic. He arranged an exchange

TABLE 13.3 *Main problems identified in a Participatory Action Plan Development Workshop in 2002 in Shuluar Beel.*

Problems	Landless women	Fishers	Landless men	Kua owners	Farmers	Total	Total Poor	Total non-poor
Natural resource related								
Natural fish declining	31	47	27	20	20	145	105	40
Lack of safe drinking water	17	3	14	8		42	34	8
Water logging		13	1	5	18	37	14	23
Siltation of canal		13		15		28	13	15
High cost of cultivation				6	22	28	0	28
Snails and aquatic plants declining		3		5	2	10	3	7
Lack of grazing land; few livestock			10			10	10	0
Lack of trees			7			7	7	0
Low crop prices			4	1		5	4	1
Encroachment on khas land by farmers			4			4	4	0
Fruit trees declining	3					3	3	0
Water pollution		3				3	3	0
Flood	1			1		2	1	1
Other problems								
Kuccha (earth) road	8	13	15	17	18	71	36	35
No electricity	5	5	7	15	18	50	17	33
Lack of homestead area	17					17	17	0
Lack of health care facility	12		4			16	16	0
Lack of school				6	2	8	0	8
Poor sanitation	6		1			7	7	0
Conflicts/lack of justice			6			6	6	0
Poverty				1		1	0	1
Total	100	100	100	100	100	500	300	200

Numbers are the number of votes placed against each problem converted into a percentage of the total votes cast by that group; each participant had 5 votes to use as he/she wished.

visit for fourteen of these villagers to Ashurar and Goakhola-Hatiara Beels to see the sanctuaries. They discussed sanctuary management and impacts with the *beel* management committees. This convinced these *kua* owners that setting aside *kuas* as fish sanctuaries would improve their fish stocks. Each of the participant *kua* owners decided to set aside one of their *kuas* for fish conservation and not to harvest it each year. In Goakhola-Hatiara Beel, the NGO paid some *kua* owners a lease fee for the *kuas* for one year, then after the year of conservation the owner regained fishing rights and harvested the fish. In Dhampara, however, a voluntary agreement was made with the *kua* owners: the project excavated two of a participant's *kuas* at a low cost (for example, 500 m³ of earth was excavated to deepen a mini-sanctuary at a cost of US$ 65) and the owner benefited from better catches in one *kua* (see

TABLE 13.4 *Local knowledge of reasons, impacts and potential solutions for main natural resource problems in Shuluar Beel (summarized across stakeholder groups).*

Problem	Reason	Impact	Solution
Natural fish declining	Fish cannot enter the *beel* because of flap gate on Sarashpur river Soluar sluice gate prevents water from entering the *beel* at proper time Catching brood fish by emptying the *kuas* in December–March Use of destructive gear Indiscriminate catch of fingerlings Fishing with fences Fish die of ulcerative syndrome (because of fertilizer and pesticide use)	Decreasing income from fishing Decreasing fish consumption due to price increase	Stop fishing in breeding period Stop emptying of *kuas* to conserve fish Let water into the *beel* at proper time Stop making embankments in river Reduce use of destructive gear
Waterlogging (filling up of river and canal)	Filling up of Soluar canal with straw and garbage Sluice gate narrow and defective Erosion of canal bank	Cannot harvest crop at proper time Polluted *beel* water	Re-excavation of canal Proper placement and use of sluice gates
Decreasing aquatic plants, snails, mussels and other aquatic animals	Overexploitation and sale Increased waterlogging Polluted water Demand for *shapla* (lilies) and snail has increased	Harmful plants increased	Preserve roots of *shapla* Stop collecting snails in breeding period
Trees and plants have decreased	Selling of trees Fewer plantations Lack of land	Disease due to lack of oxygen Storms affect homesteads Decreased income Lack of firewood Homestead loses beauty	Plant more trees More land for plantations Build awareness about plantations
Pollution of beel water	Decomposing grass Use of chemical fertilizer and pesticides Waterlogging	Skin disease Fish disease, including ulcerative syndrome	Ensure water flow from river and *khal* to *beel* Clear decomposing grass and straw from the *beel*

also Poepoe et al. and Meeuwig et al., this volume, for discussion of transfer of knowledge and applications between communities).

The project formed local advisory committees for each *beel*, with *kua* owners, their friends and relatives (co-owners) and fishers having use rights in the *beel*. They also formed an apex committee of sanctuary owners which meets every month with the Department of Fisheries sub-district officer to exchange their experiences and opinions. Sanctuaries were established in twenty-one locations with the help of *kua* owners (one *kua* as a sanctuary in six locations, two *kuas* in twelve locations, and three in three locations). The number of sanctuary *kuas* depended on the willingness of key *kua* owners and the views of the local advisory committees as to suitable ditches; there was no scientific modelling of areas of dry season water to be protected for over-wintering fish. The participants and wider community were happy with the arrangement; they have kept their *kuas* as sanctuaries for two years and plan to harvest them alternately. Except in two locations, the *kuas* are reported to have shown a satisfactory increase in the amount of fish and number of species, although some *kua* owners have since stopped protecting sanctuaries.

With knowledge gained from visiting Ashurar Beel, the Dampara project also used some *kata* as fish sanctuaries in the river and some permanent *beels*. Local people participated, but this was essentially an experiment by the project staff to modify traditional fish aggregating devices into a more effective form of sanctuary. They used different materials to assess their efficiency in conserving different fish species. To protect smaller fish from predators, the apex committee proposed placing bamboo cages of different gap sizes in the *katas* so that the smaller ones could hide where larger fish would be unable to swim.

Ashurar Beel

Dhampara is not the only example of exchange of local knowledge between fishing communities in Bangladesh. Using their experience and the lessons of an exchange visit to Hamil Beel, which is stocked each year by the community with carp, the participants from Ashurar Beel decided not to stock fish. They recognized that Ashurar Beel is much larger and fish can easily escape; moreover they feared that stocked species would cause the loss of indigenous species. This shows that fishing communities can assess experience in other locations critically to decide its relevance to themselves.

In Ashurar Beel, according to the participants, compliance with their fishing rules has been good: a few poor people catch fish during the closed season but the catch is not high and, as it is only for consumption, this is accepted by the committee. A few cases of outsiders breaking rules and facing gear confiscation or fines were reported. The fishers also reported that they all know the rules and follow them very strictly.

One important decision made by the BMC in Ashurar Beel was to harvest fish from half of the sanctuary in alternate years – half in 2000, the other half in 2002, and so on. The fishers decided this because they considered the sanctuary was overpopulated with predatory fish in the dry season and saw that they were catching fewer smaller fish than they had expected; it was also a way of earning some additional income. The first partial harvest of the sanctuary confirmed that a predatory fish, boal (*Wallago attu*), dominated in the sanctuary area and the amount of other fish caught there was low. The catch of high-value boal gave a good income. After this first harvest the fishers reported that catches and the diversity of smaller fish increased.

IMPACTS OF USE OF LOCAL KNOWLEDGE IN FISHERY MANAGEMENT

Two examples considered in this chapter are recent and their results are not fully known. In Shuluar Beel, community planning was undertaken only in 2002. The community established local fish sanctuaries late in that year, and the impacts will take at least a year to become apparent. In the Dhampara area, impact data were not collected during the short time-span of the project, and assessments are complicated by negative impacts of the embankment on surface water and the fishery. In 2002 almost all of the sanctuaries continued to function and the apex committee met each month (with virtually no external input), which indicates that the communities found some continued benefit from the sanctuaries. In Ashurar Beel, in contrast, management actions have been under way since 1997 and there has been external monitoring of fish catches since 1997 by the WorldFish Center and Department of Fisheries. Moreover there have been household socio-economic baseline and impact surveys, and regular discussions with the BMC to assess performance.

Ashurar Beel

In 1997 and 1998 there was relatively little fishing between March and May – the closed season – but from 1999 compliance with the closed season declined. Although seine-nets are not used in this season, fishers do use gillnets, which have become larger (average net length increased from 140 m in 1997 to 247 m in 1999). However, their mesh size has also increased, from 2.6 cm to 5.5 cm in the same period, indicating that the conservation arguments of the BMC have had some impact. There has been a major increase in the use of push-nets and traps – small-scale gear used by NGO participants (members of the NGO groups who fish and are represented in the BMC) and others. Total effort increased, initially in response to more water and fish in the high flood year of 1998, and continued despite fishery management; recorded catches fell in 2000 and 2001 (Figures 13.2

and 13.3). Catch per unit effort has fluctuated for the main devices used, with no clear trend.

The community said that their catches were stable in 2000 and 2001, but that there was a lack of water in the dry season. The fishers say that formal monitoring is on predefined days and misses some of the best fishing times, such as after rain, and also neglects night fishing. The Department of Fisheries regards total catch or production relative to area as the main indicator of success, while scientists are concerned to see stable and relatively high catch per unit effort. However, the fishing community sees success as having more fish to sell, catching more species

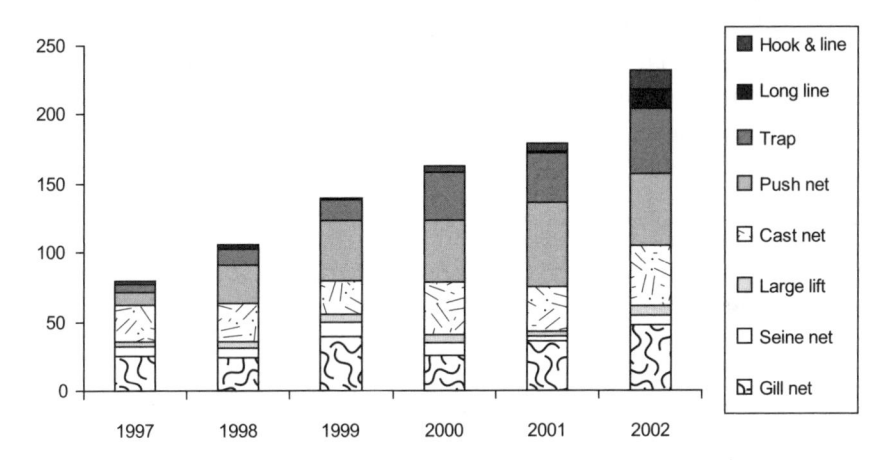

FIGURE 13.2 *Estimated total number ('00) of gear unit days operated in Ashurar Beel, 1997–2002.*

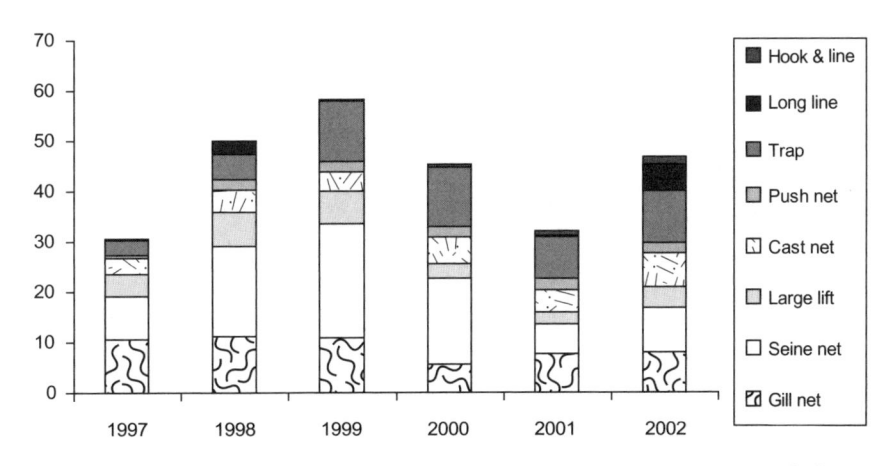

FIGURE 13.3 *Estimated volume of fish (tonnes) caught in Ashurar Beel, 1997–2002 (excluding sanctuary harvest in 1999).*

(particularly more of the higher-value species), having higher incomes, fewer conflicts over fishing and a fair distribution of benefits. In these measures they say they are generally satisfied that the fishery has improved, but they want further improvement.

Catch composition has changed over time (Table 13.5), but most of the catch is still of small fish, particularly *jat punti*. However, there are contradictory signs. The proportion of small shrimps (*gura icha*) in the catch has increased, regarded by de Graaf et al. (2001) as an indicator of overfishing, but the catch of predatory boal has also increased, which would indicate a healthier fishery. The proportion of boal caught in 2000 is understated in Table 13.5, as the table does not include another 2.7 tonnes that were caught in the sanctuary. The fishers correctly deduced from this that there would be a better catch in 2002 outside the sanctuary as there were fewer predators, and because there was a higher water level in 2002 and a good rice crop, which means that there were better feeding conditions in the flooded fields.

TABLE 13.5 *Species composition of catch (per cent by weight) from sampling on 2-4 days/fortnight in Ashurar Beel.*

Local name	Scientific name	1997	1998	1999	2000	2001	2002	% overall
Jatputi	*Puntius sophore*	23.8	25.9	32.5	35.7	20.5	19.8	27.9
Gura icha	*Machrobrachium styliferus*	14.1	14.7	15.8	24.4	29.0	18.0	18.7
Boal	*Wallago attu*	2.5	6.8	8.1	8.5	17.7	9.8	8.6
Bajari tengra	*Mystus tengara*	15.9	11.5	5.0	3.5	6.0	13.9	8.5
Guchi baim	*Macrognathus pancalus*	7.0	10.6	8.3	4.4	3.9	6.0	7.4
Ranga chanda	*Parambassis ranga*	6.1	6.9	6.8	9.0	8.5	2.2	7.0
Katari	*Salmostoma bacaila*	5.4	6.1	6.2	7.7	5.4	2.6	6.0
Mola	*Amblypharyngodon mola*	5.4	3.0	2.6	1.5	6.5	1.0	3.1
Taki	*Chana punctata*	4.1	2.0	1.0	1.0	0.8	7.8	2.1
Chapila	*Gudusia chapra*	6.0	2.2	2.0	0.3	0.0	0.0	1.7
Guzi ayre	*Aorichthys seenghala*	1.4	3.0	0.3	0.4	0.0	0.1	1.2
Common carp	*Cyprinus carpio*	1.0	1.6	1.0	1.5	0.5	0.2	1.2
Silver carp	*Hypophthalmichthys molilrix*	0.4	1.4	2.8	0.0	0.0	0.0	1.1
Shing	*Heteropneustes fossilis*	1.5	1.9	0.5	0.0	0.1	1.4	1.0
Kholisa	*Colisa fasciatus*	0.0	0.4	0.2	1.3	0.0	1.0	0.5
Kanchan puti	*Puntius conchonius*	0.0	0.0	0.0	0.0	0.0	6.3	0.5
Rui	*Labeo rohita*	0.1	0.0	2.0	0.0	0.1	0.1	0.5
Mixed fish		5.5	2.0	5.1	0.6	1.0	9.7	3.1
		100%	100%	100%	100%	100%	100%	100%
Total catch in kilos		1,038	3,678	2,532	2,265	1,442	891	

Household surveys in 1996 and 2001 indicate no significant improvement in the incomes or assets of NGO participants compared with non-participants (Table 13.6), but neither were they worse off. The NGO participants were expected to be poorer at the outset of the project than a random sample of non-participants, but were no different in this generally poor area.

TABLE 13.6 *Changes in fishing income, house construction and sanitation (1996–2001) in Ashurar Beel.*

	NGO	Non-NGO
Household net income from fishing (Tk p.a.)		
1995–96	6,830	2,360
2000–01	3,670	750
Total income (Tk p.a.)		
1995–96	14,850	13,050
2000–01	17,880	23,030*
Landholding (ha)		
1996	0.24	0.24
2001	0.28	0.31
Tin/concrete roof (% households)		
1996	28	28
2001	72	72
Water-sealed latrine (% households)		
1996	5	7
2001	14	17

NGO = members of NGO groups formed for fisheries management.
Non-NGO = not members of those groups.
Tk = Bangladesh Taka (in 1996 approx. Tk 42 = US$ 1, in 2001 approx. Tk 57 = US$ 1).
* significantly different from baseline, t-test, p<0.05.
Source: interview surveys with same households.

Self-assessment of changes in participation, well-being and fishery management in 2001 indicated that, on average, both sets of respondents believe conditions have improved significantly since 1996. On almost all indicators related to fishery participation and management, the NGO participants' increases in perceived status were significantly larger than those of the non-participants (Table 13.7). Participants have increased their active participation in fishery management and believe that fishery management has changed for the better to a greater extent than non-participants.

TABLE 13.7 *Mean difference in change in perceived indicator scores (1997 compared with 2001) between NGO participants and non-NGO respondents in Ashurar Beel.*

Indicator	NGO change	Non NGO change	t	p
General participation	0.83	0.58	0.81	>0.1
Community affairs influence	**1.28**	**0.82**	**2.00**	**<0.05**
Fisheries participation	**1.65**	**0.48**	**4.69**	**<0.001**
Fisheries influence	**1.80**	**0.63**	**4.47**	**<0.001**
Decision making	**2.32**	**1.57**	**3.83**	**<0.001**
Fishery well-being	**1.20**	**0.73**	**2.07**	**<0.05**
Household well-being	0.67	0.37	0.98	>0.1
Influence on government over common property access	0.80	0.48	1.10	>0.1
Household income	0.80	0.67	0.71	>0.1
Control over fishery resource	**1.70**	**0.42**	**5.25**	**<0.001**
Fair access right to fishery resource	1.32	1.32	0.00	>0.5
Active fishery management (sanctuary, closed season, gear limits)	**2.52**	**1.55**	**3.99**	**<0.001**
Benefits from fishery management resource	**2.22**	**1.28**	**4.31**	**<0.001**
Conflict resolving speed	**2.42**	**1.88**	**2.47**	**<0.05**
Community compliance with fishery resource	**2.68**	**1.97**	**3.19**	**<0.005**
Information flow among fishers	2.32	1.90	1.85	>0.05
Knowledge of fishery	2.42	2.13	1.28	>0.1

Note: Bold indicates that NGO participants perceived a significantly greater improvement in the indicator than did non-NGO participants.

Scores were on a 10-point scale from the worst situation imaginable to the best situation imaginable, the means of the differences between the scores given for 2001 and 1997 are shown, all changes were positive for both samples.

t = t-test statistic.

p = significance level.

DISCUSSION AND CONCLUSIONS

In Bangladesh, fishing has been a vital part of rural people's lives for generations, but governments have used fisheries administration as a means of collecting revenue. This has meant that traditional fishers often have only limited access to the most productive fisheries, and no recognized role in management decisions. Meanwhile seasonal floodplains have been converted to more intensive agriculture with the help of government flood-control and drainage projects that focus on rice self-sufficiency at the expense of fish and aquatic resources, which in the past were mostly a source of subsistence and were consequently undervalued.

Internationally, documenting and researching indigenous knowledge related to natural resources is now seen as a way of adding to scientific knowledge and understanding of complex systems. The cases discussed here are part of a trend that has seen NGOs in particular recognize the role of local knowledge and participatory planning for fisheries management. The government is changing its views as a result of demonstrated experience, for example towards proposing more fish sanctuaries. But it sees this as a mere technical fix and is more concerned about creating larger sanctuaries than encouraging small sanctuaries in the seasonal floodplains.

The move towards community-based fisheries management has drawn upon lessons from traditional management systems that regulate access, maintain fishery health and have proved resilient to external pressures. In Bangladesh, community-based management involves developing new local institutions and organizations for fisheries management. Past efforts to do this focused on technology and external interventions, but did not survive in the longer term as they were not based on local knowledge and preferences, and therefore depended on continuing external inputs. Participatory planning and community-based management are based on the knowledge of local fishers. Facilitation by NGOs has been vital in setting up local community bodies that can then take decisions that are based on their analyses of the trends and problems they face. Examples of success by one community have also encouraged others to adopt similar concepts. Often the knowledge of a few key fishers is recognized and respected by the wider community and tends to lead opinion. The planning processes and committees discussed here are a way of identifying and using local knowledge.

New participatory approaches have been criticized, for example by Cooke and Kothari (2001), for overriding existing legitimate decision-making processes and for reinforcing the voice of the powerful at the expense of the poor who were meant to be the beneficiaries. However, we argue that the participatory processes described here, for example in Shuluar Beel, fit within the range of good practice identified by Edmunds and Wollenberg (2001). Thus the processes adopted for community-based fisheries management are inclusive and adaptive, yet provide a way for each category of stakeholder to analyse natural resource problems and solutions separately and then jointly, to share their knowledge and seek a consensus in a way that the interests and voices of poorer users are heard.

The examples discussed here show that fishing communities can:

- identify and analyse their problems and plan solutions based on their local knowledge
- critically assess the experience of other communities and lessons that they have learned, and then adapt them appropriately to their own circumstances and priorities
- implement actions to restore and manage more sustainably the fish and wetland resources they depend upon.

The long-term future for these local initiatives is still uncertain. In the seasonal floodplains it depends on communities making the protection of residual water as dry-season fish habitat a norm. In the *jalmohals* it depends on government policies changing to recognize long-term community use rights, and to balance revenue demands with sustainable fish harvests and incomes for the fishers. It also depends on the government providing a framework that supports and maintains local fisher community initiatives when they are threatened by local elites. The rich and powerful previously controlled many of these fisheries, and often wait on the sidelines of a project for any opportunity to regain control of enhanced resources.

ACKNOWLEDGEMENTS

We are grateful to the Department of Fisheries, Banchte Sheka and CARITAS, and to our colleagues in WorldFish Center for their assistance. We thank Dr Munir Ahmed for sharing his experience and insight into the Dhampara Project. The original Community-Based Fisheries Management Project (CBFM-1) was supported by the Ford Foundation. CBFM-2 is supported by the UK Department for International Development. This document is in part an output from a project funded by the UK Department for International Development (DFID) for the benefit of developing countries. The views expressed are not necessarily those of DFID.

REFERENCES

AGRAWAL, A. 2001. Common property institutions and sustainable governance of resources. *World Development*, Vol. 29, No. 10, pp. 1649–72.

ALI, M.Y. 1997. *Fish, water and people*. Dhaka, Bangladesh, The University Press, 153 pp.

BERKES, F.; FOLKE, C. 1998. Linking social and ecological systems for resilience and sustainability. In: F. Berkes and C. Folke (eds.), *Linking social and ecological systems*. Cambridge, Cambridge University Press, pp.1–25.

BERKES, F.; FEENY, D.; MCCAY, B.J.; ACHESON, J.M. 1998. The benefits of the commons. *Nature*, Vol. 340, pp. 91–93.

BWDB. Undated. *Dhampara Water Management Project*. Dhaka, Bangladesh Water Development Board and Canadian International Development Agency, 16 pp.

COOKE, B.; KOTHARI, U. (eds.). 2001. *Participation: The new tyranny*. London, Zed Books, 207 pp.

DEPARTMENT OF FISHERIES. 2000. *Fish catch statistics of Bangladesh 1998–1999*. Dhaka, Bangladesh, Department of Fisheries.

DIXON, P.J.; BARR, J.J.F.; SILLITOE, P. 2000. Actors and rural livelihoods: Integrating interdisciplinary research and local knowledge. In: P. Sillitoe (ed.), *Indigenous knowledge development in Bangladesh*. Dhaka, Bangladesh, The University Press, pp.161–77.

EDMUNDS, D.; WOLLENBERG, E. 2001. A strategic approach to multistakeholder negotiations. *Development and Change*, Vol. 32, No. 2, pp. 231–53.

ISLAM, M.Z. 1999. Enhancement of floodplain fisheries: Experience of the Third Fisheries Project. In: H.A.J.Middendorp; P.M. Thompson; R.S. Pomeroy (eds.), *Sustainable inland fisheries management in Bangladesh*. ICLARM Conf. Proc. 58. Manila, International Center for Living Aquatic Resources Management, pp. 209–18.

MINISTRY OF WATER RESOURCES. 2001. *Guidelines for participatory water management*. Dhaka, Bangladesh, Ministry of Water Resources, Government of the People's Republic of Bangladesh, 76 pp.

MINKIN, S.F.; RAHMAN, M. M.; HALDER, S. 1997. Fish biodiversity, human nutrition and environmental restoration in Bangladesh. In: C. Tsai; M.Y. Ali (eds.), *Openwater fisheries of Bangladesh*. Dhaka, Bangladesh, The University Press, pp.75–88.

OSTROM, E. 1990. *Governing the commons: The evolution of institutions for collective action*. Cambridge, Cambridge University Press.

——. 1994. Institutional analysis, design principles and threats to sustainable community governance and management of commons. In: R.S. Pomeroy (ed.), *Community management and common property of coastal fisheries in Asia and the Pacific: Concepts, methods and experiences*. ICLARM Conf. Proc. 45. Manila, ICLARM, pp. 34–50.

POMEROY, R.S.; BERKES, F. 1997. Two to tango: The role of government in fisheries co-management. *Marine Policy*, Vol. 21, No. 5, pp. 465–80.

SILLITOE, P. 2000. The state of indigenous knowledge in Bangladesh. In: P. Sillitoe (ed.), *Indigenous knowledge development in Bangladesh*. Dhaka, Bangladesh, The University Press, pp. 3–20.

SULTANA, P.; THOMPSON, P.M. 2003. Methods of consensus building for community-based fisheries management in Bangladesh and the Mekong Delta. CAPRi Working Paper 30, CGIAR Systemwide Program on Collective Action and Property Rights. Washington, International Food Policy Research Institute.

THOMPSON, P.M.; SULTANA, P.; ISLAM, M.N.; KABIR, M.M.; HOSSAIN, M.M.; KABIR, M.S. 1999. An assessment of co-management arrangements developed by the Community-Based Fisheries Management Project in Bangladesh. Paper presented at the international workshop on fisheries co-management, 23–28 August 1999, Penang, Malaysia.

CHAPTER 14   The role of fishers' knowledge in the co-management of small-scale fisheries in the estuary of Patos Lagoon, Southern Brazil

Daniela C. Kalikoski and Marcelo Vasconcellos

ABSTRACT

THIS chapter analyses the ecological knowledge of small-scale fishers in the estuary of Patos Lagoon obtained from interviews and questionnaire surveys, and discusses its potential role in the local co-management of small-scale fisheries (under the auspices of the Forum of Patos Lagoon). The study demonstrates that fishers' knowledge can provide valuable information about the characteristics of practices, tools and techniques that may contribute to a more sustainable pattern of resource use. Particular attention was paid to knowledge of fishing seasons and the impact of changes in fishing technologies and practices on the resilience of estuarine resources. Such knowledge can contribute to the formulation of management plans to better adapt rules to local social and environmental conditions. However, the use of fishers' knowledge in the co-management of small-scale fisheries can be hampered by a range of factors, including: the low expectations among scientists and decision-makers of the value of fishers' knowledge for management, problems in the definition of property rights that weaken incentives for fishers to act upon their ecological knowledge, and the contradictory paradigms that at present relate to the role of scientific and local knowledge in the management of the estuarine ecosystem.

INTRODUCTION

WORLDWIDE crises in fisheries management have triggered changes in governance processes and in approaches to the study of common-property resources (CPRs). Co-management theory and common-property theory have played an important role in changing the field of fisheries CPRs management (Berkes, 1989; Pinkerton, 1989; Ostrom, 1990). The essence of co-management, as defined by Pinkerton (1989), is the involvement of fishers' organizations and fishing communities in management decision-making through power sharing, both between government and locally based institutions, and among differently situated fishers. It represents a way to decentralize decisions, delegate rights and roles to communities and move towards a joint decision-making process.

One of the strongest aspects of fisheries co-management is its capacity to access fishers' knowledge of the environment and the resources that they pursue. The term fishers' knowledge is used here interchangeably with local/traditional ecological knowledge (LEK/TEK) to refer to the cumulative body of knowledge, practice and beliefs, evolving by adaptive processes and handed down through generations by cultural transmission, about the relationship of living beings with one another and with their environment (Berkes, 1999; Neis and Felt, 2000). TEK contains empirical and conceptual aspects, is cumulative over generations, and is dynamic, in that it changes in response to socio-economic, technological and other changes (Berkes, 1999). It is well known that the knowledge held by fishers in many areas of the world, especially in small-scale traditional societies, may be extremely detailed and relevant for resource management (Berkes and Folke, 1998). In fact studies have shown that it is the complementary characteristics of local and scientific knowledge that make co-management stronger than either community-based management or government management (Pomeroy and Berkes, 1997).

Small-scale fisheries in the estuary of Patos Lagoon, located in the Southern Brazilian coastal zone (Figure 14.1), are going through a tragedy of the commons. The abundance of fisheries resources is decreasing sharply, compromising the livelihood of more than 10,000 small-scale fishers (Reis, 1999). Consensus about the failure of former institutions to protect these resources triggered the establishment of new institutional arrangements by redefining rules and rights to manage the resources (Reis and D'Incao, 2000; Kalikoski, 2002; Kalikoski et al., 2002; Kalikoski and Satterfield, 2004). A co-management forum (the Forum of Patos Lagoon) composed of different stakeholders was established in 1996 to:

- discuss and develop alternative actions to mitigate and/or resolve the problems of the fishers and the crisis in the small-scale fisheries sector
- restore the importance of small-scale fisheries
- share decisions to address problems more effectively.

The role of small-scale fishers' knowledge in this new institutional arrangement has not yet received the required attention, and the exchange of knowledge between fishers and scientists is still limited.

The main goal of this chapter is to discuss whether it is possible to identify an informal knowledge system used by small-scale fishers in the estuary of Patos Lagoon that could improve the system of co-management and so help maintain local ecosystem resilience. To this end the work analyses three questions:

1. How has the local social system developed management practices based on ecological knowledge for dealing with the dynamics of the ecosystem in which it is located?
2. How have these management practices changed over time to the present situation?
3. What are the current barriers to and opportunities for using TEK in the Forum of Patos Lagoon co-management?

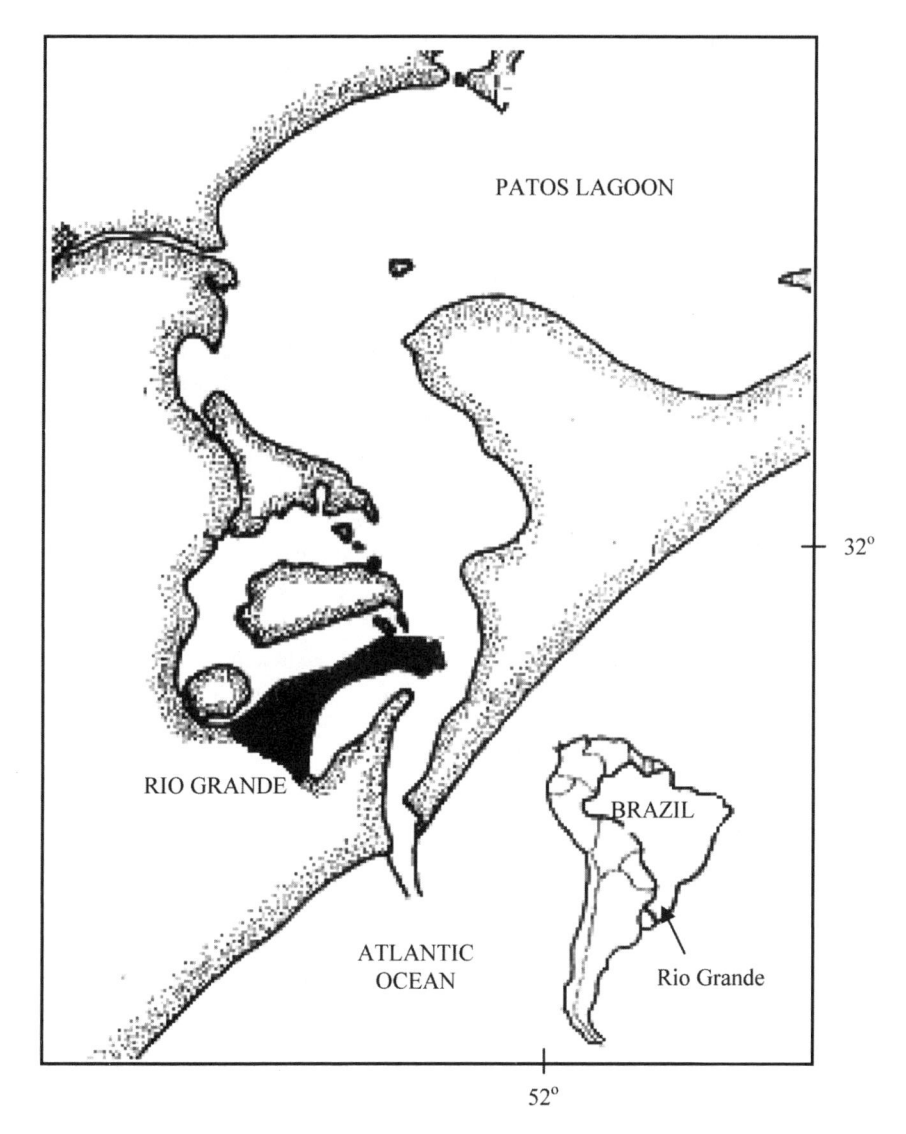

FIGURE 14.1 *Location of the Patos Lagoon estuary in Southern Brazil.*

We then discuss the role of local knowledge held by fishing communities in the estuary of Patos Lagoon and its use and relevance for the Forum of Patos Lagoon co-management scheme as a complement to scientific knowledge in devising rules and regulations for the management of small-scale fisheries.

METHODS

FIELDWORK in the estuary of Patos Lagoon was carried out from April 2000 to February 2002. Data were obtained from primary and secondary sources. The primary sources were:

- researcher observation at the Forum of Patos Lagoon meetings
- informal conversations with the members of the Forum of Patos Lagoon
- in-depth semi-structured interviewing and a questionnaire survey with fishers from the small-scale fishing communities of the estuary of Patos Lagoon.

A total of forty-eight face-to-face interviews were conducted ranging in length from forty-five minutes to three hours. Interview data were complemented by (and used to cross-validate) field observation conducted throughout the research period (Creswell, 1994; Czaja and Blair, 1996). The observational and interview data were complemented by a survey of local fishers (n = 623). Supplemental data were obtained from secondary sources including analysis of scientific publications, local newspapers, meeting minutes, laws, decrees and policy statements from national profile sources such as the Federal Institute for the Environment (IBAMA) and the Federal Sub-Secretary for Fisheries Development (SUDEPE).

Interviews and questionnaires focused on four levels of analysis, consistent with the description of TEK as a knowledge–practice–belief complex as proposed by Berkes (1999). Particular attention was paid to the first two analytical levels, although all levels were addressed intertwined. Level one relates to the local knowledge of the animals and ecosystems, such as the behaviour and habitat of fish, and the timing of fishing seasons. Such local knowledge may not, in itself, be sufficient to ensure the sustainable use of resources. Therefore, level two refers to the existence or sophistication of a resource-management system that uses local environmental knowledge to devise an appropriate set of practices, tools and techniques for resource use. However, for a group of fishers to manage resources effectively, appropriate institutions, or social organization must exist to support coordination, cooperation, rule-making and enforcement (Ostrom, 1995; Berkes, 1999).

Accordingly, the third level of analysis is about institutions: the set of rules-in-use to coordinate the management of the resources. Lastly, the fourth 'worldview' level represents the system of beliefs that 'shapes human-nature relations and gives meaning to social interactions' (Berkes, 1999). As indicated by Berkes, although the four levels of management systems and institutions are hierarchically organized, it is sometimes artificial to distinguish between them as there is frequent feedback between different levels. As a result, worldviews may be influenced by changes occurring at several levels, including the environment, the community, and also in the case of the collapse of a management system.

THE ESTUARY OF PATOS LAGOON ECOSYSTEM

With an area of approximately 10,000 km^2, Patos Lagoon is recognized as the world's largest choked lagoon, stretching from 30°30' to 32°12' S near the city of Rio Grande where it connects to the Atlantic Ocean (Figure 14.1). The estuarine region encompasses approximately 10 per cent of the lagoon, and supports a diverse and abundant flora and fauna. The estuary is shallow, with variable temperature and salinity depending on local climatic and hydrological conditions (Castello, 1985). The dynamics of estuarine waters are mainly driven by the wind and rain regime, with only minor influence of tides.

The Patos lagoon system communicates with the ocean via a channel between a pair of jetties, about 4 km long and 740 m apart at the mouth. All the estuarine-dependent marine organisms enter and leave the estuary through this channel for nursery, reproductive and feeding purposes. Of the more than 110 species of fish and shellfish species that occur in the estuary (Chao et al., 1985), four are particularly important fisheries resources, and have sustained small-scale fisheries for more than a century. They are pink shrimp (*Farfantepenaeus paulensis*), marine catfish (*Netuma barba*), croaker (*Micropogonias furnieri*) and mullet (*Mugil platanus*). The lifecycle of these species is described in Table 14.1.

Different species' life history characteristics create a well-defined seasonal variability in the diversity and abundance of resources in the estuary and in the availability of resources to small-scale fisheries (Figure 14.2). Landings in small-scale fisheries have declined steadily since the mid-1970s, to about 5,000 tonnes in the late 1990s, the lowest landings recorded in the last fifty years. Fisheries landings also present a marked interannual variability, with a periodicity that seems to be related to the occurrence of strong El-Niño Southern Oscillation (ENSO) events. Figure 14.3 uses Holling's (1986, 1992) model of the dynamics of small-scale fisheries resources, described in Table 14.1, to account for four major phases in resource lifecycles in the estuarine and coastal areas. An *exploitation* phase, where species such as mullet, catfish, croaker and shrimp enter the estuarine environment for feeding (growth) or reproduction purposes, leads to a *conservation* phase in which resources increase in size (mullet; shrimp) and/or maturity (catfish). In the *release* phase, adults leave the estuary to spawn (mullet; catfish) and recruit (shrimp) in the marine environment,

TABLE 14.1 *Summary of biology and lifecycle of main small-scale fisheries resources in the estuary of Patos lagoon.*

Pink shrimp, *Farfantepenaus paulensis*	Estuarine-dependent species. Adults spawn in shelf waters below 50 m deep, producing demersal eggs that hatch into planktonic larvae. When approaching estuaries the larvae develop a benthic habit, settling in shallow areas where they will grow for a few months until they reach the pre-adult phase, when they migrate to the ocean, reinitiating the cycle. The growing phase in the estuary may last between four and ten months when they reach about 7 cm in length. Larvae enter the estuary with varying success all year round, but mainly in the spring and summer, depending on environmental forcing of wind and freshwater outflow.
Marine catfish, *Netuma barba*	Slow-growing, anadromous species with an estimated life span of approximately 23 years, though adults may occasionally attain 36 years of age and a total length of 98 cm. At the end of the winter the species migrates into the Patos Lagoon estuary. Reproduction takes place in early spring in the estuary followed by spawning in the coastal waters. *N. barba* has low fecundity and after reproduction the males incubate the eggs for up to two months in the buccal cavity. Between spawning seasons, adults disperse over the entire shelf.
Croaker, *Micropogonias furnieri*	Species depend on the estuary of Patos lagoon as a nursery and feeding ground. Croakers spawn during spring and summer in coastal waters under the influence of freshwater runoff from the Patos lagoon. Adults normally migrate into the estuary in September–October and leave the area in December–January. Young and sub-adult croakers occur throughout the year near the coast and in the estuary of Patos Lagoon. Adults are dispersed over the shelf and migrate from Uruguay to southern Brazil during the autumn and winter and towards Uruguay in the summer.
Mullets, mainly represented by *Mugil platanus*	Mullets occur year round in the Patos lagoon and adjacent coastal waters. Juveniles are more abundant in the winter and spring in nursery areas of the lagoon. In the autumn, adult mullets leave the estuary and initiate their reproductive migration. Spawning occurs in warmer offshore waters at about 27°S between the end of the autumn and winter. Eggs and larvae are transported from spawning ground towards the surfzone, followed by longshore migration to the estuary of Patos Lagoon.

Source: Reis, 1986; D'Incao, 1991; Vieira and Scalabrin, 1991; Haimovici, 1997.

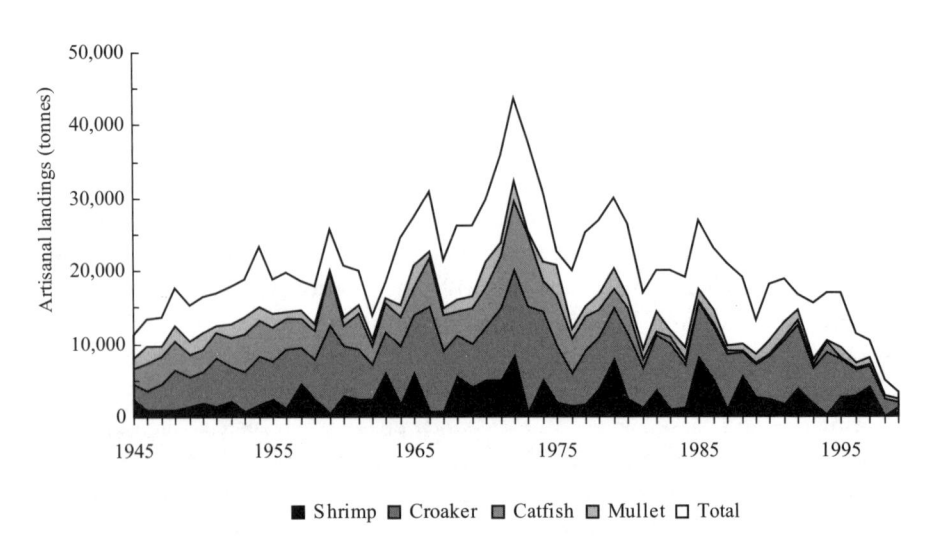

FIGURE 14.2 *Small-scale fisheries landings in the estuary of Patos Lagoon.*

closing the cycle with the *renewal* phase (Figure 14.3). The influence of climatic conditions is conspicuous in the transition from the renewal to exploitation phases and from the conservation to release phases because of the climate's effect on the recruitment success and on the migration/dispersion of resources into and out of the estuarine environment.

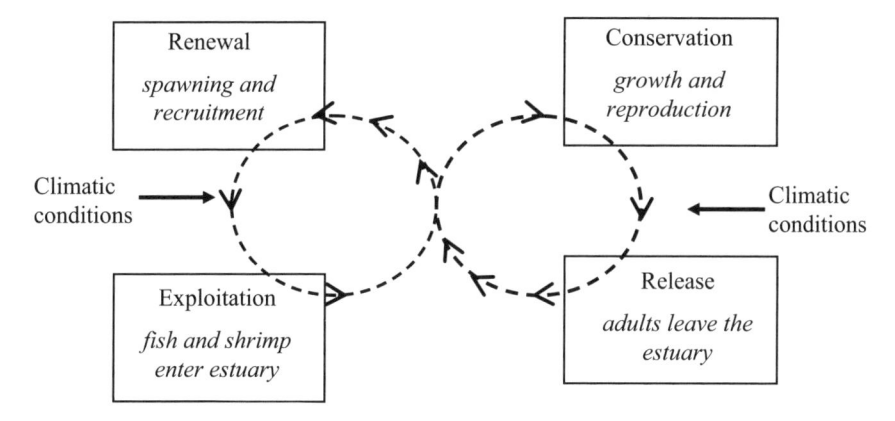

FIGURE 14.3 *Four-phase model of estuarine and coastal fisheries resource dynamics.*
Note: During the cycle of exploitation, conservation, release and renewal, biological time flows unevenly. It is normally slower from the exploitation to the conservation phase than during the transition between the release and renewal phases. Source: Adapted from Holling, 1986, 1992.

FISHING PRACTICES AND ECOSYSTEM RESILIENCE

The fishing calendar

One of the most important characteristics of estuarine small-scale fisheries is the fishing calendar. Since the time when practically no formal rules existed for fisheries management (pre-1960s), small-scale fisheries followed a calendar of activities (rules in use) determined by the abundance of different fisheries resources during the year and by the fishing technologies in use. The calendar was based on the experience of local fishers. As such it represents a form of traditional ecological knowledge with important consequences for the resilience of small-scale fisheries because, as discussed later, it created natural limits to the exploitation of CPRs (See also Poepoe et al., this volume).

From January to May, fishers captured shrimp and mullets. Mullets were fished mainly in two periods: in January, when the adults were returning from the spawning grounds in the sea, and during the spawning runs, which normally occur between the months of April and June. The catfish season normally began in July and lasted until early November. This fishery targeted large catfish entering the lagoon to reproduce, and spawning grounds in the upper estuary. This fishery captured mostly large fish with well-developed gonads. A less extensive fishery also occurred during the summer months, especially in February, when catfish migrate back to the sea, and the males incubate the young in their mouth. Few fishers were involved in this fishery because catfish were normally 'thin' and of low value; besides, fisheries such as shrimp and mullet were more attractive during the summer. The croaker season started in October, or right after the catfish season, and normally lasted until early summer.

According to fishers, the fishing calendar in the estuary of Patos Lagoon is strongly influenced by the strength of the intrusion of salt water and the rainfall regime. Many fishers consider salt water to be the single most important factor controlling small-scale fisheries activities. This influence is particularly conspicuous in the shrimp fishery, because fishers consider shrimp to be more influenced by climate than other fisheries resources. A good fishing season usually occurs if the salinity of the estuary is ideal in the period from October to December; the earlier the estuary is replenished with salt water, the earlier will be the shrimp season. Castello and Moller (1978) demonstrated a similar relationship between the rainfall regime and shrimp production. Fishers also view a warm winter as beneficial for the shrimp season.

The moon is considered an important factor in determining the timing and success of a fishery. For instance, the full moon usually produces good catches of shrimp but it is not good for the capture of croaker. One fisher explains: 'When the moon is bright the croaker is more active and difficult to catch with gillnets.' The last quarter moon is considered excellent for mullets. Fishers recognize a set of conditions, including the last quarter moon of May and the passage of cold fronts from the south, as important to trigger the schooling behaviour of mullet spawners and hence for a good fishery.

Resource use by small-scale fishers in the estuary of Patos Lagoon was, and still is to a large extent, conditioned by the availability of the resources in the estuarine environment, which is in turn controlled seasonally by the influence of the weather and also affected by the influence of the moon on the behaviour of the fish. As explained by a fisher:

> 'Nature makes its own fishing closures with the moon, the bad weather, and also the fish, because if it is too windy the fish don't move and you cannot catch them. For instance, if the mullet sees the net it does not enmesh. If it is not the right time, and the fish do not want to be captured, you cannot catch them.'

But, as will be seen in the next section, resource-use practice changed markedly as new fishing technologies were introduced and as the industrialization of fisheries brought exploitation beyond the limits of the carrying capacity of the resources.

Changes in fishing practice and resource conditions

In the past fifty years, fisheries in the estuary of Patos Lagoon and coastal areas have experienced changes in fishing technologies and materials that significantly altered resource exploitation and the sustainability of small-scale fisheries. Small-scale fisheries were initially based on a beach seine fishery at the mouth of the estuary and in other specific locations along the migratory route of the species inside the lagoon (Barcellos, 1966; Costa, 2001). The nets were approximately 300 m long and were used to encircle the schools of mullet, croaker, black drum, catfish and even shrimp, close to shore. The mullet fishery was carried out in two main places in the mouth of the estuary, one on each side of the channel. Each fisher had his turn on a specific day of the season, which was arranged among the fishers of each community. It was common to capture more than 60,000 fish (about 90 tonnes) in a single shot, and in order to handle the large catch volume, the fishery was often carried out by groups of twenty to thirty fishers.

Older fishers recall that the beach seine fishery remained important until the mid-1960s when gillnet fishing intensified (this is also confirmed by Barcellos, 1966). Gillnets were the most appropriate type of technology to be used in the large areas of the lagoon, where fish were naturally more dispersed than at the mouth of the estuary. The intensification of gillnet fishing in turn decreased the viability of the beach seine fishery.

The introduction of motors and the widespread use of gillnets allowed fishers to start fishing mullets in the lagoon as early as October. This gillnet fishery was considered unsustainable by elders, who believe the lagoon functions as a nursery area. Unlike the beach seine fishery, which captured only adult fish during a short time window, the gillnet fishery expanded the time and areas where the resource was vulnerable to exploitation, as well as targeting immature fish. Today croakers and catfish, as well as mullets, are mainly fished by means of gillnets.

Many assume that the increase in the number of small-scale fishers and the changes in fishing practice and technologies in estuarine fisheries increased the pressure on resources, which gradually became less abundant to the point of collapse in the case of some important fish resources such as catfish (Reis, 1986; Rodrigues, 1989). However, fishers and scientists agree that one of the main causes of the decline of fisheries CPRs in southern Brazil was the intensification of industrial fisheries observed during the 1960s and 1970s (Haimovici et al., 1989; Haimovici, 1997). The fishing areas and technologies employed by industrial fisheries, as viewed by fishers, have a much greater impact on resources because of the amount of fish caught and the fishing time involved. These fisheries operate in areas of the continental shelf that were formerly (and still are) inaccessible to small-scale fishers for most of the time. These fishers recall that since these industrial vessels started operations, the fish that used to enter the lagoon have been disappearing. To offset this decrease in landings, small-scale fishers, in turn, started to increase the amount of gear in the estuary and intensified their shallow coastal water fisheries (many stated that, when weather permits, the coastal area is visited regularly during the croaker fishing season, capturing the fish before they enter the lagoon). The result has been an overall decrease in fisheries abundance.

The pink shrimp fishery has also experienced marked changes in fishing technologies and fishing practices in the last decades. Shrimp fishing was initially carried out along the lagoon beaches and shallow areas using a manual trawl-net dragged by two to four people, or beach seine nets. The manual trawl nets were later (in the mid-1950s) modified into fixed nets (bag nets). Bag nets were fixed around the channels, the mouth of the net placed facing the ebb currents of the estuary, so that shrimp were caught passively through the currents. Beginning in the 1960s, otter trawling from boats became widely used in the shrimp fishery. Most of the trawling was done in the deeper waters of the estuary and in areas with 'cleaner' bottom (although fishers recognize that many of them used to trawl also in shallow nursery areas).

Stow-nets, introduced in the 1970s, are now the dominant type of gear used in the estuarine shrimp fishery. They are fixed in shallow areas of the lagoon and operate by attracting shrimp to the net with light produced by gas lamps. The stow-net fishery has changed over the years. The nets were initially placed close to small inlets, because 'shrimp was initially caught in the currents.' Now the nets are placed mostly in the shallows where, according to fishers, the young/smaller shrimp are caught before migrating from the nursery areas. Under government rules, stow-nets have become the only fishing technology allowed in the shrimp fishery. Although these rules define the number of nets and the spacing between them, the exact location of stow-nets is still informally determined among fishers on the basis of agreed fishing territories (Alumdi et al., 2006).

Fishers maintain that the introduction and widespread adoption of stow-nets impacted negatively on the operation of other types of fishing technologies (such as bag nets and trawling) because a large proportion of the shrimp are caught before they are able to migrate to the channel areas and lower parts of the estuary. It also triggered an intensification of trawling in the estuary to compensate for the decreasing yield of shrimp. The result has been an increase in fishing effort and the overexploitation of shrimp in the estuary. D'Incao (1991) estimated that the intensity of the stow-net shrimp fishery in the estuary of Patos Lagoon is so high that few shrimp leave the lagoon to complete the species lifecycle.

Fishers mentioned in the interviews that stow-nets and trawl-nets frequently produce high bycatch rates. According to them, small-scale trawling can produce little bycatch, depending on the area of the estuary and on the characteristics of the otter board and the height of the net – the higher the net is in the water column, the higher the bycatch. For practical reasons, fishers have found ways to reduce the bycatch to avoid the increased handling time on board that it entails. They have done this by decreasing the height of the net and by avoiding areas associated with high bycatch rates such as shallow estuarine waters and specific locations off the coast known to be nursery areas. Although there has been no scientific evaluation of the comparative impacts of trawl-nets and stow-nets on the estuary, all types of trawl-net fisheries are now forbidden.

MANAGEMENT LESSONS FROM TRADITIONAL PRACTICES

What can be learned from the above forms of resource use practice? When resources were still abundant, the fishing calendar worked in a way that allowed fishers to benefit from the most abundant resources in a season while limiting the amount of fishing pressure (time) on a particular species and/or during a critical period. For instance, fishing for catfish was normally discouraged during the summer months when the males are incubating the young (see 'slot limits', Poepoe et al., this volume). It was also unnecessary, given the availability of other resources such as croaker and shrimp. Similarly, the capture of large amounts of shrimp below the optimal size (between late spring and early summer) was in part prevented by the type of fishing technology in use, and by the existence of other alternative fishing resources. A failure of the shrimp fishery due to low abundance would result in a redistribution of fishing effort to the other resources available in the period, but never to the point of overexploitation, because the characteristics of the fishing practice were more compatible with the carrying capacity of the system and fewer people were involved.

An informal fishing calendar was still in place until the mid-1990s, but to a much lesser extent than in the past. Figure 14.4 shows the changes in fishing calendars for the main small-scale fisheries resources between the 1960s and the early 1990s. Species such as mullet, formerly fished mostly in late autumn (April to June) during the spawning run, were being fished throughout the year in the early 1990s. For other resources, such as catfish, the collapse of the stock brought a change in the fishing calendar from spring to winter months, when the few remaining catfish sustain a smaller-scale fishery in the upper estuary. The change in technology (from beach seines to gillnets) also made croaker and mullet more evenly vulnerable to small-scale fisheries throughout the year, because both species are present in the estuary at different life stages during the year and are susceptible to capture by gillnets.

Also, before the advent of industrial fisheries, a large proportion of the species habitat in the Patos Lagoon and in the southern Brazilian shelf worked as *de facto* spatial refugia, because small-scale fisheries were limited to specific areas of the estuary of Patos Lagoon and adjacent coastal shallow waters. Thus, the increasing competition between small-scale and industrial fisheries, and the technological

INDIGENOUS AND ARTISANAL FISHERIES

improvements in resource location and capture, undermined important factors that had made small-scale fisheries resilient, such as the limited times and areas of resource exploitation. Fishing technologies and resource-use practice in the past were intrinsically dependent upon nature, through the influence of the moon, the behaviour of the fish, and weather conditions, which created natural mechanisms for limiting excessive exploitation by small-scale fisheries.

Referring to Holling's four-phase model (Figure 14.3), small-scale fisheries were practically limited to two phases in the resource dynamics: the exploitation phase, when resources such as croaker, catfish and mullets were entering the estuary, and

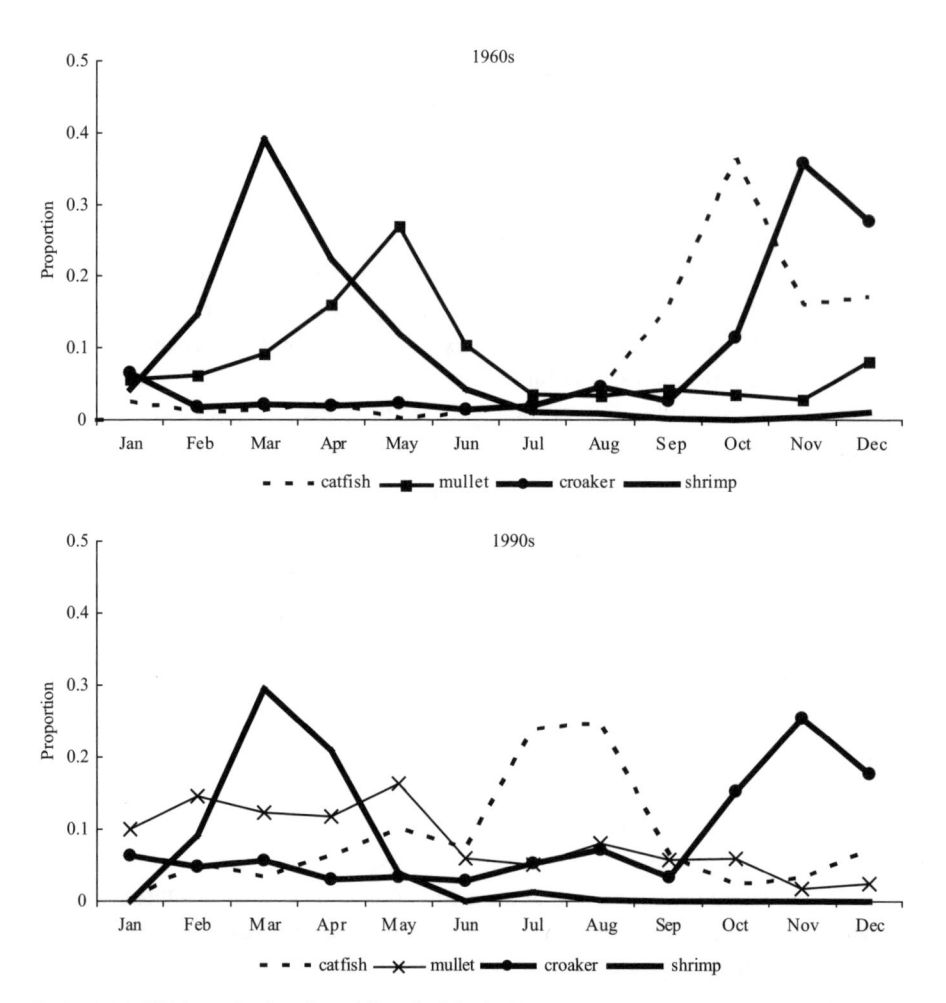

FIGURE 14.4 *Fishing calendars for small-scale fisheries in the estuary of Patos Lagoon and coastal waters during the 1960s and the early 1990s.*

Note: The lines represent the proportion of the total annual catch of each species obtained in a single month.

the release phase, when all these species and pink shrimp were leaving the estuary en route to the shelf waters. Fishers did not target fish during the other two phases (renewal and conservation) until technological advances and the industrialization of the fisheries. These, in turn, made the resources available to be exploited at any time and place. In conclusion, the hypothesis put forward here is that up to a certain point in time, the pattern of resource use by small-scale fisheries in the estuary of Patos Lagoon served conservation purposes, because it made resources less vulnerable to overexploitation while helping maintain the cycle of resource renewal. Besides serving conservation purposes, the fishing practices adopted by small-scale fishers sustained a very productive fishery from the early 1900s until practically the late 1980s (Reis, 1999). For instance, in the 1960s small-scale fisheries were responsible for over 80 per cent of the total fisheries landings in southern Brazil (about 27,000 tonnes/year; IBAMA, 1995).

The above analysis of the fishing practices adopted by small-scale fishers in the estuary of Patos Lagoon showed that indeed there is an informal knowledge system used by fishers to deal with the dynamics of the resources. These fishing practices were part of an informal resource-management system that helped maintain a productive and resilient small-scale fishery in the past. Resource-use practice in the estuary of Patos Lagoon has been changing in response to changes in technology, increasing fishing pressure and influences from internal and external (mostly government agencies) institutional transformations that shifted the management of fisheries from informal community-based, to central-government-based forms, and to the present situation of co-management (Kalikoski et al., 2002; Kalikoski and Satterfield, 2004).

FISHERIES INSTITUTIONAL ARRANGEMENTS

OVERALL, the governance of fisheries in Brazil is the responsibility of the state. Today fisheries management in Brazil is regulated by the Federal Environmental Agency (IBAMA and by the Special Secretariat for Fisheries and Aquaculture (SEAP). These agencies retain the ultimate power and control over fisheries management via the creation of laws and decrees that establish the regulations for fisheries activity within the Brazilian internal and coastal waters and within its exclusive economic zones (EEZs). Nevertheless, in the estuary of the Patos Lagoon, as in many other regions in Brazil, a certain type of devolution or delegation of power from the national government to small-scale fisheries communities or local institutions is taking place. On the one hand, government is devolving power to communities through legitimizing existing community-based management systems (examples are fishing accords in the Amazon, and extractivist reserves in the Amazon and along the Brazilian coast) (Kalikoski et al., 2006). On the other hand, power has been delegated via co-management regimes, which means that responsibilities for the design and implementation of regulations that mediate the use of the resource are shared between government and small-scale fisher stakeholders. Currently, co-management is the management regime that regulates fisheries in the estuary of the Patos Lagoon.

Historically, the management of fisheries in Brazil has been ruled by different governmental institutional arrangements that have been changing over the years. The role of the Federal Government in marine fisheries management became particularly influential in the mid-1960s with the creation of the Federal Fisheries Agency (SUDEPE) of the Ministry of Agriculture. Up to the 1960s, fisheries in the estuary were mostly informally governed within small-scale fishing communities (local communities extracted the resources by implementing a series of agreed rules and fishing practices), but these local informal governance institutions were substantially affected by the creation of this type of top-down management system. SUDEPE regulated fisheries until 1989, when it was eliminated because of its failure to manage fisheries in a sustainable fashion.

In 1989, fisheries management became one of the responsibilities of the Environmental Agency (IBAMA), a subsidiary of the Ministry of Environment. From 1998 to 2002, authority for fisheries management in Brazil was split between two agencies: the Ministry of Agriculture (Department of Fisheries and Aquaculture – DPA) and the Ministry of the Environment (IBAMA). IBAMA became responsible for conservation, enforcement and management of overfished and/or depleted resources, while the DPA was responsible for licensing and the development of fisheries regarded as underexploited. At the time of writing, DPA had been replaced by the new Special Secretariat for Fisheries and Aquaculture (SEAP) but kept the basic former DPA's fisheries management attributions with a strong policy for aquaculture development.

These changes in fisheries regulation have had many impacts on the effectiveness of fisheries management and intensified the erosion of many other identified informal systems of community-based fisheries management along the Brazilian coast (Seixas and Berkes, 2003; Pinto da Silva, 2004). In the estuary of Patos Lagoon fisheries management has shown failures in both decentralized (community-based) and centralized (government-based) forms of resource management due, to a large extent, to the mismatch between institutions crafted at the local level and the broader governmental institutions (Kalikoski et al., 2002). The local, informal, decentralized management system present until the 1960s failed to maintain its sustainable fisheries because it was unable to craft informal institutions that were able to control access by outsiders, or resist internal pressures to behave opportunistically by taking advantage of new technologies, and external pressures from market incentives generated by the governmental policies implemented by SUDEPE, IBAMA and the most recent agencies (DPA and SEAP). Attempts to control access and attenuate the overexploitation problem with locally devised rules did not reach higher levels of decision-making. This system was easily eroded by the external influence of economic development policies aimed at the industrialization of local fisheries and by a centralized and volatile management model adopted by the federal government after the late 1960s (Kalikoski et al., 2002). Relying on a system of economically driven policies, this centralized management disregarded the sustainable resource-use practices of small-scale fishers and drove many resources to overexploitation and collapse.

Due to incompatible governmental policies for achieving fisheries sustainability and the inability of locals to implement effective community-based management systems, the small-scale fisheries management situation in the estuary of Patos Lagoon called for a cross-scale linkage between local institutions and government. Steps toward this were taken in 1996 with the creation of the Forum of Patos Lagoon Co-management, as an institutional response to the crisis of estuarine fisheries. This new regime establishes rules regarding how much, when and in what way different resources can be harvested, through implementing management functions such as licensing, timing, location and vessel or gear restrictions to prevent overexploitation inside the estuary and within the three-mile zone. Government and fisher stakeholders decide in a monthly forum meeting how and where the use, extraction and management of fish resources should take place. The co-management arrangement was initiated by the Fishers' Pastoral ('Pastoral do Pescador')[1] and the Fishers' Colonies[2] ('Colônia de Pescadores') in conjunction with the local branch of the Federal Environmental Agency (IBAMA-CEPERG). Key elements to be achieved within this new fisheries management regime were a collaborative partnership among communities, governmental and non-governmental organizations, and a transition to negotiation-style, decentralized decision-making.

The Forum recognizes that fisher communities play an important role in the preservation of healthy fish stocks and that resource management will be more effective when communities are granted active participation in the management process and in turn have the potential to devise regulations that are more flexible, adaptable and appropriate to specific situations than those crafted by centralized agencies. In an attempt to include all the institutions affected by coastal resource management generally and fisheries specifically, a total of twenty-one institutions representing the principal stakeholders on coastal resource management were invited to be part of the Forum.[3] Participation in the Forum is voluntary and all representatives have the right to speak and to vote.

1. The Fishers' Pastoral is an organization that promotes small-scale fisheries communities' social and economic organization.
2. The Fisher Colony is a professional organization of fishers of a given municipality mandated by the Federal Constitution as one form of a working union.
3. Port authority; Fishers' Pastoral/CNBB; Environmental Police (PATRAM); Federal Environmental Agency (IBAMA/CEPERG); Local universities (FURG; UFPel and UCPel); State of RS (FEPAM and SAA); EMATER/ASCAR; 4 Fishers Colonies; Public Ministry; local NGO (NEMA, CEA); 4 Municipalities and Fisheries Industry syndicates.

Representatives of the Fishers' Colonies and the Fishers' Pastoral were given rights to two votes each, while the other institutions have the one vote each. Assigning more votes to the Colonies and Pastoral represents an explicit attempt to shift the locus of control to the institutions representing artisanal fishers. Other people who do not officially represent any institution, such as researchers, can participate in the meetings but do not have the right to vote. However, the interests and issues raised by all participants are accommodated as fully as possible (detailed analysis on the implementation process of the Forum of Patos Lagoon can be found in D'Incao and Reis, 2002; Kalikoski, 2002, and Kalikoski and Satterfield, 2004).

OPPORTUNITIES AND CHALLENGES FOR PUTTING FISHERS' KNOWLEDGE TO WORK

THIS study demonstrates that fishers' knowledge can provide valuable information about the relationship between fishers and the local environment, and about the characteristics of the practices, tools and techniques that underlie a more sustainable pattern of resource use. Local knowledge can broaden the knowledge base needed for management and hence improve institutions that mediate the interaction between communities and their use of resources.

The first step towards recognizing the causes of the failure to achieve sustainable fisheries management in the region and the importance of involving fishers in the whole process of decision-making occurred with the creation of the Forum. The Forum is an attempt to share responsibility and authority related to the management of fisheries resources and provides the opportunity to apply TEK in devising rules for sustainable resource use. Change towards a more inclusive process of rule making has been observed recently, and fishers' inputs have been used to revise the norms regulating resource use in the estuary (Decrees 171/98 and Instrução Normativa MMA/Seap 03/2004). The decrees established minimum mesh sizes, the number and length of fishing nets, minimum fish sizes and the calendar for each of the main resources, as well as criteria for limiting access to fisheries inside the estuary. Access is now restricted to those fishers who can prove their historical dependence on this activity for a living.

During the development of this chapter, we identified other proposed changes to the norms regulating fisheries activities in estuarine and coastal waters suggested by fishers (Table 14.2) that reflect, to a certain extent, their understanding of sustainable fishing practices. Some of them have been taken into account by the Forum (such as the restricted access of industrial purse seiners to the mouth of the estuary, under a decree approved in December 2003), while others are considered valid only if subjected to considerable scientific scrutiny (such as the adjustment of the calendar for catfish fisheries in the upper estuary).

TABLE 14.2 *Comparison between selected principles of the Code of Conduct for Responsible Fisheries (FAO, 1997) and adjustments to local fisheries management suggested by small-scale fishers during interviews and a Forum of Patos Lagoon meetings.*

Principles of responsible fisheries (FAO)	Adjustments to fisheries management indicated by fishers' knowledge in the estuary of Patos Lagoon
Control methods that damage the ecosystem	Stop industrial trawling in the coast because it kills large quantities of fish that are discarded. Replace trawling nets with gillnets with large mesh sizes, which are more selective and less damaging. Forbid or reduce small-scale fisheries in the nursery shallow waters of the estuary (such as stow-netting and trawling) because they capture large quantities of juvenile fish and shrimp. Adapt artisanal otter trawling nets to reduce bycatch (implementing bycatch reduction devices) and confine the use of artisanal trawling to the channel areas of the lagoon.
Monitoring and enforcement	Increase enforcement in the estuary all year round and not only during the shrimp season. Increase enforcement in the three-mile zone along the coast, where many industrial trawlers operate illegally.
Marine protected areas	Close the inshore area around the mouth of the lagoon, especially to industrial purse seiners. This is an area where, according to fishers, fish concentrate before entering the lagoon. If it is made a protected area, fishers believe that more fish will make their way to nursery and reproduction areas in the lagoon. The establishment of marine protected areas is also congruent with a precautionary approach to fisheries management.
Adaptive management	Adjust fishing calendars according to the environmental conditions and resource abundance. A system of time/area openings has been suggested by fishers as a way to accommodate management rules to the characteristics of the shrimp fishery.

The co-management of fisheries CPRs in the estuary of Patos Lagoon is still in its infancy. Important adjustments still need to be made before the outcomes of the Forum can be said to better reflect the interests and knowledge of fishers. It is possible to identify three interrelated factors influencing the use of local knowledge in the co-management of estuarine resources.

Illiteracy and socio-economic marginalization make scientists and decision-makers undervalue fishers' knowledge

There are many myths about small-scale fishers that still haunt management arenas and hinder a more productive interaction between scientific and local knowledge. Diegues (1995) paraphrased some of the most common myths about small-scale fishers in Brazil:

- Small-scale fishers are beach beggars; they are a social problem that needs to be treated by social aid programmes.
- Small-scale fisheries are in transition to industrial, capitalist fisheries, and therefore are doomed to disappear.
- Small-scale fishers are unintelligent and resist technological innovations.
- Small-scale fishers are predators and individualists who are unable to organize themselves.

Over time, these myths led to fishers being excluded from decision-making and consequently made them more vulnerable to the management process. As argued by Pauly (1997), the marginalization of fishers and their limited formal education have often blinded managers and scientists to their ecological knowledge, which is used in many successful common-property systems as the basis for traditional community-based management.

Despite their limited formal education, small-scale fishers developed resource-use practices that maintained a productive fishery in the estuary of Patos Lagoon until the late 1960s, when their informal systems of management practices were eroded by formal, top-down management procedures. Fishers' knowledge of sustainable fishing practices was also identified during interviews and meetings of the Forum of Patos Lagoon in the form of requests for changes in local fisheries management. Fishers' requirements mirror many of the principles one can read in higher-level environmental institutions, such as the FAO Code of Conduct for Responsible Fisheries (FAO, 1997; Table 14.2).

Misfit between institutions and the characteristics of CPRs that hinders fishers' resources stewardship and knowledge application

Although fishers' recognize the need for management, they do not comply with the management rules in place in the estuary (such as the fishing closure in the winter months and the ban on trawling). In conditions of scarcity and competition, fishers' stewardship of resources is an important yet difficult aim to achieve. Where stewardship for resources exists, it is in the best interests of those who control it not to overfish. As argued by Johannes (1981), in this case 'self-interest thus dictates conservation'. Users' interest in working towards the sustainability of a particular resource is conditioned by the benefits they expect to achieve (Ostrom et al., 1999). However, solving fisheries CPR problems involves two distinct elements that are important to the husbandry of the resources: restricting access and creating incentives for users to invest in the resource instead of overexploiting it.

Limiting access alone can be ineffective if resource users compete for shares; the resource can become depleted unless incentives or regulations prevent overexploitation (Ostrom et al., 1999). As can be observed in Table 14.2, traditional users of the estuary of Patos Lagoon feel threatened by having to share access rights with the more recent industrial users group. Resources outside the mouth of the estuary are still open to industrial fisheries. There are few rules regulating this activity on the coast, and even these are poorly enforced, despite the damage industrial fishing can cause. This creates a dilemma inside the estuary as small-scale fishers complain that the resources they do not catch today will not be available to them in the future, but rather will be fished by industrial fishers outside the estuary. Efforts to exercise stewardship in such circumstances are unlikely to succeed.

Examples of CPRs' management worldwide have shown that although the development of local ecological knowledge is a necessary condition, it is usually insufficient in itself to achieve sustainability if it is not accepted and legitimized by management institutions (Johannes, 1981; Berkes, 1999; De Castro, 2000; Seixas, 2000). A fundamental incentive to conservation involves the definition of property rights to common-property resources (Ostrom, 1990). As long as property rights to resources remain open, no one knows what is being managed or for whom, and any incentive to conserve will disappear because there is no guarantee that the benefits of any management action will accrue to the individual or group that practises conservation.

The difficult transition to a 'civic science' in the management of coastal resources

Two types of paradigms about the role of science and local knowledge are evident in local environmental management institutions. The first, which has been the dominant one, is based on the idea that scientific knowledge is objective and factual, and provides the 'truth' upon which decisions should be based (Holling et al., 1998). This paradigm has no room for local ecological knowledge, for uncertainties, or for a systemic view of the problems. This conventional way of conducting science has been shown to act against sensitive and precautionary environmental management by leading decision-makers to examine only those phenomena where cause and effect can be either proved or shown to be reasonably unambiguous (O'Riordan, 2000).

The second paradigm is based on the recognition that conventional science is value-laden, and that information and decisions can be manipulated by powerful vested interests. It acknowledges that knowledge about the ecosystem is incomplete, and that therefore uncertainties are high and surprises (when actions produce results opposite to those intended) are inevitable (Holling et al., 1998). It calls for the integration of different forms of knowledge (scientific and local) in order to better understand the nature of complex problems and to reduce uncertainties, where possible. More importantly, this paradigm recognizes that management of CPRs should not rely merely on science but on a *civic science* (Lee, 1993), that is 'deliberative, inclusive, participatory, revelatory and designed to minimise losers' (O'Riordan, 2000).

The sharing of power and responsibilities and a civic science within co-management systems can be accomplished through the use of fishers' knowledge. Using fishers' knowledge in the design of local rules that mediate the use of resources, for instance, is a mechanism to empower local communities as it gives them a voice in the decision-making process, thus providing a concrete basis for their involvement. In this sense, the use of traditional ecological knowledge becomes a strong tool for empowering small-scale communities within co-management systems. With the use of local knowledge, a greater power balance may be achieved between local small-scale fishers communities on the one hand and government, large-scale fishers and conventional, resource-management scientists on the other (Berkes, 1999; Neis and Felt, 2000).

By stimulating the exchange of information and knowledge between scientists and fishers, the Forum of Patos Lagoon is creating the conditions for a transition towards a civic science in the co-management of small-scale fisheries. One important indicator of this move is the process of defining and revising rules to regulate the fisheries of the Patos Lagoon estuary from the bottom up, with inputs from small-scale fishers. The locally devised rules were legitimized by the federal government through the creation of regulations by IBAMA (Decrees 171/98 and 144/01).

However, while Forum decisions that relate to small-scale fisheries management are triggering the transition towards a civic science paradigm, the overall process of governance of other resources and activities within the coastal zone of the Patos Lagoon are not. Instead, the overall coastal zone governance system is still locked into a top-down management system based on a conventional scientific approach (*sensu* Holling et al., 1998; Asmus et al., 1999). An example of this approach was seen in the environmental impact assessment (EIA) of the enlargement of the jetties in the mouth of the estuary of Patos Lagoon (FURG, 2000). The EIA study contained many uncertainties, which were not made explicit or communicated. The project had many outcomes that are not well defined and there are many questions that still remain unanswered, such as the ones raised within the Forum.

> 'Will the project impact the amount of shrimp entering the lagoon? What will be the impact of the project on the behaviour of the fish that migrate through the channel of Rio Grande? What will be the impact of the project on the estuarine ecosystem? How will the project affect navigation conditions for small-scale fishing boats off the mouth of the estuary?'

The above characteristics create a mix of uncertainties and ignorance about the possible consequences of the project that call for a civic science approach (*sensu* O'Riordan and Stoll-Kleemann, 2002). Contrary to civic science principles of inclusivity and participatory research, neither the small-scale fisher communities of the estuary of Patos Lagoon directly affected by the project nor the Forum of Patos Lagoon were consulted during the EIA.

Therefore, although the Forum is moving slowly towards a civic science approach to small-scale fisheries management inside of Patos Lagoon, activities in the estuary with a direct effect on small-scale fisheries are not taken into account in bottom-up or participatory approaches. However, because many of the twenty-one institutions that participate in the Forum represent interests beyond fisheries (for instance, the Federal Public Ministry[4] and the Environmental Agency), opportunities are being created for the Forum to challenge decisions that affect small-scale fisheries, thus empowering local institutions and fishing communities to call for better governance of the natural resources in the region.

The use of local knowledge in co-management systems is a way for communities to regain their right to control their resources, the right to self-determination and self-government, and the right to represent themselves through their own worldview systems (Berkes, 1999; Berkes et al., 2001). The Forum of Patos Lagoon becomes then the venue for empowering fishers to regain control over their own cultural information and reclaim their knowledge. In this sense, small-scale fishers and their knowledge – including the set of practices, tools, techniques and appropriate informal institutions embedded in a different worldview system – may represent a future-oriented concept for sustainable resource management in the estuary of Patos Lagoon.

ACKNOWLEDGEMENTS

The authors are especially thankful to the Forum of Patos Lagoon representatives and to the fisher communities of the Patos Lagoon for the support, shared knowledge and kindness that made this work so pleasant. This project was supported by the University of British Columbia (UBC, Canada) Hampton Award.

REFERENCES

ALMUDI, T.; KALIKOSKI, D.C.; CASTELLO J.P. 2006. Territorial control as a fisheries management instrument: The case of artisanal fisheries in the estuary of Patos Lagoon, Southern Brazil. In J.L. Nielsen et al. (eds.). *Proceedings of the Fourth World Fisheries Congress: Reconciling fisheries with conservation.* Symposium 49, American Fisheries Society, Bethesda, Md.

ASMUS, M.L.; CALLIARI, L.; TAGLIANI, P.R.; KALIKOSKI, D.C. 1999. Ecosystem based integrated coastal zone management in the estuary of the Patos Lagoon: Opportunities and constraints. *Proceedings of the Workshop Ecosystem Based Integrated Coastal Zone Management,* Vancouver, Canada, UBC, 10 pp.

---

4. The Federal Public Ministry is mandated to ensure that the principles of honesty, democracy and justice are respected. Accordingly, it acts as a watchdog for environmental issues and represents the interests of the society as a whole on sensitive issues such as traditional peoples, environment and human rights.

BARCELLOS, B.N. 1966. *Informe geral sobre a pesca no Rio Grande do Sul*. Porto Alegre, Rio Grande do Sul, BRDE-CODESUL, 115 pp.

BERKES, F. (ed.) 1989. *Common property resources: Ecology and community-based sustainable development*. London, UK, Belhaven, 302 pp.

——. 1999. *Sacred ecology: Traditional ecological knowledge and resource management*. Philadelphia, Taylor and Francis, 209 pp.

BERKES, F.; FOLKE, C. 1998. *Linking social and ecological systems: Management practices and social mechanisms for building resilience*. Cambridge, UK, Cambridge University Press, 459 pp.

BERKES, F.; MAHON, R.; MCCONNEY, P.; POLLNAC, R.; POMEROY, R. 2001. *Managing small-scale fisheries: Alternative directions and methods*. Ottawa, IDRC, 320 pp.

CASTELLO, J.P. 1985. The ecology of consumers from dos Patos Lagoon estuary, Brazil. In: A. Yañez-Arancibia (ed.), *Fish community ecology in estuaries and coastal lagoons: towards an ecosystem integration*. Mexico City, UNAM Press, pp. 383–406.

CASTELLO, J.P.; MOLLER, O.O. 1978. On the relationship between rainfall and shrimp production in the estuary of the Patos Lagoon (Rio Grande do Sul, Brazil). *Atlântica, Rio Grande*, Vol. 3, pp. 67–74.

CHAO, L.N.; PEREIRA, L.E.; VIEIRA, J.P. 1985. Estuarine fish community of the Patos Lagoon, Brazil: A baseline study. In: A. Yañez-Arancibia (ed.), *Fish community ecology in estuaries and coastal lagoons: Towards an ecosystem integration*. Mexico City, UNAM Press, pp. 429–50.

COSTA, J. S. 2001. *Navegadores da Lagoa dos Patos: A Saga Náutica de São Lourenço do Sul*. Pelotas, Hofstatter, 212 pp.

CRESWELL, J. 1994. *Research design: Qualitative and quantitative approaches*. Thousand Oaks, Calif., Sage, 228 pp.

CZAJA, R.; BLAIR, J. 1996. *Designing surveys: A guide to decisions and procedures*. Thousand Oaks, Calif., Pine Forge Press, 269 pp.

DE CASTRO, F. 2000. *Fishing accords: The political ecology of fishing intensification in the Amazon*. Ph.D. dissertation. Bloomington: Indiana University.

DIEGUES, A.C.S. 1995. *Povos e Mares: Leituras em Sócio-Antropologia Marítima*. São Paulo, NUPAUB, University of São Paulo, 260 p.

D'INCAO, F. 1991. Pesca e biologia de *Penaeus paulensis* na Lagoa dos Patos, RS. *Atlântica (Rio Grande)*, Vol. 13, No. 1, pp. 159–69.

D'INCAO, F.; E. G. REIS. 2002. Community-based management and technical advice in Patos Lagoon estuary (Brazil). *Ocean and Coastal Management*, Vol. 45, pp. 531–9.

FAO. 1997. *Fisheries Management*. FAO Technical Guidelines for Responsible Fisheries, 4. Rome, FAO, 82 pp.

FURG. 2000. *Relatório de Impacto Ambiental: Ampliação dos Molhes do porto de Rio Grande*. Rio Grande, Brazil, University of Rio Grande, 230 pp.

HAIMOVICI, M. 1997. *Recursos Pesqueiros Demersais da Região Sul*. Programa REVIZEE. Ministério do Meio Ambiente, dos Recursos Hídricos e da Amazônia Legal. Brasilia, FEMAR, 80 pp.
HAIMOVICI, M.; PEREIRA S.D.; VIEIRA, P.C. 1989. La pesca demersal en el sur de Brasil en el periodo 1975–1985. *Frente Maritmo,* Vol. 5, pp. 151–63.
HOLLING, C.S. 1986. Resilience of ecosystems; local surprise and global change. In: W.C. Clark and R.E. Munn (eds.), *Sustainable Development of the Biosphere*. Cambridge, Mass., Cambridge University Press, pp. 292–317.
——. 1992. Cross-scale morphology, geometry and dynamics of ecosystems. *Ecological Monographs,* Vol. 62, No. 4, pp. 447–502.
HOLLING, C.S.; BERKES, F.; FOLKE, C. 1998. Science, sustainability and resource management. In: F. Berkes; C. Folke (eds.), *Linking social and ecological systems: Management practices and social mechanisms for building resilience*. Cambridge, UK, Cambridge University Press, pp. 342–62.
IBAMA. 1995. *Peixes Demersais*. Ministério do Meio Ambiente, dos Recursos Hídricos e da Amazônia Legal. Coleção Meio Ambiente. Brasilia, Séries Estudos de Pesca, 16 pp.
JOHANNES, R.E. 1981. *Words of the lagoon: Fishing and marine lore in the Palau District of Micronesia*. Berkeley, Calif., University of California Press, 245 pp.
KALIKOSKI, D.C. 2002. *The Forum of Patos Lagoon: An analysis of institutional arrangements for conservation of coastal resources in southern Brazil*. PhD Dissertation. Vancouver, University of British Columbia, 257 pp.
KALIKOSKI, D.C.; VASCONCELLOS, M.; LAVKULICH, L. 2002. Fitting institutions to ecosystems: The case of artisanal fisheries management in the estuary of Patos Lagoon. *Marine Policy,* Vol. 26, pp. 179–96.
KALIKOSKI, D.C.; SATTERFIELD, T. 2004. On crafting a fisheries co-management arrangement in the estuary of Patos Lagoon (Brazil): Opportunities and challenges faced through implementation. *Marine Policy,* Vol. 28, No. 6, pp. 503–22.
KALIKOSKI, D.C.; SEIXAS, C.S.; ALMUDI, T. 2006. Gestão compartilhada e gestão comunitária da pesca no Brasil. Internal Report. International Development Research Center (IDRC), Ottawa, Canada, 59 pp.
LEE, K.N. 1993. *Compass and gyroscope: Integrating science and politics for the environment*. Washington, DC: Island Press, 243 pp.
NEIS, B.; FELT L. (eds.). 2000. *Finding our sea legs: Linking fishery people and their knowledge with science and management*. St. John's, ISER Books, 318 pp.
O'RIORDAN, T. (ed.). 2000. *Environmental science for environmental management*. 2nd edition. Harlow, UK, Prentice Hall, 520 pp.
O'RIORDAN, T.; STOLL-KLEEMANN, S. 2002. *Biodiversity, sustainability and human communities: Protecting beyond the protected*. Cambridge, UK, Cambridge University Press, 318 pp.

Ostrom, E. 1990. *Governing the commons: The evolution of institutions for collective action.* Cambridge, UK, Cambridge University Press, 280 pp.

——. 1995. Designing complexity to govern complexity. In: S. Hanna and M. Munasinghe (eds.), *Property rights and the environment*: Washington, DC, Beijer International Institute of Ecological Economics, World Bank, pp. 36–46.

Ostrom, E.; Burger, J.; Field, C.B.; Norgaard, R.B.; Policansky, D. 1999. Revisiting the commons: Local lessons, global challenges. *Science*, Vol. 284, pp. 278–82.

Pauly, D. 1997. Small-scale fisheries in the tropics: Marginality, marginalization and some implication for fisheries management. In: E.E. Pikitch, D.D. Huppert and M.P. Sissenwine (eds.), *Global trends in fisheries management.* Bethesda, Md., American Fisheries Society, pp. 40–49.

Pinto Da Silva, P. 2004. From common property to co-management: Lessons from Brazil's first maritime extractive reserve. *Marine Policy*, Vol. 28, pp. 419–28.

Pinkerton, E. (ed.). 1989. *Co-operative management of local fisheries: New directions for improved management and community development.* Vancouver, University of British Columbia Press, 297 pp.

Pomeroy, R.S.; Berkes, F. 1997. Two to tango: The role of government in fisheries co-management. *Marine Policy*, Vol. 21, pp. 465–80.

Reis, E.G. 1986. Reproduction and feeding habits of the marine catfish *Netuma barba* (*Siluriformes, Ariidae*) in the estuary of Lagoa dos Patos, Brazil. *Atlântica, Rio Grande*, Vol. 8, pp. 35–55.

——. 1999. Pesca artesanal na Lagoa dos Patos: história e administração pesqueira. pp. In: F.N. Alves (ed.), *Por uma História Multidisciplinar do Rio Grande*, pp. 81–84. Rio Grande, Fundação Universidade Federal do Rio Grande, 241 pp.

Reis, E.G.; D'Incao, F. 2000. The present status of artisanal fisheries of extreme southern Brazil: An effort towards community based management. *Ocean and Coastal Management*, Vol. 43, No. 7, pp. 585–95.

Rodrigues, G. 1989. *A atividade pesqueira no estuário da Lagoa dos Patos.* Subprojeto a pesca artesanal na Lagoa dos Patos. Projeto Lagoa dos Patos. Rio Grande, University of Rio Grande, 27 pp.

Seixas, C. 2000. State-property, communal property or open-access? The case of Ibiraquera Lagoon, Brazil. *Proceedings of the Eighth Conference of the International Association for the Study of Common Property*, Bloomington, Indiana, US, 31 May to June. Available online at: http://dlc.dlib.indiana.edu/documents/dir0/00/00/10/28/index.html (last accessed 26 June 2005).

Seixas, C.S.; F. Berkes. 2003. Dynamics of social-ecological changes in a lagoon fishery in southern Brazil. In: F. Berkes, J. Colding and C. Folke (eds.), *Navigating social-ecological systems: Building resilience for complexity and change.* Cambridge, UK, Cambridge University Press.

Vieira, J. P.; Scalabrin, C. 1991. Migração reprodutiva da tainha (*Mugil platanus*, Gunther 1980) no sul do Brasil. *Atlântica (Rio Grande)*, Vol. 13, No. 1, pp. 131–41.

CHAPTER 15   The value of local knowledge
in sea turtle conservation
A case from Baja California, Mexico
Kristin E. Küyük, Wallace J. Nichols and Charles R. Tambiah

ABSTRACT

IN many coastal communities around the world, hunting of sea turtles is still part of local traditions and culture despite evidence of decreasing turtle numbers and strict laws prohibiting their exploitation. While the ways that fishers have negatively affected sea turtle populations are well documented, what is often overlooked is how these same individuals can contribute to their conservation. A major goal of community-based efforts in sea turtle conservation is to develop practices which, while protecting sea turtle populations and habitats, are also compatible with the socio-economic system and cultural ecology of local resource-dependent communities. Within a conservation mosaic, the incorporation of both biological and social research methods as well as effective communication are critical. This chapter presents a case study of sea turtle recovery efforts within the Baja California peninsula, Mexico. It shows that community-based research can result in an improved knowledge base with benefits for both long-term conservation and resource users. Partnership development through local education, informal conversations and community meetings is shown to be a fundamental part of sea turtle conservation. By combining the knowledge gained through scientific investigations with the insights of the local population, we stand a much better chance of succeeding in recovery efforts, particularly those that rely on adaptive management techniques designed through community-based research and action.

INTRODUCTION

COASTAL communities throughout the world continue to utilize sea turtles according to their traditions and culture despite evidence of decreasing turtle numbers and strict laws against turtle hunting and use (for some examples see Parsons, 1962; Tambiah, 1989; Frazier, 1995; King, 1995; Kowarsky, 1995; Nietschmann, 1995; Tambiah, 1995). In north-western Mexico, and specifically the Baja California

INDIGENOUS AND ARTISANAL FISHERIES

peninsula, sea turtles were first exploited in subsistence hunting, but over time this use broadened into a directed fishery (Caldwell, 1963; Clifton et al., 1995). Although legislation is now in place to protect Mexican sea turtles, laws and enforcement have not adequately abated hunting and related declines in their populations, especially in rural areas like Magdalena Bay where there has been a long history of use. In this region, turtle meat is viewed at times as a tasty delicacy, and at others as a necessary means of feeding a large family. Sea turtles are consumed for a variety of reasons, ranging from special events and holidays to medicinal uses (Figure 15.1). In an area where the average monthly income is less than the equivalent of US$ 500 and the average number of dependants in a household is four to six people, there is also a strong economic motive for fishers to supplement their income by the sale of harvested sea turtles on the black market (Bird, 2002).

Of the five threatened or endangered sea turtle species known to inhabit the coastal waters of Pacific Mexico, two species most commonly frequent the waters within and adjacent to Magdalena Bay: the East Pacific green turtle, also known as the black turtle, (*Chelonia mydas*) and the loggerhead turtle (*Caretta caretta*) (Clifton

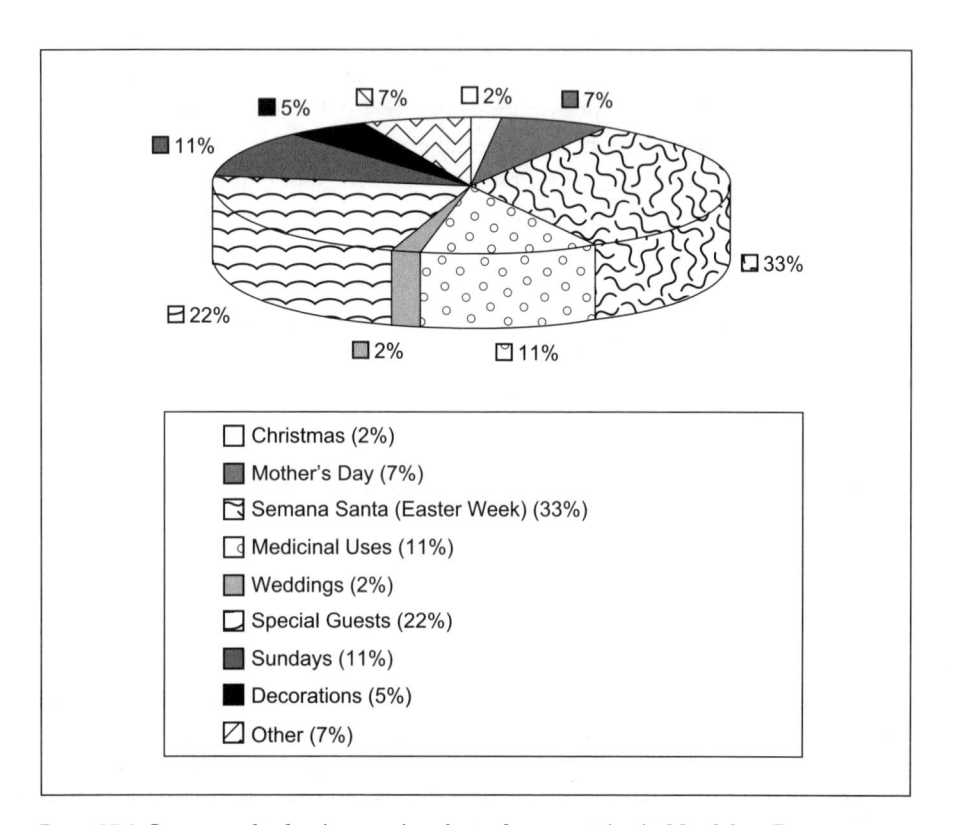

FIGURE 15.1 *Survey results showing occasions for turtle consumption in Magdalena Bay.*

et al., 1995; Nichols et al., 2001). These are also the species that are most commonly caught by the fishers of Puerto San Carlos, Puerto Magdalena and Lopez Mateos, the largest communities on the shores of Magdalena Bay (Gardner and Nichols, 2001). The coastal waters around the Baja California peninsula serve as critical feeding and developmental habitat for these and other sea turtles after they migrate from as far as Michoacan, south-west mainland Mexico (Nichols et al., 1998) and Japan (Nichols et al., 2000b).

SITE DESCRIPTION

THE Baja California peninsula, which extends about 1,000 miles (~1,600 km) into the Pacific Ocean south of the US State of California, comprises two Mexican states, Baja California and Baja California Sur. Magdalena Bay, a large mangrove estuarine complex on the Pacific side, is one of the largest bays in the entire peninsula and is bordered by several barrier islands (Figure 15.2). Due to its location between Pacific and Californian ocean currents, which allows for a mixture of both warm and cold water species, and the relative protection provided by the barrier islands, Magdalena Bay is a highly productive ecosystem which boasts enormous biodiversity. The mangroves of this bay are at the northernmost reaches of their range; their presence is a unique feature of the coastal ecology which contributes to the high productivity of a bay that has been called 'the Chesapeake of the Pacific' (Dedina, 2000).

Many of the towns on the shores of Magdalena Bay were settled by *rancheros* (ranchers) from the Santo Domingo valley and surrounding inland areas. While Magdalena Bay was first discovered by *conquistadores* (conquerors) in the sixteenth century, migration to this region did not get underway until the 1920s when inland agricultural projects began to fail and new means of subsistence – shell and finfish – were sought (Dedina, 2000). More permanent settlement began in the late 1950s when the cannery and deep-water port projects were initiated in Puerto San Carlos. Since that time people have continually been migrating to the town. Though many who currently inhabit Puerto San Carlos have lived there for a number of years and consider themselves residents of the area, their roots may lie in other states in mainland Mexico (Bostrom et al., 1999). Today, migrant fishers continue to come from the mainland and other parts of the Baja California Peninsula in order to exploit the seasonal resources.

Currently, numerous fish camps are scattered along the coastline of Magdalena Bay, many of which are only occupied seasonally. There are also a few permanent settlements, most notably the towns of Puerto Adolfo, Puerto Lopez Mateos, Santo Domingo, Puerto Magdalena and Puerto San Carlos, which is the largest settlement on the bay. The population of Puerto San Carlos varies seasonally with the fisheries,

INDIGENOUS AND ARTISANAL FISHERIES

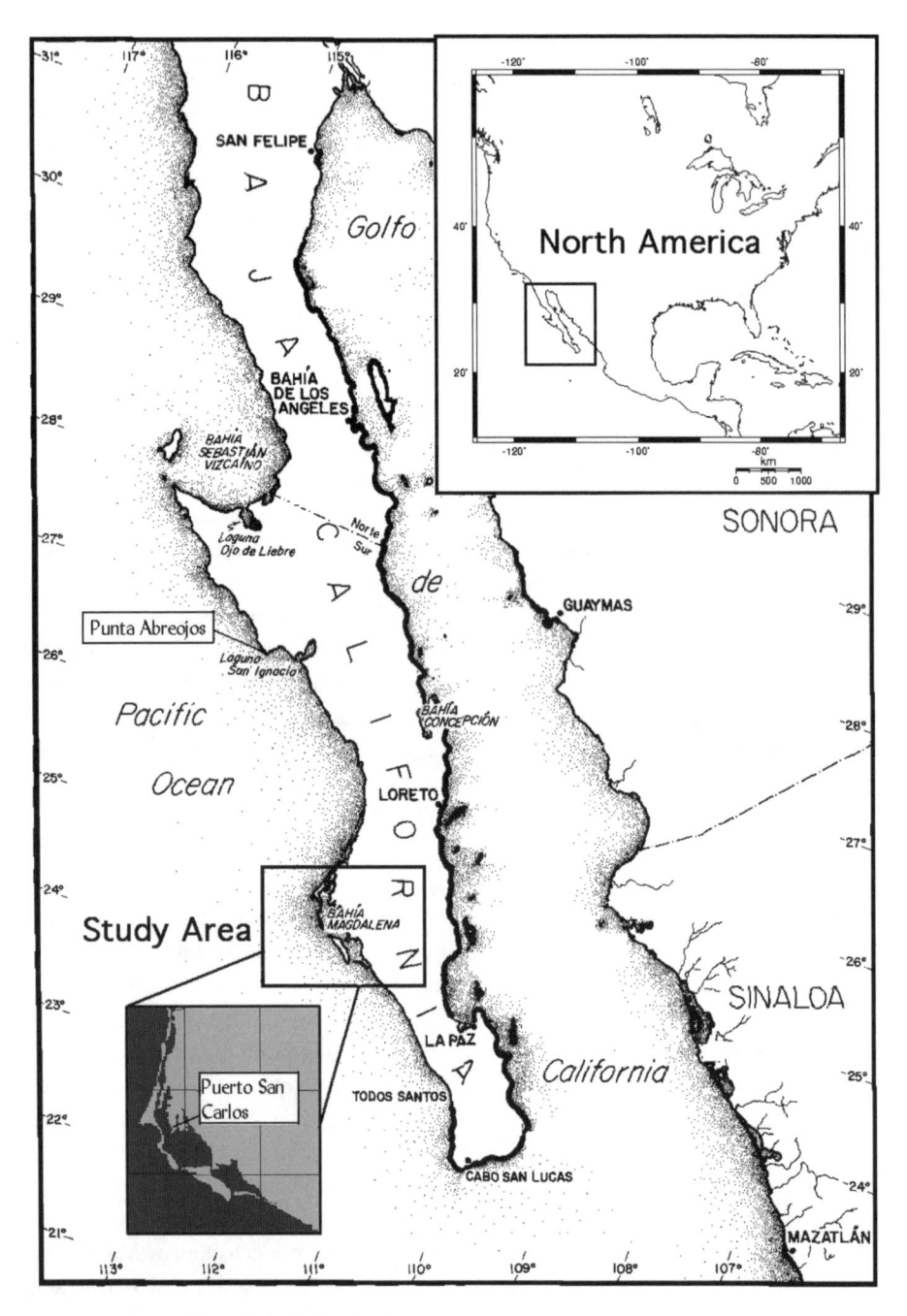

FIGURE 15.2 *Map of the Baja California Peninsula.*
Source: Adapted from Thompson et al. (2000).

and ranges between 3,000 and 5,000 people. Its people have been called 'the people of the mangroves'; they form a resource-dependent community (Serge Dedina, Executive Director, Wildcoast International Conservation Team, personal communication), reliant on marine and coastal ecosystems for their livelihood and survival. While there is a cannery, port and large-scale commercial fisheries, as well as a thermoelectric plant in the area, the community and character of Puerto San Carlos rest on the shoulders of small-scale artisanal fishers and their families. These fishers may be members of a fishing cooperative or among the many *pescadores libres* (independent fishers) in the region.

THE CONSERVATION MOSAIC

FRAZIER (in press) posed the question: 'Is increased scientific production conserving turtles?' stating that 'we are learning more and more about what is becoming less and less'. Frazier's comments suggest that despite scientific progress in this field, many sea turtle populations are becoming increasingly endangered. Unfortunately increased 'scientific' understanding does not always translate into 'conservation' on the ground.

Throughout the world, fishers have been blamed for declining sea turtle populations, even in areas with sparse populations and inadequate utilization assessments (Caldwell, 1963; Parsons, 1962; Clifton et al., 1995; King, 1995; Tambiah, 1995). In addition, local knowledge has historically been excluded from sea turtle research and from the conservation process. Furthermore, the active participation by fishers in sea turtle conservation initiatives has rarely been considered (Nader, 1996). Placing value on the opinions, experiences, and knowledge of local fishers, and involving them directly in a conservation initiative from the design through the implementation and evaluation phases can contribute to the development of strong conservation alliances (Bird, 2002; Bird and Nichols, in press; Peckham et al., in press; Pesenti et al., in press). A developed awareness and understanding of local cultures and values is essential to the success of such initiatives. Within a *conservation mosaic*, the incorporation of both biological and social research methods and communication are critical (Nichols, 2003).

Local involvement in turtle conservation has been increasing over the past decades, but has tended to be guided by an outside 'expert' organizing and/or overseeing community work and selecting appropriate conservation techniques. Community-based approaches are also not new to sea turtle conservation. Existing approaches include community monitoring of lighting practices on nesting beaches, community-based stranding networks and beach patrols, self-enforcement by fishing communities, formal sharing of traditional knowledge (Nabhan et al., 1999) and the systematic consideration of results of interviews with fishers (Tambiah, 1999). While such practices are increasing, community-based efforts are still not widely accepted as a valid conservation approach (Frazier, 1999; Tambiah, 2000).

A major goal of community-based efforts in sea turtle conservation is to develop practices that will protect sea turtle populations and habitats but that are also compatible with the socio-economics and cultural ecology of local resource-dependent communities (Bird and Nichols, 2002; Tambiah, 2000). In many of the 'community-based conservation' cases documented in the literature, external researchers have initiated conservation projects that have included local community participation in their design (Govan, 1998; Hackel, 1999; Tambiah, 1995).

Projects that integrate local science into the conservation initiative are less common. External researchers often only have the time and resources to make a snapshot assessment. In addition, research projects are sometimes designed to allow researchers to operate with complete autonomy from the community in terms of food and equipment and to permit them to 'get in and get out', gathering as much data as possible as efficiently as possible. All too often, once the data are collected researchers never return. An alternative approach aims to strengthen integration into the community and to promote dependence on local hosts as a means to begin building trust and partnerships that may be critical to the long-term success of conservation initiatives.

RESEARCH APPROACH AND METHODOLOGY

Our interdisciplinary project on sea turtle conservation relied on two main tools: conservation research and active community involvement. Qualitative research in Puerto San Carlos yielded some important primary data related to the cultural and socio-economic factors that affect a fisher's decision to capture a turtle, or whether to keep or throw back one that is captured incidentally (Bostrom et al., 1999). To build on these data we carried out socio-economic studies of current and historic sea turtle utilization within Baja California Sur, particularly in the Magdalena Bay region (Bird, 2002). We are also carrying out ongoing biological monitoring and ecological studies (Brooks et al., in press b; Garcia-Martinez and Nichols, 2000; Gardner and Nichols, 2001; Hilbert et al., in press; Nichols et al. 2001, Peckham and Nichols, 2003), including the deployment of radio and satellite transmitters to monitor the distribution, movements and long-distance migratory patterns of sea turtles (Brooks et al., in press a; Nichols et al., 1998; Nichols et al., 2000b).

A series of surveys and interviews were also conducted in order to document local knowledge related to sea turtles and other marine species. Fishers were asked questions about the types of species captured in their gear, the frequency and/or seasonality of capture, as well as the specific locations they viewed as productive for setting their nets. They were also asked questions related to use and conservation of sea turtles and other marine species so that researchers could better understand the values, both economic and cultural, associated with the species they hunt. The results of such surveys and interviews have been factored in alongside the data gathered through biological research and monitoring in order to provide more complete information on the effectiveness of conservation efforts, while also validating the interview results.

Fishers, including turtle poachers, provided the insight and guidance that led the research team to study sea turtles in Estero Banderitas, an estuary in the northern reach of the Magdalena Bay Complex (Figure 15.3). After hearing from fishers that this area was a particularly productive place to capture turtles, our research team decided to conduct an informal mapping exercise. Fishers who were willing to share their knowledge were given a map of the Magdalena Bay area and asked to mark the locations where they regularly encountered turtles, using different symbols and colours for different species and different times of the year. After reviewing the results of these maps, as well as information gathered through conversation with other local fishers, our research team began mark-recapture studies in Estero Banderitas. A number of fishers have continued to help us track more than thirty turtles over four years. Results of this type of systematic monitoring are validating what the fishers had been telling us initially.

Local fishers from the community have been involved in all aspects of data collection related to biological monitoring and ecological research. Thus they helped identify optimal locations and times to set nets, assisted in captures, measurements and marking, as well as informally monitoring turtle movements while fishing on the bay (Nichols et al., 2000a). By sharing their detailed knowledge about the ecology of the bay, including the seasonal movements of marine species and the daily movement of the currents, fishers have contributed immensely to our work, improving the accuracy of the information collected and providing a more complete picture of the sea turtle's natural history.

The partnerships formed with individual fishers through this collaborative work have been integral to other aspects of our research in the area. Conversations about sea turtle ecology have led to reports of findings in the bay of flipper tags on sea turtles tagged as far away as Japan and southern Mexico. Local members of the research team have initiated sea turtle monitoring at new locations within the bay and have identified areas where incidental catch of turtles may be of concern. In the latter case a sea turtle monitoring project conducted by fishers and community members in Puerto Lopez Mateos has elucidated mortality trends, and potential solutions, of global significance (Koch et al., 2006).

Our research approach seeks to utilize local knowledge and to foster partnerships which facilitate the exchange of information and active community participation. We used a number of steps in our approach to outline general research considerations for the integration of local science into conservation initiatives.

INDIGENOUS AND ARTISANAL FISHERIES

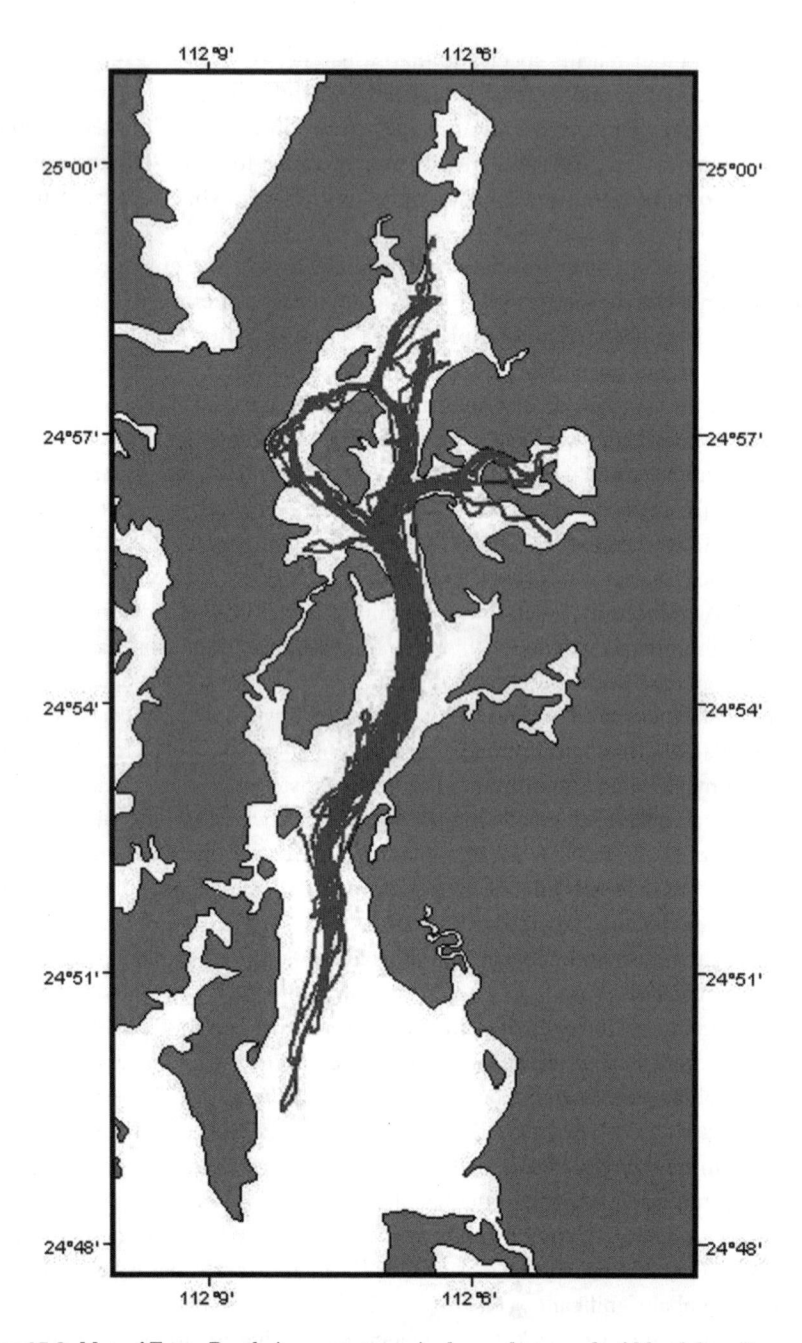

FIGURE 15.3 *Map of Estero Banderitas, an estuary in the northern reach of Magdalena Bay identified by fishers as a productive area for sea turtles. This map shows movements of turtles tracked over several years.*
Source: Adapted from Brooks et al. (in press b).

The first step involved researchers and fishers getting to know each other, an ongoing process now reaching the end of its first decade. We sought to build trust through friendships and partnerships within the local community and to show respect in our interactions with all individuals.

Since our initial visit to the community, we have used informal conversation, surveys and semi-structured interviews to learn about the community issues relevant to local conservation efforts (Delgado et al., in press). We worked within the existing socio-economic framework by paying attention to cultural norms and beliefs, personal needs and politics. For example, visits during the mid-afternoon to a fisher's home were timed in accordance with their fishing schedule. Visits by male researchers to fishers' homes were avoided during times when boats were at sea.

We shared the knowledge we possess with local fishers (specifically when asked) and also sought to learn from the fishers. Both 'outsiders' and 'insiders' were engaged in participant observation, sharing with and learning from each other. For example, when we set out to investigate new areas of the bay for the presence of sea turtles, fisher-led inquiries in the community, focused on the most knowledgeable individuals, helped refine our initial reconnaissance trips. At sea, the research team continuously communicated regarding research goals, adapting them as needed. Methodologies related to net management and other technical aspects were closely guided by local knowledge.

We integrated local knowledge and information with external and local science, and used them to outline general and/or specific action plans that were implemented with the support, knowledge and active participation of the local population. Fishers produced maps of the bay and the distribution of sea turtles based on years of fishing. Public meetings were held to share the results of cooperative research and planning, with findings presented by fishers.

The progress of the conservation initiatives, which include social marketing campaigns to reduce turtle hunting, the establishment of a community-based sea turtle sanctuary (Brooks, Nichols et al., in press b; Marsh et al., 2003), expansion of sea turtle eco-tourism, and a natural history museum, is monitored on an ongoing basis through observation, interviews and surveys and mark-recapture monitoring studies. The initiatives are also adjusted on the basis of adaptive management strategies. For example, residents told the conservation team which communication channels are most resonant; these included local radio, comics, flyers and festivals, rather than newspaper and television (Peckham et al., in press). This has resulted in a series of popular sea turtle comics, an annual sea turtle festival organized by a local committee, announcements using simple informational flyers and regular discussion of the conservation programme on local radio stations (for some examples, visit www.grupotortuguero.org).

OUTCOMES AND LESSONS LEARNED

THROUGHOUT the course of this research, the value of local knowledge has become increasingly obvious. Local technical knowledge, provided by experienced fishers about appropriate gear, fishing locations and timing, helped increase our turtle capture success rate for the research. By taking the time to ask questions and to observe the community, we learned about the values and needs associated with the use of sea turtles and other marine species in this area, and about how to conduct our research in the least obtrusive way. This approach contributed to the development of respect between researchers and community members, supporting productive communication and collaboration among diverse parties.

A number of meetings have been held within various communities in Baja California and Baja California Sur, particularly in the Magdalena Bay area, in order to identify community issues and generate conservation strategies related to sea turtle recovery efforts. Through both formal meetings and impromptu discussions aboard *panga*s (small fishing boats) and in the back of pickup trucks, local fishers and outside researchers have learned from each other and learned how to incorporate local and outside science into their daily activities (Bird and Nichols, 2002).

Over the past several years of involvement, interest in sea turtle conservation has been increasing due to informal education and outreach initiatives, initially implemented by outside researchers from the United States and Mexico. More recently, we have witnessed changing values and attitudes in some of the local fishers who have been involved in the biological research. Some fishers are now taking on their own educational pursuits within the region, leading discussions or simply setting examples by releasing turtles that were accidentally entangled in their nets. Fishers such as Juan Sarrabias, Rodrigo Rangel, Julio Solis and Miguel Lizarraga are leading efforts in their community to protect and restore sea turtle populations. They present results of sea turtle monitoring at regional and international meetings, conduct educational workshops and inform other community members, on a casual basis, of the need to allow sea turtles to recover. The research conducted in the Estero Banderitas reach of Magdalena Bay was informed by fishers' knowledge and has produced convincing data to support what the fishers already knew. This is the best tool we have to promote the region as a sea turtle reserve, an idea that many local fishers are now espousing. In fact, members of the local fishing community and fishing cooperative, hopeful that ecotourism in the area will continue to increase, are actively promoting the establishment of a protected area for sea turtles at Estero Banderitas. This grassroots effort, initiated by involving fishers in research, has put pressure on the state and local officials to comply with their mandate to protect endangered species. Local members of the conservation team frequently report poaching activities, using their own boats to conduct enforcement patrols with federal agents.

Cross-regional communication is important in the development of successful sea turtle conservation initiatives (Trono and Salm, 1999). There is a growing interest in collaboration and the sharing of information reflected, for example, in the recovery of flipper tags placed on sea turtles locally and at distant locations. As word has spread and fishers have become increasingly aware of sea turtle conservation initiatives, flipper tag returns have also increased. An organized network of sea turtle conservation and monitoring has been created spanning the Baja California peninsula from the Pacific coast to the Gulf of California, including both Baja California and Baja California Sur. Through the annual meetings of the Grupo Tortuguero or Sea Turtle Conservation Network – STCN (www.grupotortuguero.org) – started in 1998 and held in Loreto (Baja California Sur, Mexico), several fishing communities have indicated their interest in contributing more towards sea turtle conservation efforts through systematic monitoring (Nichols and Arcas, 1999). Fishers know the general movements and distribution of the turtles. Now, through the coordinated efforts of seven dedicated communities, monthly monitoring enables fishers to attach quantitative weight to their observations (Bird and Nichols, in press). The results of these studies are shared between communities year round, with additional formal reports at the annual Grupo Tortuguero meetings. These meetings have yielded important collaborative results, which are becoming increasingly embedded in regional and local management decisions (Pesenti et al., in press).

RESEARCH IMPLICATIONS AND COMMUNITY BENEFITS

An interdisciplinary approach allows for the utilization of many 'sciences' and provides the basis for a more holistic understanding of the relationship between sea turtles and local communities, which is essential for effective conservation that does not undermine local cultures and practices. The inclusion of local people and their knowledge in the development of conservation initiatives can provide many benefits. Stronger conservation alliances based on the mutual construction and sharing of knowledge, along with the combination of local science and structured monitoring, may produce the greatest conservation benefits. The integration of knowledge generated through quantitative approaches with the qualitative knowledge of local fishers based on daily observations over years, and sometimes generations, can produce an account of turtle behaviours and movements day-by-day throughout the year, as well as the threats to sea turtles. Recognizing that outsiders and locals share the goal of conserving sea turtles, we recognize that all involved have a right to be, and must be, part of the solution.

'Western science' does not have all the answers, nor can it routinely collect all the information required for effective, long-term conservation (Nader, 1996; Johannes and Neis and others, this volume). By looking to local communities to provide the 'missing links' within the data, the time needed to develop the biological and social pieces of the conservation mosaic can be tremendously reduced. Fishers and other members of local host communities will more readily share their intimate knowledge of their environment, including information on the daily movements and distribution of sea turtles, when friendship and trust are fostered through partnerships. Once the value of local fishers' knowledge is recognized, the next step is the active integration of that knowledge into marine conservation planning and management by ensuring that fishers feel empowered to participate. In this way, the fishers are viewed, and view themselves, as an integral part of the conservation team, contributing valuable knowledge and ideas and sharing the benefits: in this case the potential for enhanced eco-tourism opportunities associated with the establishment of a sea turtle sanctuary.

ACKNOWLEDGMENTS

WE thank the communities of Puerto San Carlos, Puerto Magdalena, and Lopez Mateos for their immense contributions to this work. Many thanks to the staff and students at the SFS Centro para Estudios Costeros in Puerto San Carlos, especially Carlos de la Alba, Salvador Garcia-Martinez, Volker Koch, Pamela Kylstra and Rodrigo Rangel for their ongoing support. Also, thanks to Hoyt Peckham, Hector Lozano and two anonymous reviewers for valuable comments and critique on this manuscript. We are particularly grateful for the support shown by all of the individuals who granted interviews and wish to acknowledge the fishing cooperatives of Puerto San Carlos and Punta Abreojos for inviting us into their communities. We also thank members of other local communities from different countries who have shared their knowledge and enriched our collaborations.

REFERENCES

BIRD, K.E. 2002. *Community-based sea turtle conservation in Baja Mexico: Integrating science and culture*. MAIS Thesis. Department of Anthropology. Oregon State University, Corvallis. 128 pp.

BIRD, K.; NICHOLS, W.J. 2002. Community-based research and its application to sea turtle conservation in Bahía Magdalena, Baja California Sur, Mexico. In: A. Mosier; A. Foley; B. Brost (eds.), *Proceedings of the Twentieth Annual Symposium on Sea Turtle Biology and Conservation*. NOAA Technical Memorandum NMFS-SEFSC-477, pp. 339–40.

——. In press. The evolution of community-based conservation action: formation of the Comite Para Protecion de las Tortugas Marinas, Bahía Magdalena, Baja

California Sur, Mexico. In: *Proceedings of the Twenty-first Annual Symposium on Sea Turtle Biology and Conservation.* Philadelphia, Pennsylvania. February 2001. NOAA Technical Memorandum.

BOSTROM, L.; CRAIG, E.; DONLOU, E.; FONG, C.; FORST, C.; GARCIA DE LEON FERRER, A.; SMITH, A. 1999. *Assessing the demand for sea turtles in Bahia Magdalena, Baja California Sur, Mexico.* Puerto San Carlos, Baja California Sur, México, The School for Field Studies Center for Coastal Studies. (Unpublished document.)

BROOKS, L.B.; HARVEY, J. T.; NICHOLS, W. J. In press a. From turtles to the moon: the importance of tides for movements, home range, and marine protected areas. In: *Proceedings of the Twenty-fifth International Symposium on Sea Turtle Biology and Conservation.* Savannah, Georgia. January 2005.

BROOKS, L.B.; NICHOLS, W.J.; KOCH, V.; HERNANDEZ, A. In press b. Preliminary results on the distribution and movement of green turtles (*Chelonia mydas*) in Estero Banderitas, Baja California Sur, Mexico. In: *Proceedings of the Twenty-first Annual Symposium on Sea Turtle Biology and Conservation.* Philadelphia, Pennsylvania. February 2001. NOAA Technical Memorandum.

CALDWELL, D.K. 1963. The sea turtle fishery of Baja California, Mexico. *California Fish and Game,* Vol. 49, pp. 140–51.

CLIFTON, K.; CONREJO, D.O.; FELGER, R.S. 1995. Sea turtles of the Pacific coast of Mexico. In: K.A. Bjorndal (ed.), *Biology and conservation of sea turtles. Revised edition.* Washington, Smithsonian Institution Press, pp.199–209.

DEDINA, S. 2000. *Saving the gray whale: People, politics and conservation in Baja California.* Tucson, University of Arizona Press, 186 pp.

DELGADO, S.; LAUDINO SANTILLÁN, J.; OCHOA DÍAZ, R.; RANGEL ACEVEDO, R.; MONTAÑO MEDRANO, B.; MALDONADO, D.; PECKHAM, S.H.; NICHOLS, W.J. In press. Local perceptions and ocean conservation: A study of human consumption, exploitation, and conservation of endangered sea turtles in Baja California Sur, Mexico. In: *Proceedings of the Twenty-fifth International Symposium on Sea Turtle Biology and Conservation.* Savannah, Georgia. January 2005.

FRAZIER, J.G. 1995. Subsistence hunting in the Indian Ocean. In: K.A. Bjorndal (ed.), *Biology and conservation of sea turtles. Revised edition.* Washington, Smithsonian Institution Press, pp. 391–6.

——. 1999. Community-based conservation. In: K.L. Eckert; K.A. Bjorndal, F.A. Abreu-Grobois; M. Donnelly (eds.), *Research and management techniques for the conservation of sea turtles.* IUCN/SSC Marine Turtle Specialist Group Publication No.4, pp. 15–18.

——. In press. Science, conservation and sea turtles: What is the connection? *Proceedings of the Twenty-first Annual Symposium on Sea Turtle Biology and Conservation.* Philadelphia, Pennsylvania. February 2001. NOAA Technical Memorandum.

GARCIA-MARTINEZ, S.; NICHOLS, W.J. 2000. Sea turtles of Bahia Magdalena, Baja California Sur, Mexico: Demand and supply of an endangered species. In: *Proceedings of the Tenth Biennial Conference of the International Institute of Fisheries Economics and Trade*. Oregon State University, Corvallis.

GARDNER, S.; NICHOLS, W.J. 2001. Assessment of sea turtle mortality rates in the Bahía Magdalena Region, B.C.S., Mexico. *Chelonian Conservation and Biology*, Vol. 4, No. 1, pp. 197–9

GOVAN, H. 1998. Community turtle conservation at Rio Oro on the Pacific coast of Costa Rica. *Marine Turtle Newsletter*, Vol. 80, No. 10, p. 11.

HACKEL, J.D. 1999. Community conservation and the future of Africa's wildlife. *Conservation Biology*, Vol. 13, No. 4, pp. 726–34.

HILBERT, S.C.; GARDNER, S.C.; RIOSMENA RODRIGUEZ, R.; NICHOLS, W.J. In press. Diet composition of east pacific green turtles (*Chelonia mydas*) in Bahia Magdalena, Baja California Sur, Mexico. In: *Proceedings of the Twenty-first Annual Symposium on Sea Turtle Biology and Conservation*. Philadelphia, Pennsylvania. February 2001. NOAA Technical Memorandum.

KING, F.W. 1995. Historical review of the decline of the green turtle and the hawksbill. In: K.A. Bjorndal (ed.), *Biology and conservation of sea turtles. Revised edition*. Washington, Smithsonian Institution Press, pp. 183–8.

KOCH, V.; NICHOLS, W.J.; PECKHAM, H.; DE LA TOBA, V. 2006. Fisheries mortality and poaching of endangered sea turtles in Bahía Magdalena, Mexico. *Biological Conservation*. Vol. 128, No. 3, pp. 327–34.

KOWARSKY, J. 1995. Subsistence hunting of turtles in Australia. In: K.A. Bjorndal (ed.), *Biology and conservation of sea turtles. Revised edition*. Washington, Smithsonian Institution Press, pp. 305–13.

MARSH, J.C.; NICHOLS, W.J.; GARCIA-MARTINEZ, S.; PALACIOS-CASTRO, E.; BIRD, K.E. 2003. Community-based coastal resource management: success factors and the sea turtle reserve in Estero Banderitas, BCS, Mexico. In: J.A. Seminoff (ed.), *Proceedings of the Twenty-Second Annual Symposium on Sea Turtle Biology and Conservation*. Miami, Florida, NOAA Technical Memorandum NMFS-SEFSC-503, pp. 102–3.

NABHAN, G.; GOVAN, H.; ECKERT, S.A.; SEMINOFF, J.A. 1999. Sea turtle workshop for the indigenous Seri tribe. *Marine Turtle Newsletter*, Vol. 86, p. 44.

NADER, L. 1996. *Naked science: Anthropological inquiry into boundaries, power and knowledge*. New York, Routlege.

NICHOLS, W.J. 2003. *Biology and conservation of the sea turtles of the Baja California peninsula, Mexico*. PhD Dissertation. Tucson, Department of Wildlife and Fisheries Science, University of Arizona.

NICHOLS, W.J.; ARCAS, F. 1999. First meeting of the Baja California Sea Turtle Group held in Loreto, Mexico. *Marine Turtle Newsletter*, Vol. 85, p. 19.

NICHOLS, W.J.; BIRD, K.E.; GARCIA, S. 2000a. Community-based research and its application to sea turtle conservation in Bahía Magdalena, BCS, Mexico. *Marine Turtle Newsletter,* Vol. 89, pp. 4–7.

NICHOLS, W.J.; BROOKS, L.; LOPEZ, M.; SEMINOFF, J.A. 2001. Record of pelagic east Pacific green turtles associated with *Macrocystis* mats near Baja California Sur, Mexico. *Marine Turtle Newsletter,* Vol. 93, pp. 10–11.

NICHOLS, W.J.; RESENDIZ, A.; SEMINOFF, J.A.; RESENDIZ, B. 2000b. Transpacific migration of the loggerhead turtle monitored by satellite telemetry. *Bulletin of Marine Science,* Vol. 67, pp. 937–47.

NICHOLS, W.J.; SEMINOFF, J.A.; RESENDIZ, A.; DUTTON, P.H.; ABREU, A. 1998. Using molecular genetics and biotelemetry to study life history and long distance movement: a tale of two turtles. In: S.P. Epperly and J. Braun (eds.), *Proceedings of the Seventeenth Annual Symposium on Sea Turtle Biology and Conservation.* NOAA Technical Memorandum NMFS-SEFSC-415.

NIETSCHMANN, B. 1995. The cultural context of sea turtle subsistence hunting in the Caribbean and problems caused by commercial exploitation. In: K.A. Bjorndal (ed.), *Biology and conservation of sea turtles. Revised edition:* Washington, Smithsonian Institution Press, pp. 439–45.

PARSONS, J. 1962. *The green turtle and man.* Gainesville, Florida, University of Florida Press, 126 pp.

PECKHAM, S.H.; W.J. NICHOLS. 2003. Why did the turtle cross the ocean? Pelagic red crabs and loggerhead turtles along the Baja California coast. In: J.A. Seminoff (ed.), *Proceedings of the Twenty-Second Annual Symposium on Sea Turtle Biology and Conservation.* NOAA Technical Memorandum NMFS-SEFSC-503, pp. 47–48.

PECKHAM, S.H.; D. ASH, J. LAUDINO-SANTILLÁN; W.J. NICHOLS. In press. Necessary antecedents made plain: Empowering *costeños* to conserve their marine resources through targeted outreach initiatives. 2004 CfAO Maui High Tech Industry Education Exchange, May 19, 2004.

PESENTI, C.; W.J. NICHOLS; R. RANGÉL-ACEVEDO; J. LAUDINO-SANTILLÁN; B. MONTAÑO MEDRANO; M.C. LÓPEZ CASTRO; S.H. PECKHAM. In press. Grupo Tortuguero: open networks as models for conservation. In: *Proceedings of the Twenty-fifth International Symposium on Sea Turtle Biology and Conservation.* Savannah, Georgia, January 2005.

TAMBIAH, C.R. 1989. Status and conservation of sea turtles in Sri Lanka. In: K.L. Eckert; S.A. Eckert; T.H. Richardson (eds.), *Proceedings of the 9th Annual Workshop on Sea Turtle Conservation and Biology.* NOAA Technical Memorandum NMFS-SEFSC-232.

———. 1995. Integrated management of sea turtles among the indigenous people of Guyana: planning beyond recovery and towards sustainability. In: *Proceedings of the Twelth Annual Workshop on Sea Turtle Conservation and Biology.* NOAA Technical Memorandum. NMFS-SEFSC-361.

——. 1999. Interviews and market surveys. In: K.L. Eckert, K.A. Bjorndal, F.A. Abreu-Grobois and M. Donnelly (eds.), *Research and management techniques for the conservation of sea turtles.* IUCN/SSC Marine Turtle Specialist Group. Publication No.4.

——. 2000. Community participation in sea turtle conservation: Moving beyond buzzwords to implementation. In: H. Kalb and T. Wibbels (eds.), *Proceedings of the Nineteenth Annual Symposium on Sea Turtle Conservation and Biology.* NOAA Technical Memorandum NMFS-SEFSC-443, pp. 77–9.

THOMPSON, D.A.; FINDLEY, L.T.; KERSTITCH, A.N. 2000. *Reef fishes of the sea of Cortez: The rocky-shore fishes of the Gulf of California. Revised Edition.* Austin, University of Texas Press, 407 pp.

TRONO, R.B.; SALM, R.V. 1999. Regional collaboration. In: K.L. Eckert; K.A. Bjorndal, F.A. Abreu-Grobois; M. Donnelly (eds.), *Research and management techniques for the conservation of sea turtles.* IUCN/SSC Marine Turtle Specialist Group, Publication No. 4, pp. 224–7.

For more information and on-going progress on sea turtle conservation activities discussed in this case study see web-site <www.grupotortuguero.org> or <www.wildcoast.net> (Last accessed 17 May 2006).

CHAPTER 16 Can historical names and fishers' knowledge help to reconstruct the distribution of fish populations in lakes?

Johan Spens

ABSTRACT

RECONSTRUCTING the historical distribution of local brown trout populations is of great importance. Information about what has actually been lost and why, is necessary for rebuilding natural lake ecosystems and recreational fisheries, as well as for monitoring future changes. I interviewed older fishers and local fishing-right owners in sixty-three private fishery management organizations (FMOs) in northern Sweden, focusing on current species distribution, stocking, introductions and extinctions in 1,509 lakes. Names were collected for each lake from modern and historical maps. Historical archival information about fish species' distribution and stocking was also compiled. Brown trout lake candidates were surveyed with multi-mesh-sized gillnets or other methods. Chemical, physical and biological anthropogenic impacts were assessed by means of archival data and limnological surveys. Information was obtained from a number of sources and methods, which allowed for comprehensive validation of lake name evidence and interviews. A third of all lakes with historical or present brown trout populations had *Rö* or other dialectal terms commonly used for brown trout included in their names. In these '*Rö*-lakes', there was at least a 92 per cent chance of finding a historical or present brown trout population, compared with 11 per cent when lakes were randomly chosen. Data suggest that the distribution of brown trout lakes under pre-industrial natural conditions was stable until the 1930s when extinctions became evident. Lake names were shown to be strongly associated with details regarding the fish fauna as well as the habitat. Historical names, fishers' knowledge and documentary evidence combined with limnological data proved powerful in revealing the past.

INTRODUCTION

MOST marine and freshwater ecosystems around the world are being degraded, and fish species pushed towards extinction (Moyle and Leidy, 1992; Maitland, 1995; Pitcher, 2001). European inland waters are subjected to chemical, physical and biological anthropogenic disturbances that are leading to extinction of local fish populations (Lelek, 1987; Maitland and Lyle, 1991; Bulger et al., 1993; Crivelli and Maitland, 1995). Knowledge about such basic questions as which populations have survived and which have been lost is fundamental for practical conservation and management.

Anthropogenic impacts are eradicating or reducing brown trout (*Salmo trutta* L.) populations all over the species range (Laikre et al., 1999). Reconstructing the historical distribution of local brown trout populations is of great importance, because information about what has actually been lost is essential for rebuilding natural lake ecosystems and recreational fisheries, as well as for monitoring future changes. Prior to this study, no scientific investigations had addressed the problem of reconstructing the historic distribution of local brown trout populations in any country in Europe, or assessing what has actually been lost to date or the extinction rate. Laikre et al. (1999) strongly recommended that such studies of local brown trout populations be carried out both on a national and international level. Empirical studies that include the historical dimension are needed to provide insight into conservation and management on a wider scale.

Spatial dimension

THOUSANDS of lakes, covering large areas, may need to be surveyed to achieve a wider landscape-scale study of fish species presence or absence. Conventional scientific methods involving use of multi-mesh-sized gillnets (Appelberg, 2000) by skilled personnel would be too time consuming, labour intensive and costly if every lake is to be sampled. Larger-scale studies can be conducted with less effort by gathering local fishers' knowledge through interviews which, if properly validated, can produce valuable data (Hesthagen et al., 1993).

Historical dimension

IN the absence of palaeontological methods, the sources of information on historical distribution of species are limited to interviews and rare, fragmented archival records, where they exist. With first-hand interviews it might be possible in some cases to extend our perspective eighty or so years back in time, and perhaps even longer with the aid of some rare archival data. A few studies have suggested that maps showing place names that may be many hundreds of years old can be useful historical sources of information on different species occurrence and habitat. Place-name evidence for the former distribution of beaver, wolf, crane and pine-marten has been presented in three studies in Britain (Aybes and Yalden, 1995; Boisseau and Yalden, 1998; Webster, 2001). Wallace (1998) mentions the use of mapped place names as historical sources indicating the past occurrence of halibut, sturgeon and whale. The feasibility of using place names as indicators of original landscapes has been tested and verified in a recent study (Sousa and Garcia-Murillo, 2001). Lake names with species terms are potentially valuable historical records of fishers' knowledge that can take us back to pre-industrial periods. As such, they may be among the few pre-industrial sources of information about fish species for many lakes.

In the present study, I aim to show that historical lake names from maps can be useful indicators of past and present fish distribution if properly validated. To my knowledge, this is the first published scientific attempt to employ lake names in investigating fish species distribution. The main objective is to demonstrate how fishers' knowledge gathered from interviews and historical fishers' knowledge stored in maps and archives, together with limnological surveys, can be used to elucidate the past and present distribution of fish species. This is illustrated by identifying brown trout lakes among 1,509 lakes in northern Sweden. I tested the following hypothesis: the proportion of 'historical brown trout term' lakes (or *Rö*-lakes) with/without brown trout populations is the same as for other lakes with/without such populations. Making use of fishers' knowledge, it is intended that results from this study will serve as a template for ecosystem reconstruction as well as helping management develop policies and actions to prevent present populations becoming extinct.

MATERIAL AND METHODS

Study area

The present study focused on one geographic region rather than selecting a random sample of lakes. This strategy was designed to increase the likelihood of detecting phenomena in local dialects relevant to the distribution of brown trout. The study area, with its centre situated near 63°32′N 18°12′E, extended over roughly a third of northern Västernorrland and parts of Västerbotten in the northern boreal region of Sweden (Figure 16.1). The investigation covered 1,509 lakes and was delimited to the lake watersheds that extend to over 700,000 hectares. The region is sparsely populated, averaging eight inhabitants per km^2, primarily concentrated in a few population centres. A majority of the lakes belong to sixty-three privately owned fishery management organizations (FMOs). These FMOs are associations of private and company landowners that sell licences to the public and manage the waters, as well as providing information about the fisheries (Figure 16.2).

Methods

I conducted face-to-face, in-depth interviews with older fishers and 250 local fishing-right owners in FMOs between 1985 and 2001, focusing on current spe-

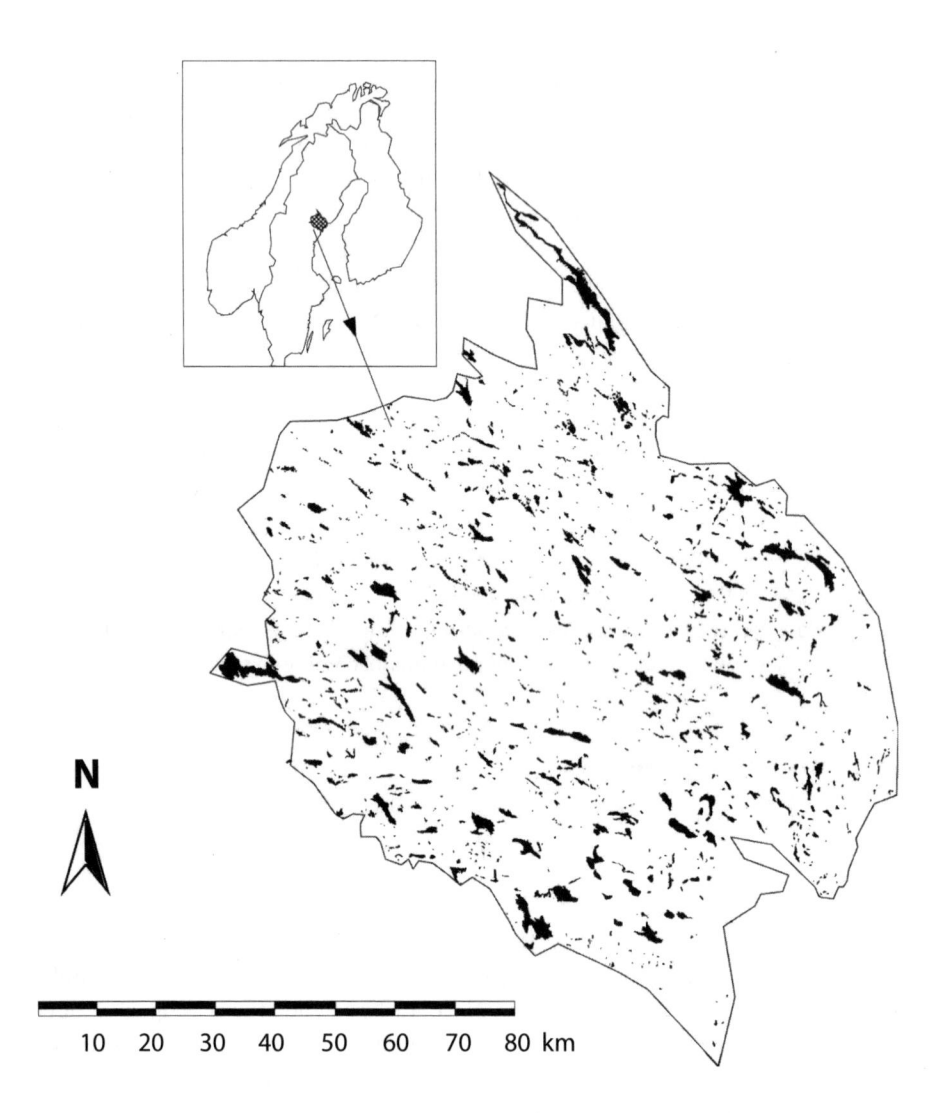

FIGURE 16.1 *The 1,509 lakes within the study area, the northern boreal region of Sweden.*
Note: Upper box illustrates position of study area within Sweden and Nordic countries.

cies distribution, stocking, introductions and extinctions in all lakes. I also collected similar data from local fishers in remaining areas not organized by FMOs. Interviews generally began in a structured manner with specific questions about key issues such as fish species distribution, spawning areas and stocking. A less structured, more in-depth part of the interview explored the informants' general knowledge and elicited additional contacts who could provide knowledge about specific areas,

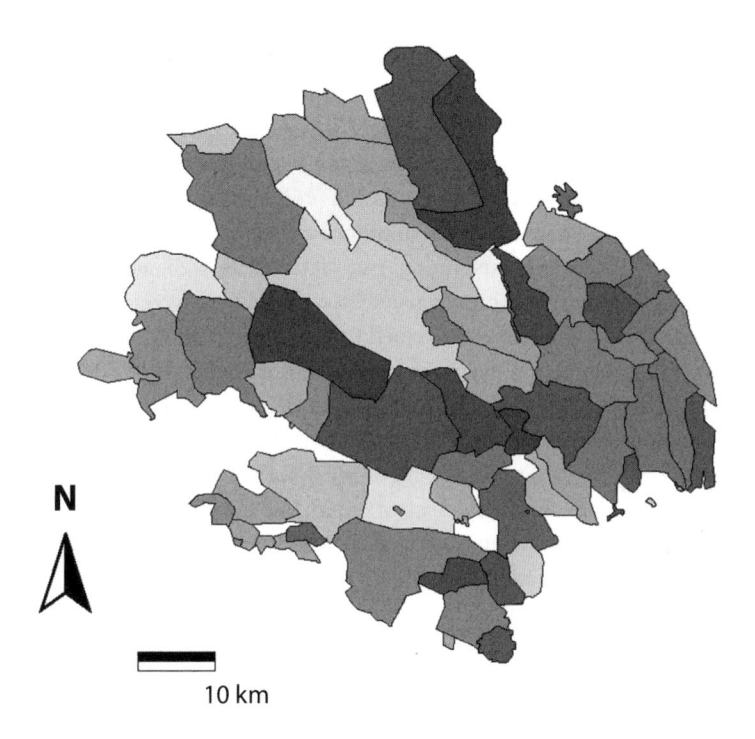

FIGURE 16.2 *Coverage of sixty-three fishery management organizations (FMOs) within the study area. Shading indicates area boundaries, 10 km scale bar.*

fish species or historic events concerning the fisheries. In return, fishers were given information on management and conservation, contributing to a comprehensive exchange of information about the water bodies being studied. Formal meetings were held indoors, often with the aid of maps for proper orientation and to avoid any mix up of lakes. In most FMOs, additional field meetings were combined with observations of essential features of their waters. Relationships were established with most interviewees, leading to additional contact over the years. Data were sought from at least two primary sources that confirmed each other when evaluating fish species presence/absence records from interviews. I collected discrete presence/absence data less prone to impacts of ordinary natural sweeping cyclic environmental change so as to make it possible to compare data that were collected using different methods and sources, as well as to avoid subjective personal opinion. I also investigated archived audio recordings and written records in local dialects of fishers born in the nineteenth century from the region of interest, dealing with fish species. Scientific papers, encyclopaedias and archives with dialectology, onomastics and folklore research in Scandinavian languages were explored, focusing on lake names and historical brown trout names.

Historical documents concerning fish species distribution and stocking between 1872 and 2000 were collected from three major forest companies, county and municipality administrations, FMOs, the National Board of Fisheries and other sources. Approximately nine months were spent in archival research work, collecting hard-to-access fisheries-related information concerning the waters of interest. I evaluated stocking data in concert with other investigations to discriminate between native and introduced self-sustaining populations as well as non-reproducing populations. The majority of brown trout lake candidates were inventory sampled with multi-mesh-sized gillnets as described by Appelberg (2000) or with a somewhat modified stratification. A few were surveyed with other methods such as trapping, rod or single-pass electrofishing with a (LUGAB Inc.) backpack unit in the inlets and outlets. A population was considered extinct when sampling efforts using 0.5–2 multi-mesh-sized gillnets per hectare/night plus electrofishing in potential spawning areas did not generate any fish.[1] I also tested the classification of each lake for consistency with limnological survey data and interviews.

Lake tributaries and outlets were classified as suitable for brown trout spawning and early growth on the basis of the stream size, calculated from hydrological data and field studies. Visual qualitative observations of bottom substrate confirmed or ruled out the existence of proper habitat conditions for spawning of salmonids, determining the capability of lakes to hold self-sustaining populations of brown trout. In the current study, waters were considered to lack a suitable spawning substrate for brown trout if the bottom material consisted entirely of sand or organic fine material (< 1 mm). The presence of a suitable spawning substrate was confirmed if gravel, pebble or cobble-sized particles (Bain, 1999) could be found in patches of a minimum length that varied according to the particle size. (See Witzel and MacCrimmon, 1983, and Crisp, 1996, for formulas on critical minimum sizes of spawning substrate.)

Natural fish migration barriers up and downstream from brown trout lakes were identified, thus showing the possibility of access to spawning grounds as well as the progeny's ability to return to the lake. I assessed chemical, physical and biological anthropogenic impact by archival data from 1925 to 2000, and by limnological surveys for 1985–2001. Names were collected for each lake from 1:50,000 topographic maps (The Swedish National Land Survey, 1961–1967). Additional names from county, parish, ordinance or village maps (The Swedish National Land Survey, 1672–1908) were also collected. The production date of each map provided a minimum age for every lake name. All data were temporally as well as geographically referenced and stored in a GIS-linked database referred to as the LIMNOR database. With modern tools like GIS systems and database software, it was possible to store and access large amounts of information and achieve a

---

1. In pike-invaded lakes, extinction classification did not consistently include electrofishing.

wider view of both space and time. Having access to a number of sources and methods on species presence and absence, such as fishers' knowledge, archival data, historical names from maps and limnological surveys, allowed validation of data concerning each lake. The hypothesis that the proportion of 'historical brown trout term' lakes (or *Rö*-lakes) with/without brown trout populations is the same as for other lakes with/without such populations if any lake is randomly chosen, was tested with Pearson Chi-square ($p < 0.001$).

I provided feedback to FMOs on the preliminary results generated in this study in an effort to make use of the knowledge gained, to help management, and in some cases to initiate lake restoration.

Quality control of presence/absence

I utilized face-to-face in-depth interviews that gave an understanding of the informants' area of knowledge, and allowed for collection of data that matched their expertise in order to generate more reliable data. In *Rö*-named lakes, presence/absence data from interviews were validated with the combined data from test-fishing results, stocking records and other archival data as well as habitat surveys. In this respect, interviewees succeeded in identifying all lakes with past and present self-sustaining brown trout populations, although two extinct populations were reported to be still present in these lakes. Archival data confirmed the interview results except for two cases where non-brown trout lakes had been stocked with this species, and a brown trout lake that was identified as a single-species perch lake. Further validations were made to verify the informants' ability to identify non-brown trout lakes. An additional sixty lakes described in interviews as non-brown trout waters were confirmed to be free of brown trout by multi-mesh-sized gillnet surveys. One possible brown trout-term lake was not classified as present or extinct in this study because of insufficient data and was excluded from all the results and evaluations.

RESULTS

Fishers' knowledge gathered from interviews and historical documents suggested several hundred brown trout lake candidates among the 1,509 lakes in the study area. Some lakes were eliminated when surveys found no suitable brown trout habitat, for example where there was no spawning substrate. Stocking data and other investigations revealed a number of introduced, self-sustaining populations as well as non-reproducing populations that were totally dependent on hatcheries. These translocated brown trout populations were also excluded from further evaluation. Finally, multi-mesh-sized gillnets and other methods verified that 162 lakes – the majority of the remaining brown trout lake candidates – represented past or present self-sustaining local brown trout populations. If a lake was randomly chosen in this area, there was an 11 per cent chance of it being a brown trout lake (Figure 16.3). In addition to the *Rö*-named lakes considered in this chapter, the entire set of lakes will be reported on elsewhere.

My interviews with an elderly fisher revealed an old oral traditional term for brown trout – *Rö* – which is not a recognized term for this species in modern language but is a common prefix of lake names in modern and historical maps. Furthermore, several records relating to the name form *Röa* in local dialect were found in archives. The following excerpts are from part of the interviews conducted in Norrland around forty or fifty years ago, freely translated: 'Röding, i.e. brown trout, we call it rödingen' (Dahlstedt, 1956). *'Röa* is a large kind of brown trout with red meat, not Arctic charr (*Salvelinus alpinus*)' (Dahlstedt, 1961). The term had also been dealt with in onomastic papers that referred to this geographic area; for example, the *'Rö*-lake is characterized by its richness in *röa*, i.e. brown trout' (Edlund, 1975). However the linkage of the term *Rö* to brown trout was not known among other fishers interviewed in the study area.

Of all lakes with historic or present brown trout populations, 29 per cent had *Rö* as part of their name. In a further 4 per cent of brown trout lakes on the outskirts of the study area, the names included two other vernacular terms commonly used for brown trout (Figure 16.3). Hence, at least one-third of all brown trout lakes in the study area had been named after brown trout. When sampling *Rö*-lakes, there was a minimum 92 per cent chance of finding a historic or present brown trout population. The hypothesis that the number of *Rö*-lakes with/without this species is proportionate to the number of brown trout populations if any lake is randomly chosen was rejected by a Pearson Chi-square test ($x^2 = 365.2$; $p < 0.001$). Thus it can be concluded that *Rö*-lakes were associated with historic or present self-sustainable brown trout populations. The *Rö*-name indicated that natural good habitat conditions for this species could be found in these lakes (Table 16.1):

- Ninety-six per cent had outlet or inlet streams of sufficient size for brown trout spawning and early growth.
- Ninety-six per cent had outlet or inlet streams with proper spawning substrate.
- Ninety-six per cent had no natural barriers to potential spawning areas.
- Ninety-six per cent were free of indigenous severe brown trout predators such as northern pike (*Esox lucius*).
- One hundred per cent were isolated by natural barriers from several fish species downstream.

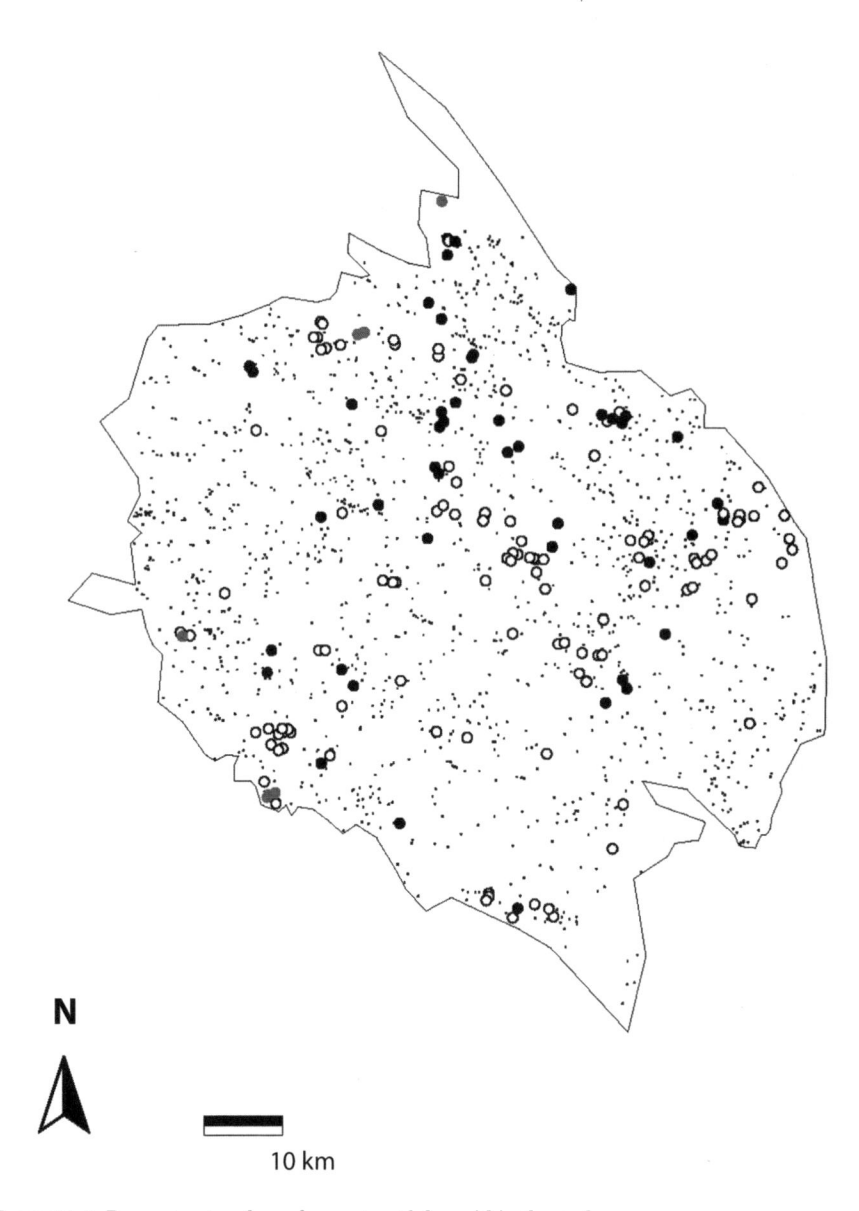

FIGURE 16.3 *Brown trout and non-brown trout lakes within the study area.*
Filled circles = brown trout lakes with *Rö*-names (n = 47).
Grey circles = brown trout lakes with common brown trout names (n = 7).
Open circles = brown trout lakes with other names (n = 108).
Small black dots = Non-brown trout lakes.

INDIGENOUS AND ARTISANAL FISHERIES

TABLE 16.1 Rö-named lakes and methods elucidating past and present brown trout populations.

Lake names	Lat° Long° (WGS 84)	Self-sustaining brown trout population	Sufficient spawning stream size and substrate	Earliest records (A.D.) Rö-names	Archive	Interviews[5]	Test-fishing[8]
Hattsjö-Röjdtjärnen	63°37′02′′N 18°59′15′′E	Present	Yes	1852	1961	1930	2001
Hemling.-Rödtjärnen	63°37′20′′N18°29′55′′E	Extinct 1930s	Yes	1766		1930[7]	1995[9]
Inner-Rötjärnen	63°25′05′′N 18°40′13′′E	Present	Yes	1799		1950	2001
Inre Rödingträsksjön	63°59′33′′N 18°12′14′′E	Present	Yes	1792		1980	1999
Lill-Rödtjärnen	63°42′55′′N 18°21′38′′E	Present	Yes	1837		1930	2000
Lill-Rödtjärnen	63°45′30′′N 18°38′29′′E	Present	Yes	1837		1920	1998
Lill-Rödtjärnen	63°14′46′′N 17°59′18′′E	Never existed	No	1824		1950[6]	
Lill-Rödvattenssjön	63°50′15′′N 17°36′04′′E	Present	Yes	1758[3]	1958	1930	2001
Lill-Rödvattnet	63°46′13′′N 18°10′10′′E	Present	Yes	1865	1955	1940	2000
Lill-Rötjern[1]	63°45′12′′N 18°40′23′′E	Present	Yes	1864		1920	2000
Lill-Rötjärnen	63°45′31′′N 18°41′50′′E	Present	Yes	1961		1920	2001
Norra Rötjärn[1]	63°54′40′′N 18°08′33′′E	Present	Yes	1837		1930	1998
Rödingtjärnen	63°55′14′′N 18°33′51′′E	Present	Yes	1856		1940	2001
Rödtjärnarna	63°50′21′′N 18°15′56′′E	Present	Yes	1961		1930	2000
Rödtjärnarna	63°50′34′′N 18°16′12′′E	Present	Yes	1961		1930	2000
Rödtjärnen	63°38′30′′N17°48′12′′E	Extinct 1930s	Yes	1968		1920	2000[9]
Rödtjärnen	63°39′16′′N 17°58′22′′E	Extinct 1920s	Yes	1961		1920[7]	2000[9]
Rödtjärnen	63°36′35′′N 18°06′54′′E	Present	Yes	1707	1930	1970	2001
Rödtjärnen	63°42′01′′N 18°08′39′′E	Present	Yes	1830		1950	1995
Rödtjärnen	63°35′33′′N 18°28′48′′E	Extinct 1990s	Yes	1766	1940	1930	2000[9]
Rödtjärnen	63°29′52′′N 18°37′15′′E	Present	Yes	1961		1950	1990
Rödtjärnen	63°26′27′′N 18°08′59′′E	Present	Yes	1961		1940	1999
Rödtjärnen	63°26′43′′N 17°51′09′′E	Present	Yes	1680	1958	1940	1990
Rödtjärnen	63°45′00′′N17°31′36′′E	Never existed	No	1820		1940[6]	
Rödtjärnen	63°07′54′′N18°20′21′′E	Extinct 1950s	Yes	1762		1940[7]	2001[9]
Rödtjärnen	63°25′24′′N 17°53′08′′E	Present	Yes	1776		1920	1999
Rödtjärnen	63°45′25′′N 18°20′19′′E	Present	Yes	1837		1930	2000
Rödtjärnen	63°19′34′′N 17°46′58′′E	Extinct 1970s	Yes	1755	1961	1940	2000[9]
Rödtjärnen	63°58′16′′N 18°12′09′′E	Present	Yes	1886		1930	2001
Rödvattensjön	63°47′05′′N 17°54′22′′E	Extinct 1980s	Yes	1752	1967	1930	2000[9]

CAN HISTORICAL NAMES AND FISHERS' KNOWLEDGE HELP TO RECONSTRUCT THE DISTRIBUTION OF FISH POPULATIONS IN LAKES?

Rödvattnet	63°28'23''N 17°38'47''E	Extinct 1990s	Yes	1856	1943	1990	2000[9]
Röftierna[1]	63°28'30''N 18°48'06''E	Present	Yes	1711		1970	1998
Röjdtjärnen	63°35'58''N 18°53'37''E	Present	Yes	1774		1960	2000
Röjtjärnen	63°43'36''N 18°51'50''E	Present	Yes	1790		1990	1998
Rörsjötjärnen	63°45'07''N 18°09'45''E	Reintroduced	Yes	1825	1959	1940	2000
Rötenburstjerna[1]	63°23'24''N 18°37'07''E	Extinct 1950s	Yes	1676	1953	1940[7]	2001[9]
Rötjern[1]	63°44'50''N 18°41'57''E	Present	Yes	1864		1920	2000
Rötjärnen	63°26'39''N 17°37'53''E	Extinct 1970s	Yes	1804	1951	1930	2001[9]
Rötjärnen	63°34'03''N 18°45'53''E	Present	Yes	1705		1940	1999
Rötjärnen	63°53'24''N 18°10'42''E	Present	Yes	1837		1930	2000
Rötjärnen	63°21'13''N 19°04'57''E	1930s[2]	Yes	1902	1958[4]	1930[6]	2000[9]
Stor-Rödtjärnen	63°43'22''N 18°23'32''E	Present	Yes	1844		1930	2001
Stor-Rödtjärnen	63°14'44''N 18°00'22''E	Present	Yes	1672	1958	1930[7]	2001
Stor-Rödvattenssjön	63°49'49''N 17°36'42''E	Present	Yes	1758[3]	1937	1930	2000
Stor-Rödvattnet	63°46'54''N 18°12'45''E	Present	Yes	1865	1955	1940	2000
Stor-Röjdtjärnen	63°38'19''N 18°58'28''E	Extinct 1960s	Yes	1901	1940	1960	2000[9]
Stor-Rötjärnen	63°45'21''N 18°42'40''E	Present	Yes	1837	1960[4]	1920	1997
Södra Rötjern[1]	63°41'31''N 18°09'21''E	Present	Yes	1830		1970	2001
Västergiss.-Rötjärnen	63°33'35''N 18°47'25''E	1920s[2]	Yes	1901		1920[6]	1996[9]
Ytter-Rötjärnen	63°24'25''N 18°40'53''E	Present	Yes	1799		1940	1998
Yttre Rödingträsksjön	63°59'22''N 18°13'36''E	Present	Yes	1792		1980	1999

1. Modern maps list different name.
2. Possibly extinct before indicated decade or no population ever existed.
3. reproductive area called Rö-.
4. historical document does not mention brown trout.
5. Fishers' earliest recollection of brown trout population (decade).
6. brown trout not found from listed decade until present.
7. several concordant 2nd hand sources.
8. brown trout population sampled year.
9. no brown trout caught.

Temporal perspective

I found all types of lake names to be 'evolutionarily' conservative, and most remained virtually unchanged through the centuries. A few *Rö*-lakes, however, had been renamed with terms unrelated to brown trout. Many older fishers used an older form of pronunciation not found in modern maps, thus providing evidence of names being passed on in a conservative oral tradition. Detailed maps over 100 years old were scarce and did not cover all the heart of the study region; they were generally too coarse to include the small lakes discussed here. Even so, forty-four *Rö*-names were found dating back between 100 and 330 years, most to pre-industrial times (Table 16.1 and Figure 16.4). It was also assumed that the remaining seven smaller *Rö*-lakes only found on maps produced in the 1960s were initially named more than a hundred years ago. This is because the historical *Rö*-term nearly vanished as a species word during the nineteenth century, and because the smaller size of these lakes could explain their absence from the coarse and simple maps produced in this area more than 100 years ago. Archival sources referring to brown trout presence in lakes were found dating back 129 years. First-hand interviews had a maximum scope of eighty years back in time with a median of fifty-six years.

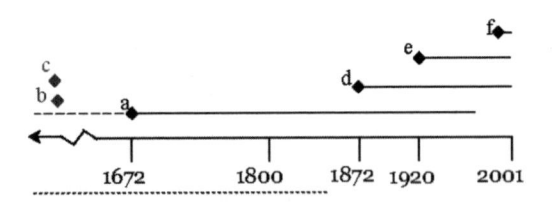

Figure 16.4 *Scale bar (A.D.) illustrating temporal range of methods to reconstruct brown trout distribution in lakes within the current study.*

a. Lakes names.
b. Palaeontology: Lack of fish fossil evidence makes reconstruction impossible for individual lakes.
c. Models are not yet developed for reconstruction of fish fauna.
d. Archival data.
e. First hand Interviews.
f. Field Surveys.
Black dotted line = pre-industrial times.

ANTHROPOGENIC PERMANENT EXTINCTIONS (1920s–1990s)

Interviews identified ten of the *Rö*-lakes with brown trout as having lost their populations during the last eighty years (Figure 16.5). Archival data confirm that the majority of these were historical brown trout lakes, and two independent test-fishing results showed that the lakes no longer harboured this species. Two additional recent extinctions were discovered by test fishing, giving a total of twelve (25.5 per cent) lost in eight decades. The average anthropogenic extinction rate during this time was estimated to exceed 3 per cent per decade. Insight into possible explanations for these eradications was gained by limnological surveys and from archival data.

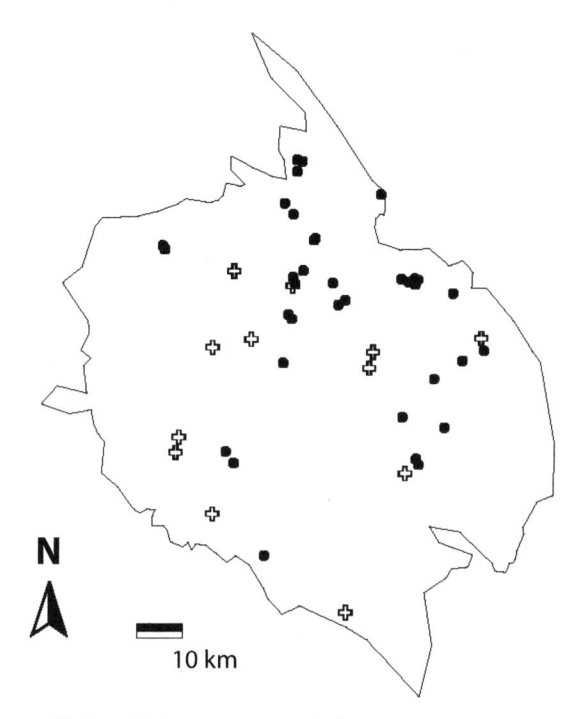

FIGURE 16.5 Rö-*named lakes with brown trout populations.*
Filled circles = brown trout populations present.
Crosses = brown trout populations extinct.

TABLE 16.2 *Estimation of maximum (E_{MAX}) permanent extinctions 1672–1920, from lake names (before the scope of possible detection by interviews and historical documents).*

Self-sustainable brown trout populations	Number of lakes
Present [a]	35
Extinct [a]	12
Possibly extinct or never existed [P]	2
Max. number of *Rö* brown trout lakes [M]	49
Never existed (impossible habitat)	2
Total number of Rö-lakes	51
Non-*Rö* brown trout lakes [a]	115
Total number of brown trout lakes [a]	162
E_{MAX} (1672-1930) [P/M] (4.2%)	2/49
E_{MAX} (1672-1930) Estimated No. of pop.	7
Max. brown trout lakes (1672-1920)	169

a = Brown trout confirmed 1920–2001.

All *Rö*-lakes where brown trout populations were classified as extinct had experienced major anthropogenic impacts, which in many cases were decisive for the survival of populations (Table 16.3). Such anthropogenic impacts were not observed in any other *Rö*-lake with brown trout present (except for brook charr (*Salvelinus fontinalis*) at a few spawning areas), strengthening the suggestion from interview and archival data that affected lakes once possessed self-sustaining populations. The lakes (n = 12) that had lost populations were more stricken by anthropogenic impact than lakes (n = 35) where populations still existed: Fisher's exact test (Systat 10.2) (anthropogenic impact p < 0.001), (brook charr in spawning areas p < 0.05). Feedback to local fishing-right owners of the preliminary results generated in this study led to action by FMOs to restore *Rö*-lakes with self-sustaining populations.

TABLE 16.3 *Factors associated with the extinction of brown trout populations in* Rö-*named lakes*.

Anthropogenic Impact		Brown trout habitat	L[a]
Biological	Brook char	Spawning area overtaken	5[b]
	Pike introduced	Strong predation	3
Chemical	Acidification	Impossible	2[c]
	Rotenone	Impossible	1[d]
Physical	Barrier	Impossible	2[b]

a. Number of extinct brown trout lakes (n = 12) affected by specific impact.
b. One lake was classified in two categories.
c. Permanently acidified pH = 4.7 to 4.9.
d. Once impossible, now brown trout has been reintroduced.

Maximum natural or anthropogenic permanent extinctions (1672–1930)

A total of 47 out of 51 *Rö*-named lakes still harboured self-sustaining brown trout populations in the twentieth century (Tables 16.1 and 16.2). Interviews with a maximum historical scope of thirty to eighty years back in time, suggested that four of the *Rö*-lakes did not contain self-sustaining brown trout populations during this time. Habitat surveys in the same four lakes determined that reproducing brown trout populations could never have existed in two of them. The remaining two lakes were found to have historically suitable habitat conditions for brown trout, although an artificial barrier prevented reproduction in one of these lakes. Test fishing confirmed that these lakes did not hold brown trout. Since all but these two out of forty-nine lakes with natural potential conditions for brown trout were confirmed brown trout waters, 2/49 was found to be the maximum potential fraction of lakes suffering permanent extinction that were not captured in interviews and historical documents. If *Rö*-lakes represented a non-biased sample of all brown trout lakes in the study area (there are no indications to the contrary), then we may estimate that from none out of 162 up to possibly seven brown trout lakes out of 169 in the whole study area would have suffered permanent extinction prior to 1930. It was concluded that the pre-industrial distribution of brown trout was 11 per cent across all lakes in the study area, and remained so until the 1930s.

Possible misinterpretations of the *Rö*-term

I excluded three *Rö*-term lakes from the current study because the earliest name forms in older maps made it plain that these names were originally derived from *Ry*, meaning something other than brown trout. One explanation of the *Rö*-term in lake names, red water colour, was refuted during field visits since none of the waters were more reddish in colour than other lakes in general. Another possible confusion of the *Rö*-term meaning was suggested to be Arctic charr (*Salvelinus alpinus*), called *röding* in Swedish. However, the *Rö*-name was not an indication of suitable Arctic charr habitat. The majority of *Rö*-lakes did not contain spawning grounds for Arctic charr and could never have harboured self-sustaining charr populations. This species was only found in three out of fifty-one *Rö*-lakes, and there were too few in all lakes to provide any statistical evidence of an association with the name. Since repeated stocking of charr had been performed in all three lakes, I could not rule out the possibility that these populations were non-native to these lakes. Nothing in all of the data collected indicated that Arctic charr could historically have had a wider distribution in *Rö*-lakes. It was an uncommon species in the whole study area and was only considered possibly indigenous in one other lake out of the 1,509 investigated.

DISCUSSION

Historical names

The results allow some general conclusions to be drawn. For instance, lake names reveal details about the fish fauna as well as habitat in these lakes. The historical records of fishers' knowledge in the form of lake names on maps can communicate valuable information on environmental history, which can in turn have an impact on management and conservation. Danko (1998) recommends collecting ecological data from the regions studied to increase the reliability of fish-terms used as evidence of past occurrence. The present study used a number of sources and methods, thus allowing comprehensive validation of lake name evidence. When lake names are verified as being positively (or negatively) associated with certain species, the spatial and temporal data linked to the name can then be used in a variety of ways. This study verifies that *Rö*-named lakes are associated with past or present self-sustaining brown trout populations. Thus, lakes with species-associated names can help identify habitats suitable for deeper investigations or restoration.

Could landscape-scale inventories of certain fish species benefit from selecting lakes from names in maps instead of performing a random survey? An inventory in the present study area using knowledge of local dialect and the deciphered *Rö*-term could provide wide spatial coverage with less effort. A simple overview of local maps can identify at least a third of all brown trout populations among 1,509 lakes. To pick out the same amount of brown trout lakes by random sampling with multi-mesh-sized gillnets (Appelberg, 2000), would take approximately five years of full-time fishing by two persons during the ice-free season. The gillnet inventory would, moreover, have missed all extinct populations and would also lack the temporal perspective that lake names provide.

Another useful feature of lake names is that historical anthropogenic impact or past natural disturbances may be discovered and further investigated where lake names do not correspond with the species currently living in lakes. The remaining two *Rö*-lakes (4.1 per cent) that cannot be confirmed by interviews or archival data as brown trout waters in spite of being historically suitable habitat might have harboured populations now lost both in nature and in local collective knowledge. In that case, the populations became extinct long before the scope of possible detection by interviews or archival data. However, it is predicted that one of these lakes will be colonized in the near future from a downstream population, once an artificial migration barrier discovered in this study is removed. Other essential ecological information such as details about habitats and fish communities are also associated with these lake names. Inlet or outlet streams of a specific minimum size with spawning gravel suitable for brown trout are found in 96 per cent of *Rö*-lakes, and in the same proportion of *Rö*-lakes we also find that the original fish communities are not exposed to large predators like pike; 100 per cent of the lakes are isolated by natural barriers that stop the upward migration of various fish species downstream of the lakes. *Rö*-lakes can thus be considered as refuges protected from major predators.

Pike are present in most lakes elsewhere in the study area, and studies indicate that predation by pike limits brown trout distribution in slow-flowing streams (Näslund et al., 1998) and in lakes (Went, 1957; Toner, 1959). Thus, with the *Rö*-names, fishers from hundreds of years back in time are communicating to us and saying: 'This lake is characterized by its richness in brown trout. There are good habitat conditions for this species here'. The past distribution of fish populations in a given area can be estimated from the wide temporal and spatial data generated from historical lake names associated with fish species, providing that associations are properly validated. This is demonstrated in the present paper by utilizing occurrences of lake names fixed in time by historical maps. Most *Rö*-names are found to be more than 160 years old, revealing a pre-industrial perspective on brown trout distribution. All types of lake names on maps were found to be 'evolutionarily' conservative and most meanings or core structures were virtually unchanged through the centuries. This is further supported in this study by findings that *Rö*-lake names are being passed on in a conservative oral tradition, even though the historical species name *Röa* has disappeared from the common language. For this reason, it is proposed here that there is little chance the core structure will change once a lake has been named.

Edlund (1997) suggests that prehistoric fishers and trappers developed a fixed onomastic system for lakes and rivers and gives examples together with C_{14}-dating of settlements, isostatic uplift and other data implying the genesis of a fisheries-related name-complex in the heart of the study area 1,900 years ago. It is possible that *Rö*-lakes were named during this prehistoric period. Since all but two out of forty-nine *Rö*-lakes with possible brown trout populations are accounted for in interviews and archival data, it is highly unlikely that extensive permanent extinctions of brown trout took place prior to the 1930s. Therefore the entire data supports the idea of long-term, stable brown trout lake distribution under pre-industrial natural conditions. The past distribution of brown trout was thus 11 per cent of all lakes in the study area, and remained so until the 1930s when extinctions started to become evident.

Interviews

The use of fishers' knowledge obtained from interviews can also provide wide temporal and spatial insight into the past and present distribution of fish populations. This is demonstrated in the present paper by utilizing fishers' knowledge gathered from in-depth interviews and validated by a number of methods. Interviews result in a temporally and geographically more extensive picture of the fish fauna distribution than could ever be achieved through conventional scientific methods with the same effort. No populations 'new' to the informants were discovered by test-fishing among the *Rö*-lakes. However, interviewees were slightly over-optimistic about the existence of self-sustaining populations. Masking of abundance by stocking activities was discussed in Hesthagen et al. (1993), who reported that interviewees assessing the status of fish-populations in Norwegian acid lakes were too optimistic. They also suggested that bias might result from a time lag before anthropogenic damage becomes evident to fishers. This might be the case for one *Rö*-lake, where fishers were clearly unaware of a recent extinction. Another *Rö*-lake was restocked annually, masking the extinction of the original population.

Apart from these two examples, fishers' knowledge obtained from in-depth interviews regarding the *Rö*-lakes was totally reliable, matching the test-fishing results and consistent relative to habitat surveys. Discussing the future of fisheries science,

Mackinson and Nøttestad (1998) emphasize that it is imperative for scientists to use diverse data sources for maximum potential, and advocate making greater use of local fishers' knowledge. Face-to-face interviews are claimed to be most effective. This view is supported by the findings in this study. The accumulated interviews reveal that the great majority of brown trout population extinctions occurred during the last eight decades. Archival data can confirm that most of these extinct populations once existed, while their current absence is shown by a range of test-fishing methods. More than a quarter of the populations are lost. We need to understand the cause of this wave of extinctions if these lakes are to be restored. Limnological surveys demonstrate that all the cases are associated with severe anthropogenic impacts. Extinctions of brown trout populations caused by acidification of Scandinavian lakes during the twentieth century are reported in several papers (Bergquist, 1991; Bulger et al., 1993; Lien et al., 1996) as well as in this study. Local extinction of fish species caused by anthropogenic biological impact is reported on by, for example, Nilsson (1985), Crivelli (1995), Lassuy (1995) and Townsend (1996). Similarly, historical records and present data in this study led to an estimate that at least 95.7 per cent of all brown trout populations survived until the twentieth century, when successful colonization by the introduction of fish species new to the lakes resulted in the extinction of the original trout populations.

Before this study, the methods available to collect historical data on fish species distribution in northern lakes were limited to interviews and archival data. Integrating the use of historical names and historical fishers' knowledge into fisheries science will enable investigations to move from brief snapshots at a local scale to the context of the wider landscape and to incorporate the historical dimension.

In conclusion, historical names, fishers' knowledge and documentary evidence combined with limnological surveys have proven useful in revealing the past natural distribution of brown trout in northern Sweden. Many of these populations are now long gone and forgotten, but the names of the lakes remain and, once deciphered, help to remind us of all that is lost. In part owing to the *Rö*-names, people are now motivated to restore *Rö*-lakes with self-sustaining local populations of brown trout.

ACKNOWLEDGEMENTS

THIS research has been made possible by the environmentally engaged and fisheries management-committed municipality of Örnsköldsvik, together with FMOs that provided parts of data. I gratefully acknowledge the financial support for this work of the Shwartz foundation and the Carlgren foundation. Finally I thank the Center for Fish and Wildlife Research Sweden, for travel funds to the Fishers' Knowledge Conference in Vancouver.

REFERENCES

APPELBERG, M. 2000. Swedish standard methods for sampling freshwater fish with multi-mesh gillnets. *Fiskeriverket information*, 2000, No. 1, pp. 1–32.

AYBES, C.; YALDEN, D.W. 1995. Place-name evidence for the former distribution and status of wolves and beavers in Britain. *Mammal Review*, Vol. 25, No. 4, pp. 201–26.

BAIN, M.B. 1999. Substrate. In: M.B. Bain and M.J. Stevenson (eds.), A*quatic Habitat Assessment: Common Methods*. Bethesda, American Fisheries Society, pp. 95–103.

BERGQUIST, B. C. 1991. Extinction and natural recolonization of fish in acidified and limed lakes. *Nordic Journal of Freshwater Research*, Vol. 66, pp. 50–62.

BOISSEAU, S.;YALDEN, D.W. 1998.The former status of the Crane *Grus grus* in Britain. *Ibis*, Vol. 140, No. 3, pp. 482–500.

BULGER, A. J.; LIEN, L.; COSBY, B.J.; HENRIKSEN, A. 1993. Brown trout (*Salmo trutta*) status and chemistry from the Norwegian Thousand Lake Survey: Statistical analysis. *Canadian Journal of Fisheries and Aquatic Science*, Vol. 50, No. 3, pp. 575–85.

CRISP, D.T. 1996. Environmental requirements of common riverine European salmonid fish species in fresh water with particular reference to physical and chemical aspects. *Hydrobiologia*, Vol. 323, No. 3, pp. 201–21.

CRIVELLI, A.J. 1995. Are fish introductions a threat to endemic freshwater fishes in the northern Mediterranean region? *Biological Conservation*, Vol. 72, No. 2, pp. 311–19.

CRIVELLI, A.J.; MAITLAND, P.S. 1995. Future prospects for the freshwater fish fauna of the north Mediterranean region. *Biological Conservation*, Vol. 72, No. 2, pp. 335–7.

DAHLSTEDT, K. H. 1956. Taped and written record number DAUM 3994 Tape 133, Department of Dialectology, Onomastics and Folklore Research: Box 4056 904 03 Umeå, Sweden. (In Swedish.)

——. 1961. Written record number ULMA 25260 p.16, Institute for Dialectology, Onomastics and Folklore Research: Department of Onomastics Box 135, 751 04 Uppsala, Sweden. (In Swedish.)

DANKO, P.D. 1998. Building a reliable database from a native oral tradition using fish-related terms from the Saanich language. In: D. Pauly, T.J., Pitcher and D. Preikshot (eds.), *Back to the future: Reconstructing the Strait of Georgia ecosystem*. Fisheries Centre Research Reports, Vol. 6, No. 5, pp. 29–33. Vancouver, University of British Columbia.

EDLUND, L.E. 1975. *Namn på vattensamlingar i Trehörningsjö och Björna socknar, Ångermanland. Bildning med särskild hänsyn till efterlederna*. Masters thesis, Department of Nordic languages, Umeå University, Umeå. (In Swedish.)

——. 1997. Från Gene till Myckelgensjö: Kring ett hydronymkomplex i norra Ångermanland. In: S. Strandberg (ed.), *Nomina Germanica*, Vol. 22, pp. 85–106. Uppsala, Uppsala university. (English summary.)

HESTHAGEN, T.; ROSSELAND, B.O.; BERGER, H.M.; LARSEN, B.M. 1993. Fish community status in Norwegian lakes in relation to acidification: A comparison between interviews and actual catches by test-fishing. *Nordic Journal of Freshwater Research*, Vol. 68, pp. 34–41.

LAIKRE, L.; ANTUNES, A.; APOSTOLIDIS, A.P.; BERREBI, P.; DUGUID, A.; FERGUSON, A.; GARCIA-MARIN, J.L.; GUYOMARD, R.; HANSEN, M.M.; HINDAR, K.; KOLJONEN, M. L.; LARGIADER, C.R.; MARTINEZ, P.; NIELSEN, E.E.; PALM, S.; RUZZANTE, D.E.; RYMAN, N.; TRIANTAPHYLLIDIS, C. 1999. *Conservation genetic management of brown trout (*Salmo trutta*) in Europe*. EU FAIR Report by the Concerted Action on Identification, Management and Exploitation of Genetic Resources in the Brown Trout (*Salmo Trutta*). ('TROUTCONCERT'; EU FAIR CT97–3882.)

LASSUY, D.R. 1995. Introduced species as a factor in extinction and endangerment of native fish species. In: H. L. Schramm, Jr. and R. G. Piper (eds.), *Uses and effects of cultured fishes in aquatic ecosystems*. Bethesda, *American Fisheries Society Symposium*, Vol. 15, pp. 391–6.

LELEK, A. 1987. Threatened fishes of Europe. *The Freshwater Fishes of Europe*, Vol. 9 (ed European Committee for Conservation of Nature and Natural Resource). West Germany, Aulag-Verlag, 343 pp.

LIEN, L.; RADDUM, G.G.; FJELLHEIM, A.; HENRIKSEN, A. 1996. A critical limit for acid neutralizing capacity in Norwegian surface waters, based on new analyses of fish and invertebrate responses. *Science of the Total Environment*, Vol. 177, pp. 173–93.

MACKINSON, S.; NØTTESTAD, L. 1998. Points of view: Combining local and scientific knowledge. *Reviews in Fish Biology and Fisheries*, Vol. 8, No. 4, pp. 481–90.

MAITLAND, P.S. 1995. The conservation of freshwater fish: Past and present experience. *Biological Conservation*, Vol. 72, No. 2, pp. 259–70.

MAITLAND, P.S.; LYLE, A.A. 1991. Conservation of freshwater fish in the British Isles: The current status and biology of threatened species. *Aquatic Conservation: Marine and freshwater ecosystems*, Vol. 1, No. 1, pp. 25–54.

MOYLE, P.B.; LEIDY, R.A. 1992. Loss of biodiversity in aquatic ecosystems: Evidence from fish faunas. In: P.L. Fiedler and S.K. Jain (eds.), *Conservation biology: The theory and practice of nature conservation, preservation, and management*. New York, Chapman and Hall, pp. 127–69.

NÄSLUND, I.; DEGERMAN, E.; NORDWALL, F. 1998. Brown trout (*Salmo trutta*) habitat use and life history in Swedish streams: Possible effects of biotic interactions. *Canadian Journal of Fisheries and Aquatic Science*, Vol. 55, No. 4, pp. 1034–42.

NILSSON, N.A. 1985. The niche concept and the introduction of exotics. *Report of the Institute of Freshwater Resources*, Vol. 62, pp. 128–35.

PITCHER, T.J. 2001. Fisheries managed to rebuild ecosystems? Reconstructing the past to salvage the future. *Ecological Applications*, Vol. 11, No. 2, pp. 601–17.

SOUSA, A.; GARCIA-MURILLO, P. 2001. Can place names be used as indicators of landscape changes? Application to the Doñana Natural Park (Spain). *Landscape Ecology*, Vol. 16, No. 5, pp. 391–406.

THE SWEDISH NATIONAL LAND SURVEY. 1672–1908. The general staffs' maps of Sweden 1:100 000 and 1:200 000; county, parish, ordnance or village maps surveyed 1672–1908. Generalstabens karta övfer Sverige 1:100 000 and 1:200 000; läns-, församlings-, stabs- and by-kartor rekognocerade 1672-1908.

THE SWEDISH NATIONAL LAND SURVEY. 1961–1967. The general staffs' topographic maps of Sweden 1:50 000 surveyed 1961–1967. Generalstabens Topografiska karta 1:50 000 Rekognoserad 1961-1967.

TONER, E.D. 1959. Predation by pike (*Esox lucius*) in three Irish lakes. *Report of Sea and inland Fisheries, Eire*, pp. 67–73.

TOWNSEND, C.R. 1996. Invasion biology and ecological impacts of brown trout *Salmo trutta* in New Zealand. *Biological Conservation*, Vol. 78, No. 1–2, pp. 13–22.

WALLACE, S.S. 1998. Sources of information used to create past and present ecosystem models of the Strait of Georgia. In: D. Pauly; T.J., Pitcher; D. Preikshot (eds.), *Back to the future: Reconstructing the Strait of Georgia ecosystem*. Fisheries Centre Research Reports, Vol. 6, No. 5, pp. 19–21. Vancouver, University of British Columbia.

WEBSTER, J.A. 2001. A review of the historical evidence of the habitat of the Pine Marten in Cumbria. *Mammal Review*, Vol. 31, No. 1, pp. 17–31.

WENT, A.E.J. 1957. The pike in Ireland. *Irish Naturalists' Journal*, Vol. 12, pp. 177–82.

WITZEL, L.D.; MACCRIMMON, H.R. 1983. Embryo survival and alevin emergence of brook charr, *Salvelinus fontinalis*, and brown trout, *Salmo trutta*, relative to redd gravel composition. *Canadian Journal of Zoology*, Vol. 61, No. 8, pp. 1783–92.

PART III: COMMERCIAL FISHERIES

CHAPTER 17   Putting fishers' knowledge to work
Reconstructing the Gulf of Maine cod spawning grounds on the basis of local ecological knowledge
Ted Ames

ABSTRACT

IN today's fisheries and centralized management strategies, fishers' knowledge often gets dismissed as subjective, anecdotal and of little value. Yet, fishers have spent much of their lives accumulating intimate, fine-scale ecological information that is not otherwise available to the scientific community. Accessing this wealth of fisher-based knowledge, however, is not without its pitfalls. This chapter reviews problems encountered while accessing information during the mapping of historical cod and haddock spawning grounds in the Gulf of Maine, and discusses the strategies developed to overcome them. Current and future roles for fishers' knowledge in managing coastal fisheries are examined. Various ways to integrate the local place-based information of fishers into current management strategies and the potential for introducing a new local management paradigm are explored.

INTRODUCTION

IN New England, fishers' local ecological knowledge (LEK) has often been dismissed as subjective, anecdotal, and dealing only with local situations. In addition, it often relates to stocks that were fished out decades ago, leading some to suggest that since these fish no longer exist, the fishers' accounts should only be used as historical footnotes.

I tend to disagree. I have used LEK often in my life, not only in order to catch fish, but also as an important source of ecological information about a fishery. From this perspective, the accuracy and breadth of knowledge shared by fishers is very impressive. Fishers and their descriptions have a pivotal role to play in the development and functioning of sustainable fisheries.

Whether LEK gets integrated into mainstream science so that it can influence management will ultimately depend on the ways it is used. Fishers and their vessels are currently being used to develop 'real time' catch data for faster, ongoing stock assessments. Though useful in bolstering the status quo, this approach tends to employ fishing vessels rather than fishers' knowledge, which deals with local populations and their seasonal habitats.

Fisheries science, involved as it is with the study of large population units, has not focused on local-level phenomena such as the changes in behaviour and distribution of local populations associated with the collapse of a stock that are so often described by fishers. The preoccupation of fisheries science with system-wide characteristics has left it without the historical parameters needed to interpret fine-scale changes in stock distribution, behaviour, or migration patterns over time. Consequently, management has lacked the ability to detect or interpret these changes in abundance.

A NEW ROLE FOR FISHERS' ECOLOGICAL KNOWLEDGE

This lack of historical perspective may have aggravated attempts to manage New England's commercial fisheries. We have all been so preoccupied by the depressed state of our fisheries that we may have missed some of the root causes of their depletion. If we are to develop sustainable fisheries, we must at the very least understand how and why the stocks collapsed in the first place. While fishers and scientists acknowledge that many stocks have declined because of high catch rates, the problem is far more complex than the simplistic rationale of 'too many fishers chasing too few fish' (National Research Council, 1999). Declines in abundance have consistently been accompanied by local changes in distribution, migration patterns and species assemblages. Clues abound about the disruption of local interrelationships and changes associated with this. But fine-scale changes cannot be detected by today's system-wide fisheries assessments.

It is here that fishers' knowledge can play an important and perhaps critical role. Fishers are, in fact, the only available source of local, historical, place-based fisheries information. Just to survive, let alone succeed, each fisher has to become proficient at figuring out how local changes in a fish stock affect distribution and abundance. This creates a pool of people who have unique experiences of local marine ecology.

Not only do fishers have special knowledge about what is presently there, but each generation has developed its own particular fishing patterns that are attuned to the stock migrations and behaviour present during that period. With a little effort, information can be retrieved about such factors as distribution, behaviour and species assemblages that are unique to those periods.

Information collected from different generations of fishers can be used to create a series of historical windows into a fishery's local ecology that can be used to identify long-term processes in the fishery. Compiling a historical database forms a timeline that allows those processes to be studied. If a relatively short time-span is used to capture changes occurring before, during and after the depletion of a fishery, the sequential effects of its depletion on the marine ecosystem can be analysed. Linking the intimate, place-based knowledge of fishers with that of scientists would help in understanding how highly productive coastal ecosystems functioned when they were more robust. This would also provide historical perspective into the fine-scale details so lacking in the analysis of commercial stocks.

The value of fishers' historical insights into fisheries ecology goes beyond its benefit to research. Fishers' knowledge may be most effective when applied to fisheries management because it offers management a new paradigm. For the first time, long-term trends, seasonal, site-specific habitats, and species interactions will be available to management. With this knowledge, alternative approaches such as area-based management using local knowledge and local participation could be used to protect reproduction and juveniles as part of the local fishery. This would enhance the possibility of consistent local reproduction while, at the same time, surveys and assessments of larger population units would be continued.

THE GULF OF MAINE COD SPAWNING GROUNDS PROJECT

A good example of the use of traditional fishers' information surfaced during efforts in New England to revitalize the collapsed inshore cod (*Gadus morhua*) fishery. Two fishing associations, the Maine Gillnetters Association and Maine Fisherman's Co-op, successfully petitioned the Maine State Legislature to form a groundfish hatchery commission to study the feasibility of establishing one or more groundfish hatcheries. The hatcheries were funded by raising the groundfish licence fee for commercial fishers. The commission found large areas of groundfish habitat along the coast that used to be highly productive, but were now abandoned. They concluded that, if hatchery production could be used to increase the number of active spawning sites along the coast by reintroducing groundfish into these areas, the resulting spawning success would drastically reduce the time depleted stocks would need to recover. The commission recommended that young cod and haddock (*Melanogrammus aeglefinus*) be released near once-productive spawning grounds and nursery areas in an attempt to jump-start the process. Releasing juveniles in the right habitats would be a critical step.

Unfortunately, most of the inshore grounds that were suitable for such a project had been fished out decades before and had long been abandoned and forgotten by today's fishers. With cod and haddock stocks collapsed, scientists were unable to locate spawning areas by conventional methods. Despite the fact that the Gulf of Maine had maintained a directed cod fishery for more than three centuries, few spawning grounds were known to science. Most of the spawning areas suitable for such a project were abandoned and forgotten, having been 'fished out' decades earlier. Few current fishers were even aware of their existence.

A study was funded to locate and interview the few remaining fishers who had fished those areas and could identify coastal spawning and nursery areas of cod and haddock. It became my privilege and great pleasure to interview these older fishers and to draw the spawning ground maps on the basis of their knowledge.

Prior to the fisher-based spawning ground study, very few coastal spawning locations for cod and haddock were known, causing researchers to raise important questions about whether either species had actually been year-round coastal residents. As the interviews proceeded, the number of confirmed spawning sites mounted. It soon became clear that both cod and haddock once had spawning areas along the entire length of the Gulf of Maine's coast. By the time the study was over, more than 2,800 km^2 of spawning grounds for cod and haddock had been identified, and numerous questions had been raised about what actually precipitated the collapse of those coastal fisheries. The contributions of these fishers have provided new insights into the causes of the collapse of Atlantic cod in the study area. (Ames et al., 2000)

An accompanying study using side-scan sonar confirmed the substrates and depths of the spawning locations given by fishers, indicating that their descriptions were exceptionally accurate (Barnhardt et al., 1996). This reinforced general acceptance of the locations identified by fishers as coastal New England's historical spawning grounds for Atlantic cod.

PITFALLS TO AVOID WHEN INTERVIEWING FISHERS

COLLECTING fisheries information about commercial stocks does not come without its own set of hurdles. Simply interviewing some fishers and then cleaning up the data to make it presentable to the scientific community is only a small part of what has to be done to interview fishers effectively. The process of figuring out who can best provide the information you seek can be formidable. Just any old fishers will not do.

In addition, the majority of interviewers confirm that fishers can be difficult to interview, their information is difficult to verify and, once verified, is very difficult to integrate into conventional fisheries information. A well-defined strategy for surmounting these hurdles is essential for good results. It is especially important to obtain ethical clearance for LEK interviews, for it may involve proprietary information and cultural issues. A brief, concise form disclosing who will have access to their information and how it will be used can dispel the concerns of many fishers, while simultaneously avoiding any misunderstanding.

Also be aware that different gear types may give quite different types of information. What is observed by one fishing technique alone can be very misleading. For example, an overview of coastal New England shows that hook fishers caught cod in their feeding areas. Since fish feed less when they are spawning, hook fishing may not provide good information about spawning locations. Otter trawlers and gillnetters caught fish whether or not they were feeding and so became a prime source for spawning ground information.

A brief description of problems that emerged during the spawning ground project and the strategies used to resolve them is provided below. It is hoped that this summary will be of use to others.

1. When we started, we did not know the names or addresses of the fishers who were part of the collapsed coastal fishery for cod and haddock. Most of them were retired and had not fished for decades. We asked Maine's two coastal groundfish organizations to help us identify older fishers to interview. Their members prepared a list of older fishers for us who were well known locally and respected for their skill at catching cod and haddock in coastal waters.

    The fishers interviewed during the project were selected from a potential list of several hundred groundfishers. They were retired captains who averaged about 65 years of age and had been very effective in Maine's inshore cod and haddock fisheries. All had been lifelong fishers with at least thirty years experience on small and medium-sized boats engaged in otter trawling or tub trawling/longlining. Many had started out as handliners or lobster (*Homarus americanus*) fishers and shifted to various technologies as opportunities appeared.

2. Fishers generally mistrusted fisheries researchers and managers. Countering this was the credible fishing history of my family and myself. In addition, a local fisher accompanied me, introduced me, and participated in most sessions. This effectively put everyone at ease. The fishers who accompanied us during the interviews were younger, active fishers whom I knew personally or by their reputation and who were members of the two fishing associations supporting the project. They were unpaid, untrained, and became involved because of a collective desire to rebuild the fishery for their communities.

3. In general, fishers are not inclined to hand over hard-won knowledge that could threaten the livelihood of friends, family, and self by inviting competition or closures. However, this difficulty was not often encountered because the fishers being interviewed were older and had little motivation to safeguard or falsify information. In addition, the interviews focused on coastal spawning areas that had been fished out years ago, rendering their location relatively worthless. Notably, information about current fishing areas was not forthcoming.

4. Fishers are often reluctant to answer questions if they perceive the interviewer to be collecting information simply for the sake of collecting it, or worse yet, for management purposes they do not support. The survey addressed this concern by explaining that its purpose was to rebuild the fishery for the benefit of fishers. The few remaining fishers who had taken part in the fishery were the only ones left who knew where the spawning grounds were located.

I stated that if we could find where the grounds were, funding would be available to support an effort to rebuild the stocks. In the end, fishers themselves were to be the beneficiaries. All recognized that restoration efforts were a long shot at best, but felt that it was worth talking with us anyway. And, if all went well, fishers in their area would regain a fishery.

5. Fishers feel especially threatened when asked to share information that may become public, and often refuse to talk. Interviewers should recognize the economic consequences fishers may face when fishing secrets are revealed. These are not trivial issues. Once published, facts affecting the fishers' landings that were casually shared with the interviewer become available to competitors and anti-fishing interests. An important step includes thoughtful decisions about what to ask and how to handle such information. Only then does a strategy to persuade fishers to share their knowledge become realistic. In the spawning ground study, questions were deliberately limited to depleted coastal grounds no longer used by local fishers.

PITFALLS TO AVOID WHEN PROCESSING FISHERS' INFORMATION

TRADITIONALLY, many fisheries scientists have brushed fishers' information aside because it is so difficult to integrate into the world of high-tech, statistics-based research. Even when fishers' subjective observations can be confirmed, they lack the reproducibility and precision of carefully controlled experiments. Given these concerns, controlling data quality becomes critical. Researchers who find ways to accommodate these limitations by developing ways to validate fishers' knowledge, however, may find a great deal of site-specific information about fisheries ecology.

The strategies developed in the spawning ground study for validating data included requiring that each spawning ground and its location be independently verified by two or more fishers, and that the depth and substrate present at the site should agree with known spawning ground preferences. In addition, the exact location of the site described by fishers required validation. Two or more independent identifications by fishers were needed when spawning grounds were identified directly on nautical charts. Most, however, preferred to simply name a fishing ground in an area, or gave marks and bearings leading to the bottom they had once fished. The location of specific grounds had to be corroborated by interviews with additional fishers or historical references, while spawning areas identified by sets of landmarks had to be plotted and their location independently confirmed by other fishers. Once identified, the site then had to agree with the bottom types reported on nautical charts and, where possible, confirmed by side-scan sonar.

Of all parameters encountered in the study, timelines were perhaps the most difficult to establish and verify. Fishing information collected during the spawning ground study was, by necessity, decades old. Even though fishers were quite sure of the season or month they had caught ripe fish, they often could not recall the exact year when it happened. In such cases, supporting information occurring during the same period was used to identify and then determine the approximate year when the fish were caught.

For example, when a participant was unsure of when he had found ripe cod on a particular ground, questions such as 'Was it before or after the war?, Were you married then?', 'What grade in school was your oldest boy then?' were used to bracket the period and eventually allowed the date to be identified.

EPILOGUE TO THE SPAWNING GROUND PROJECT

A unique aspect of the spawning ground study was that all the participants involved were attempting to rebuild the fishery, even though retired fishers had no interest in returning to the sea and younger fishers knew their efforts might be for naught. This idealism was undoubtedly the key to the project's success. All wanted local fishers from coastal fishing communities to continue harvesting cod in a limited, hook fishery once the fishery recovered. As events unfolded, however, this was not to be. The depleted groundfish stocks precipitated management regulations that eliminated most of the active fishers involved in the study, even though they were instrumental in efforts to improve the fishery through spawning season closures.

It seems ironic that nearly all the fishers involved in the project have now lost access to the fishery, an outcome that was once inconceivable to Maine fishers. Six years after the study, the eastern two-thirds of Maine's long coastline has but three active groundfish permits left among the 10,000-odd fishers who live there, and those three will disappear with Amendment 13, leaving many embittered and frustrated fishers with few business alternatives, and Maine's coastal fishing communities disenfranchised.

Perhaps the most grievous insult came as the aquaculture industry consumed US$2 million of Federal groundfish assistance in a three-year period to grow and release 450 fingerling cod. Much of the funding disappeared in their efforts to commercially grow pen-raised haddock, rather than cod.

New applications for fishers' knowledge

The mapping project of cod and haddock spawning grounds displays only a fraction of the potential value found in fishers' knowledge. It has since been use to build a prototype LEK database for Atlantic cod to analyse stock structure in the Gulf of Maine during the 1920s, a period when the population was more robust (Ames, 2004). The historical spawning grounds were used as points of origin for tracking the cod's seasonal movements within a spatial plot of fishing grounds and were instrumental in determining movement patterns.

By mapping the distribution of cod for each season of the year on a geographic information system (GIS), and then displaying the seasons sequentially, fine-scale details of movements could be tracked. From this, the location of sub-populations and their spawning components and/or local populations were tentatively identified. Recent discoveries show cod returned to specific spawning grounds for reproduction (Wroblewski, 1998; Green and Wroblewski, 2000). Concentrations of cod were tracked from spawning areas to bordering fishing grounds and then back to the same spawning ground through each season of the year. When viewed in their entirety, the collective movements of Atlantic cod among fishing grounds in the Gulf of Maine followed seasonal migration corridors associated with three sub-populations, and local spawning components made local, circular movement patterns between feeding areas and their spawning ground.

Many of the historical cod spawning grounds could be verified by recent cod egg distribution surveys (Berrien and Sibunka, 1999), confirming that not only had fishers identified the right spawning areas, but that historical spawning components still used the same grounds (Figure 17.1). Many abandoned spawning areas were also found. The absence of recent spawning activity and cod landings near those sites identified them as spawning areas used by extinct spawning components or local stocks.

A new paradigm for management?

Today's fisheries managers and fishers are trapped in a management system dependent on system-wide stock assessments that are not designed to detect local depletions (Frank et al., 1994; Sinclair et al., 1997; Smedbol and Stevenson, 2001). All have been helpless in avoiding the depletion of valuable fisheries that are now diminished to a fraction of their historical productivity.

The linking of fishers' ecological knowledge (LEK) with current fisheries reports, however, offers fishers, managers and environmentalists a new paradigm that can be used to identify and evaluate temporal changes in fine-scale population structure. Ames (2004) used LEK to create an overarching framework of historical stock structure and behaviour patterns as part of an analysis of Gulf of Maine Atlantic cod. The distribution of historical spawning components within the Gulf of Maine grouping was described and their interactions were summarized, on the basis of seasonal movements to and from specific spawning grounds. The results were then compared to recent fisheries surveys and studies, first to validate the methodology used, and then to evaluate changes that have occurred in the disposition of today's spawning components. Such insights are pivotal if the reproductive capacity of non-panmictic populations such as cod and herring are to be maintained and if

PUTTING FISHERS' KNOWLEDGE TO WORK IN THE GULF OF MAINE COD SPAWNING GROUNDS

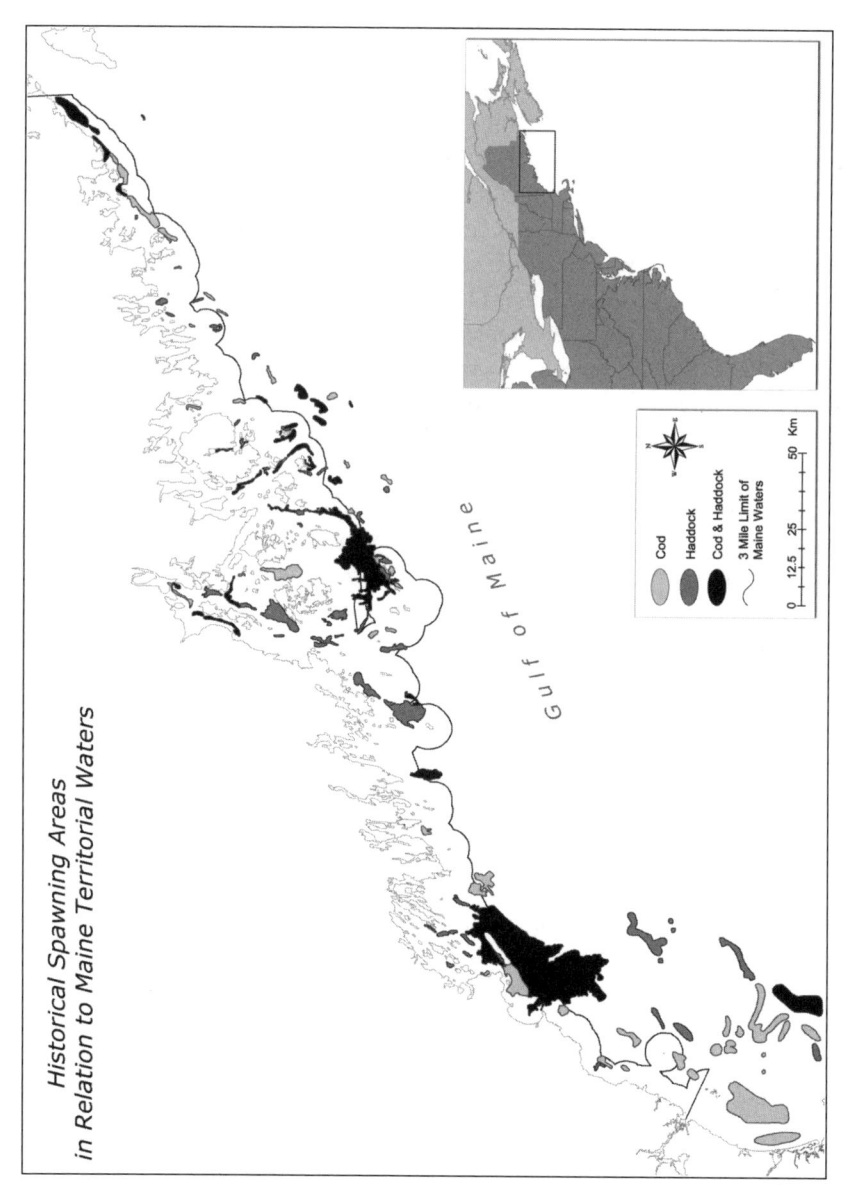

FIGURE 17.1 *Historical cod spawning grounds and recent distribution patterns of cod eggs in the northern Gulf of Maine.*
Source: Ames, 1997.

functional ecological boundaries for fisheries management areas are to be defined. The information derived from fishers' local, fine-scale knowledge can facilitate strategies to improve reproduction and recruitment, and protect critical habitats.

The New England Fisheries Management Council (NEFMC) recently considered the Gulf of Maine Conservation and Stewardship Plan, which would have been used to manage three sub-populations of Atlantic cod spawning components along the US coastal shelf of the Gulf of Maine. The proposed plan would have created three ecologically-discrete subdivisions on the coastal shelf, accessible only to fishers who agreed to fish in one of the areas for five years, making it imperative that they develop a good rebuilding programme to protect spawning aggregations, juveniles, nursery habitats and forage stocks. Harvesting was to be restricted to modest levels that allow development of a sustainable fishery that provides long-term economic benefits to local economies in the area.

The NEFMC was to delegate local management plans for each area to a committee, pending the council's approval. The plan proposed a committee chaired by the NEFMC, with a scientist-advisor, area fisher delegates of each gear type, fishing community delegates and environmentalists. The committee was to be patterned after the State of Maine's Lobster Zone Councils where consensus building and peer-group pressure could be used to support an ecosystem-based recovery plan for area fishers, who would be the principal beneficiaries.

Several reports identify a need to manage cod stocks at finer scales (Frank and Brickman, 2001; Smedbol and Stevenson, 2001). One way to accomplish this would be by adding area management units for rebuilding sub-populations. The spawning ground project succeeded because inshore fishers chose to be stewards of their local fishery in an attempt to improve it. This exemplifies a practical form of stewardship shared by many coastal fishers who could be enlisted in innovative, area-based management plans to rebuild individual coastal spawning components in order to establish sustainable fisheries. Improvements in component abundance should be detected adequately by improvements in the current larger-scale assessment surveys.

The success of such an approach, of course, would depend on creating management units that were predisposed to support rebuilding programmes for depleted coastal stocks. The Gulf of Maine Conservation and Stewardship Plan's strategy proposed to do that by restricting access to fishers who were willing to be dependent on the area's local stocks and by focusing peer-group pressure to improve stewardship efforts through participation in the management process.

REFERENCES

AMES, E.P. 1997. Cod and haddock spawning grounds of the Gulf of Maine from Grand Manan to Ipswich Bay. In: I. Hunt von Herbing; I. Kornfield; M. Tupper; J. Wilson (eds.), *The implications of localized fish stocks*. Ithaca, NY, NRAES-118, pp. 55–64.

———. 2004. Atlantic cod structure in the Gulf of Maine. *Journal of the American Fisheries Society*, Vol. 29, No. 1, p. 10–27.

AMES, E.P.; WATSON, S.; WILSON, J. 2000. Rethinking overfishing: Insights from oral histories of retired groundfishermen. In: B. Neis and L. Felt (eds.), *Finding our sea legs*. St. Johns, ISER Press, pp. 153–64.

BERRIEN, P.; SIBUNKA, J. 1999. Distribution patterns of fish eggs in the U.S. NE Continental Shelf ecosystem 1977–1987, Woods Hole, Massachusetts, *NOAA Technical Report, NMFS*, Vol. 145.

BARNHARDT, W.A.; BELKNAP, D.F.; KELLY, A.R.; KELLY, J.T.; DICKSON, S.M. 1996. Surficial geology of the inner continental shelf of the northwestern Gulf of Maine. *Maine Geological Survey*, Geologic Maps 96-6, 96-7, 96-8, 96-10, 96-11, and 96-12.

FRANK, K.T.; BRICKMAN, D. 2001. Contemporary management issues confronting fisheries science. *Journal of Sea Research*, Vol. 45, pp. 173–87.

FRANK, K.T.; DRINKWATER, K.F.; PAGE, F.H. 1994. Possible causes of recent trends and fluctuations in Scotian Shelf/Gulf of Maine cod stocks (ICES). *Marine Symposia*, Vol. 198, pp. 110–20.

GREEN, J.M.; WROBLEWSKI, J.S. 2000. Movement patterns of Atlantic cod in Gilbert Bay, Labrador: Evidence for bay residency and spawning site fidelity. *Journal of Marine Biological Assessment*, UK, Vol. 80, No. 3675, pp. 1–9.

NATIONAL RESEARCH COUNCIL. 1999. *Sharing the fish: Toward a national policy on individual fishing quotas*. Washington, DC, National Academy Press, 164 pp.

SINCLAIR, M.; O'BOYLE; R. BURKE; D.L.; PEACOCK, G. 1997. Why do some fisheries survive and others collapse? Developing and sustaining world fisheries resources: the state of science and management. *Proceedings of the Second World Fisheries Congress*. Melbourne, CSIRO, pp. 23–35.

SMEDBOL, R.K.; STEVENSON, R. 2001. The importance of managing within-species diversity in cod and herring fisheries of the northwestern Atlantic. *Journal of Fish Biology*, Suppl. A, pp. 109–28.

WROBLEWSKI, J.S. 1998. Substocks of northern cod and localized fisheries in Trinity Bay, Eastern Newfoundland and in Gilbert Bay, Southern Labrador. In: I. Hunt von Herbing, I. Kornfield, M. Tupper and J. Wilson (eds.), *Proceedings from the implications of localized fish stocks*. Ithaca, NY, NRAES, pp. 104–16.

CHAPTER 18   Integrating fishers' knowledge
with survey data to understand the structure,
ecology and use of a seascape off
south-eastern Australia

Alan Williams and Nicholas Bax

ABSTRACT

AUSTRALIA involves fishers at all stages of the fishery assessment and management process. A key factor in the success of this approach is using fishers' information to supplement and interpret standard fisheries data. From 1994, we collected fishers' information on fishing grounds and habitats as part of a five-year study of a continental shelf fishery. We met regularly with experienced fishers during port visits, commercial fishing operations at sea and in formal (management) meetings. This pattern of liaison enabled us to build relationships and a level of trust that facilitated a two-way sharing of knowledge. We integrated the ecological knowledge of fishers with scientific survey data to map and understand the seascape (seabed landscape) in a way that would not have been possible from scientific data alone. Fishers provided detailed information on the fishery, navigation, fishing effort distribution, individual species, fish behaviour, productivity, seabed biology, geology, and oceanography. A key result was an interpreted seascape map incorporating geomorphological features and biological facies at a variety of spatial scales of resolution from tens to hundreds of kilometres. Supported by the industry, we have extended the mapping project to the entire shelf and slope of the South East Fishery region. Fishers believe the project provides them with the opportunity for input to developing spatial management under Australia's 'Oceans Policy', and guarantees their involvement in a developing programme of 'regional marine planning'. However, they also fear that their information will be used against them – especially to close off valuable fishing areas. We discuss the importance of fishers' knowledge – interpreting scientific data, and the need for an ongoing dialogue between the fishing industry, scientists and managers. Only this ongoing dialogue will ensure that fishers' knowledge is used appropriately and, as importantly, that fishers' concerns are addressed in developing management options for this area.

INTRODUCTION

MANAGEMENT of the world's oceans has typically been driven by single issues – for example, how many fish to catch, where to discard waste, where to mine, dredge or drill for oil, and more recently which areas to protect (McNeill, 1994; Allison et al., 1998). At its simplest, single-issue management can be achieved with specific and limited information and by ignoring many of the potential interactions with other issues or aspects of the marine environment. However, coincident with our increasing awareness of the ecosystem services provided by the marine environment (Norse, 1993) is an increasing recognition of the limitations of single-issue management (Sainsbury et al., 1997), especially as our use of the oceans continues to increase.

It is no longer sufficient to manage a fishery solely on the basis of the number of fish removed; instead, where and how fishing occurs, and with what impacts, have become equally important questions. To answer these questions requires first that we define the management units we are dealing with (Langton et al., 1995). In particular, and as been the case on the land for centuries, spatial attributes of the marine environment have become increasingly important for effective management. This requires that we understand the ecological patterns at regional and local scales, and integrate over these scales to provide a 'seascape' perspective (Garcia-Charton and Perez-Ruzafa, 1999).

Australia is developing integrated management of its marine resources through Australia's Oceans Policy, launched in December 1998. Principal drivers for the policy are: ecosystem-based management, integrated oceans planning and management for multiple use, promoting ecologically sustainable marine-based industries, and managing for uncertainty (Commonwealth of Australia, 1998). It is recognized that real success of the plan will depend on all Australians gaining an appreciation and understanding of both the complexity of the ocean environment, and the interaction of humans within that environment (Sakell, 2001).

The marine environment off south-east Australia is the test case for 'regional marine planning' in Australia as it forms the first of thirteen 'large marine domains' (LMDs) that will eventually be covered by management plans. While there are some spatial data relevant to fishery management available for this area, in general they are either of low resolution (for example, the start and end positions of commercial fishing operations from fishery logbook records), or lack ecological interpretation (as in the case of bathymetric and geological maps from geoscience sampling). Until recently, little was known about the spatial organization of habitats (substrata, biota and adjacent water column) or the ways in which the seabed is used as fishing grounds. Seabed habitat in the South East Fishery (SEF) was mapped for the first time as part of a five-year study to interpret the ecological processes contributing to the productivity of the shelf fishery ecosystem – 'the ecosystem project' (Bax and Williams, 1999). The SEF is a complex, multi-species, multi-sector fishery (Tilzey and Rowling, 2001) that operates in a large fraction of the South East LMD adjacent to mainland Australia. The mapped area was ~24,000 km^2 of the continental shelf (~25–200 m depths) adjacent to the coastline between Wilsons Promontory in eastern Victoria and Green Cape in southern NSW – the south-eastern point of the Australian continental margin where east and south coasts meet (Bax and Williams, 2001: Figure 1). In that study, survey data provided the means to determine the structure of the seabed and its association with biological communities and environmental factors at particular scales in space and time (Bax and Williams, 2001; Williams and Bax, 2001). The addition of fishers' ecological knowledge aided the interpretation of those associations, as well as enabling an understanding of the ways in which the seabed is used by the commercial fishing fleet. As it turned out, fishers' information was so useful that we developed a second study – 'the mapping project' – using fishers' information on habitat types and distribution (interpreted through scientific knowledge and ground-truthing) as the primary data source to develop fine-scale maps of the south-east Australian seascape.

In this chapter, we first describe how fishers' knowledge contributed to the ecosystem project and explain why this provided a better understanding than a study based on scientific survey data alone. Second, we provide an overview of our methodology for collecting and integrating fishers' knowledge in the follow-up mapping project. Finally, we draw attention to the benefits of combining fishers' ecological knowledge with scientific survey data to provide a seascape perspective of the marine environment, and stress that this combination requires an ongoing dialogue between the fishing industry, scientists and managers. The direct benefit of combining our knowledge in this way is an improved understanding of the seascape. An indirect benefit is that it empowers fishers with the opportunity to be actively involved in developing management options for the marine environment with which they are most familiar.

THE SOUTH EAST FISHERY

THE continental shelf and slope off south-eastern Australia is the area of greatest fishing effort within the South East Fishery (SEF) – Australia's largest scalefish fishery, and the most important source of scalefish for domestic markets (see also Baelde, this volume). Trawling started in the early 1900s, and by 1999 the SEF fleet was made up of eighty-nine operating otter-board trawlers (draggers) and twenty Danish seiners (the 'trawl sector') (Tilzey and Rowling, 2001), as well as a smaller number of demersal longliners, dropliners, mesh-netters and trappers (the 'non-trawl sector'). More than 100 species form the commercial catch of the fishery, but eighteen species or closely-related species-groups managed by a system of catch-quotas make up the bulk (> 80 per cent) of the catch. Annual total allowable catches of individual species range from a few hundred to a few thousand tonnes, generating a total value for the fishery of about AUS$ 70 million.

OVERVIEW OF THE 'ECOSYSTEM' AND 'MAPPING' PROJECTS

THE ecosystem project was designed to consider the ways in which management intervention, beyond the established single-species fisheries management, could have a direct effect on the long-term productivity of this fishery ecosystem (Bax et al., 1999). Production was taken to mean both the production of fish and the factors that determine their availability to the fishery, while our concept of 'ecosystem management' was tied strongly to the notion of needing to manage peoples' interactions with ecosystem components (Bax et al., 1999). Engagement with the fishing industry was desirable to understand how fishers viewed the ecosystem, how they interacted with it, and how to best target our limited survey time. Accordingly, we initiated a two-pronged industry liaison programme when the project started. Depending on individual skills and experience, members of the project team became involved in formal fishery management and assessment meetings, and/or spent time in the two big ports in our study area (Eden and Lakes Entrance) and made trips to sea on fishing boats (several trips in the first year, then only one or two per year). A particularly useful feature of our sampling programme was using industry vessels for specialized fishing. Collectively, these interactions enabled us to establish contact with a range of industry personnel from working skippers to association executives. This gained us the support (and data) of individual operators and, in addition, the endorsements of the executive to further develop the project.

We maintained fairly regular contact with a core group of operators (about a dozen experienced working skippers) and were able to build up a level of trust and dialogue with this group as the project developed. Our findings were reported back to individuals and the major industry associations on an ad hoc basis during the course of the project. So, in summary, our approach to industry involvement evolved naturally during the ecosystem project – importantly, it lacked systematic planning or protocols, and there were no obvious benefits for the industry.

The contacts with industry members and associations that we developed during the ecosystem project proved crucial in mustering support for the second project – the mapping project – which makes extensive use of industry information and has explicit benefits (and risks) for the industry. In this partnership project, we are extending the seascape mapping to the entire continental shelf and upper slope (to ~1,300 m depth) of the SEF region. In contrast to the ecosystem project, the mapping project has a planned methodology for collection, review and release of industry data.

However, our approach will need to be adaptive as the scale and detail of outputs is realized, and as the industry responds to a rapidly evolving environmentally-focused fishery management regime. Key elements of the methodology are discussed in the final part of this chapter.

VALUE OF FISHERS' KNOWLEDGE FOR NAVIGATING AND MAPPING

When we started the ecosystem project our means of navigating around the fishery seabed was limited to what could be gleaned from third-party, coarse-scale bathymetry data and navigation charts – primarily point-source depth soundings, the approximate positions of key depth contours, including the continental shelf edge at ~200 m, and the positions of some near-surface rocky banks identified as shipping hazards (Table 18.1). This information, in combination with some prior survey data and some rapid exploration by echosounding during survey, enabled us to fix a set of transects and sampling sites, stratified by depth and latitude (Bax and Williams, 2001: Figure 1). These were used for a broad-scale coverage of the area during four seasonal trawl surveys – by definition on sediment substrata. But to meet the core aim of the project, which was to understand the importance of habitat to fisheries productivity, we needed both to survey a range of characteristic rocky reef habitats in the study area and understand the spatial context of habitats, such as patch sizes, boundary types and distributions.

TABLE 18.1 *Sources and types of information used to describe the continental shelf seascape in the south-eastern South East Fishery during the 'ecosystem project'.*

Information	Project surveys	Fishers' knowledge
Navigation over seabed	Navigational charts, depth contours	Accumulated maps in charts and plotters; names for features
Fishery	Fish species and size composition (quantified seasonal catches – trawl, trap, mesh-net)	Fish species and size composition (unquantified daily catches – trawl, mesh-net)
Fish behaviour (use of grounds)	Seasonal, diel (at times of surveys)	Time scales from days to decades
Fishing effort distribution	Logbooks (aggregated start position data)	Detailed tracks and marks of individual vessels
Productivity	Detailed energy flows at set points in time	Stability of fishing grounds over decades
Seabed biology	Fish and invertebrate communities (quantified, but few samples from nets, sleds, and photography); detailed species information	Dominant fish and invertebrate types (unquantified, but numerous net catches); local species-mixes or 'taxonomies'
Seabed geology	Rock type and geological history (dredge rocks); sediment classification (grab samples); depth contours (echo soundings from survey track lines)	'Ground-type' classification (gear damage/ wear, bycatch of rocks, mud etc.); depth contours (echo soundings accumulated over years of exploration)
Oceanography	Regional surface currents (SSTs; sea surface height) and local vertical structure (CTDs); bottom currents (sediment modification in photographs)	Local surface and bottom current direction and speed (gear/ vessel behaviour)

This is where we really started to benefit from our dialogue with fishers – they told us where to look. At an early stage we were able to build a focused study of habitats into the field surveys to intensively sample at a relatively small number of sites (Bax and Williams, 2001: Figure 1). This enabled us to understand the ecological roles of particular features, and their often small spatial scales (ranging from hundreds of metres to a few kilometres), for example the use of prominent reef edges by commercially important semi-pelagic, feature-associated species. Fishers' knowledge (Table 18.1) enabled us to progressively build a spatial framework on which to interpret the range of information we were collecting during our surveys. For example, by providing information on the boundaries of rocky reefs we were able to produce thematic maps of underlying geology (Bax and Williams, 2001: Figure 3). Over the course of the project we collected sufficient spatial information from fishers to put together what we called our 'fishers map' (Figure 18.1). In many ways it is a coarse-scale map of habitats, although its units – fishing grounds – are actually a hybrid mix of geomorphological features such as sediment plains and rocky banks, together with biological facies or biotope types – patches of substratum dominated by one particular community or animal. In summary, fishers contributed unique mapping knowledge, such as ground types, boundaries and names, that enabled us to understand the make-up of the seascape at variety of spatial scales – from small-scale features through to a regional overview.

VALUE OF FISHERS' INFORMATION FOR UNDERSTANDING SPECIES' ECOLOGY AND THEIR ENVIRONMENT

Two fundamental differences between observations made by fishers during commercial fishing and by scientists during surveys relate to the timing and frequency of sampling – the temporal and spatial resolution (Table 18.1). While time spent at sea by skippers varies considerably, some average over 200 days per year and sustain this for many years, building on the experience of their parents or other older skippers. In addition to learning where to fish, their mode of operation often includes searching and watching to enable precise target-fishing of fish 'marks' seen on echosounders. For example, the first shot of the day is often delayed until the 'feed layer' (or acoustic scattering layer) descends to the bottom – around first light (Prince et al., 1998).

In contrast, our survey samples (a combination of randomly directed and targeted) were fixed on the calendar, but essentially random in time as they took no account of the annual variability in seasonal progression (Bax et al., 2001) or of fine-scale patterns of fish movement. Sampling was only regulated (standardized) to either day or night, but not by season, or by considering a site–season interaction. Relative to the high number and frequency of commercial samplings, surveys

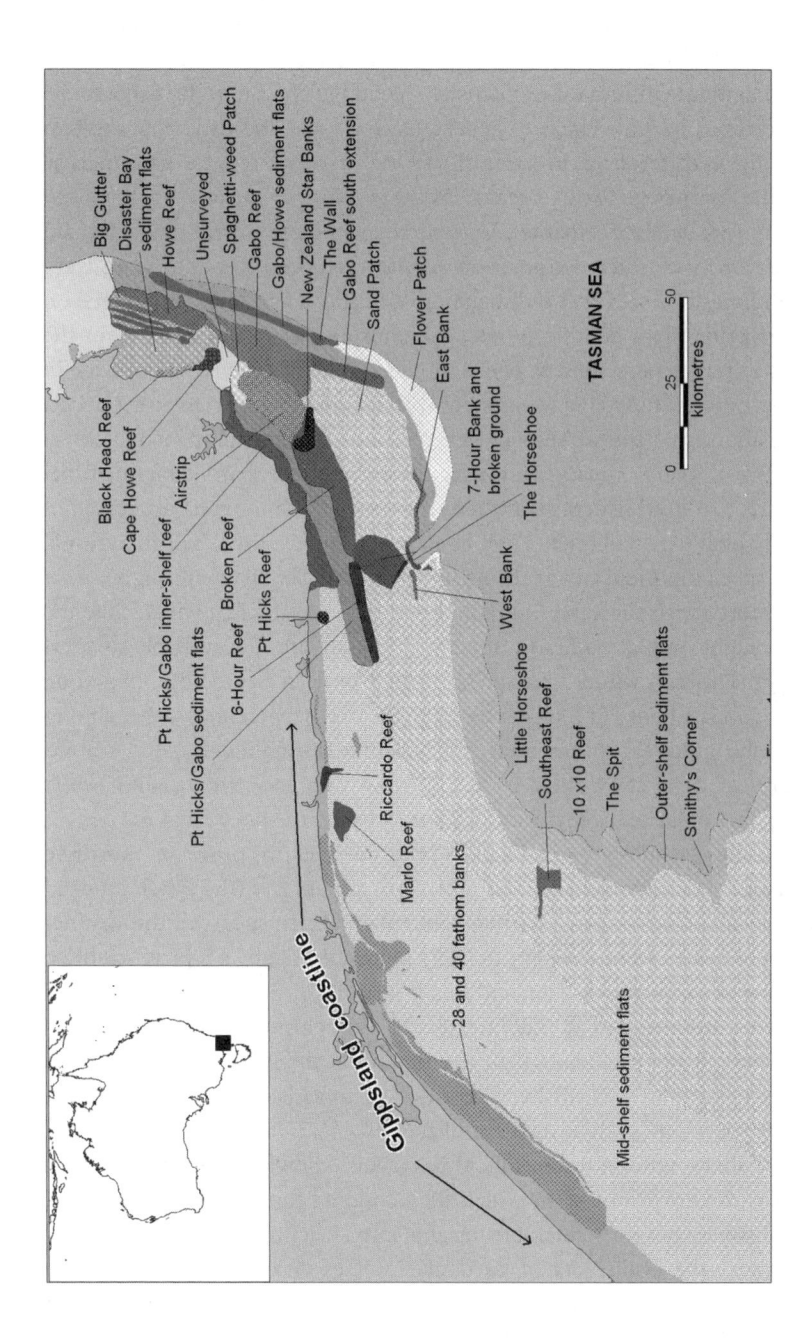

INTEGRATING FISHERS' KNOWLEDGE WITH SURVEY DATA IN AUSTRALIAN SEASCAPE MANAGEMENT

FIGURE 18.1 *A coarse-scale map of habitats – the 'fisher map' – made for the 'Twofold Shelf Bioregion' an area of the continental shelf off SE Australia. The map is a mix of fisher-delineated geomorphological features (mostly sediment plains and rocky banks) ground-truthed with physical samples and photographs from surveys.*

Source: Bax and Williams, 2001: Figure 4.

represent very brief snapshots in time and space. In the year when we sampled most intensively (two surveys in 1996) we completed fewer than 100 trawl tows on the continental shelf (< 250 m depth) while the trawl fleet completed over 10,000 – a two orders of magnitude difference in intensity spread widely across the fishery.

What differences in knowledge of species ecology and the fishery ecosystem resulted from these differences in sampling? One of many species examples is illustrated by the morwong, or sea bream (*Nemadactylus macropterus*), a mainstay quota species on the domestic market. Our survey sampling – including targeted sampling based on prior information from fishers – showed that morwong were associated with limestone reef and sediment substrata, and had high abundance on reef edges. It is primarily a benthic feeder, and presumably moves away from the shelter of reefs to forage on sediment plains. It had a generally higher abundance in the southern part of the study area (consistent with its broad temperate distribution) and was most abundant (in our seasonal trawl samples from sediment plains) in spring and autumn. Catch rates were higher during the day than at night in diel gillnet samples. Local trawl fishers report that movements of morwong are linked to season, depth, habitat type and time of day in a more complex way. Thus, in autumn, they catch this species in the south of the area, but catches are taken in progressively shallower and more northward areas over a period of weeks, during which time the morwong are caught only at night (in other words, they are not available to trawl during the day). Through winter and spring, with a peak in September, morwong move onto the elongate banks of limestone reef to the north where they are caught in what are called the 'gutters' between reefs, but now only during the day.

Our scientific data show this is not a spawning movement, and while oceanographic data indicate a general correlation between the horizontal movement of fish and opposing seasonal flows of warm and cool currents, the processes that drive the depth-related, substratum-associated and vertical patterns (the latter inferred from variable availability to trawl) remain unexplained. Irrespective, the distinct patterns known to fishers would be very unlikely to be detected by a typical scientific survey or by analyses of logbook data, and this is just one of the many examples for individual species. Information at this fine spatial and temporal resolution, unless provided by fishers, is not available to survey design, for the interpretation of catch per unit effort (CPUE) or other fishery statistics, or to assist an understanding of individual species' ecology such as habitat utilization.

Although fishers tend not to talk about their knowledge of the fishery 'ecosystem', it is the environment in which they conduct the business of catching fish. Successful fishers have considerable insights into structures and processes that affect production – the availability of particular species or species-groups, of the right size and in commercial quantities. In our region, fishers know that production is concentrated at the shelf break and on the upper slope (~150–700 m) particularly around canyon heads. Successful fishing depends on knowing when and where the right combinations of depth, bottom types, currents and good feed marks occur together. There are hotspots, but they are dynamic over periods of days, weeks or years – for example, with hydrodynamic climate being influenced by daily tide and the moon, they include episodes of upwelling, wind-driven currents, as well as 'long-term' seasonal events. Fishers may not be aware of the movement of the eddies of the East Australian Current onto the shelf, but their observations of how the fish catch changes with 'clean' or 'dirty' water matches the movement of these eddies. The extent to which hotspots can be detected or predicted is closely linked to the degree of success in fishing over time (see also Worm et al., 2003, for discussion of predator diversity hotspots in open water).

We were able to explain some of the patterns known to fishers by identifying food webs and sources of primary production from analysis of diets, stable isotopes and pigment breakdown products in survey data (Bax and Williams, 1999; Bax et al., 2001). Oceanic production (food) is highly important, while terrestrial or nearshore inputs are relatively trivial. Commercial shelf fishes, including many traditionally viewed as demersal or 'bottom dwelling', prey heavily on the animals that form 'feed layers' in the oceanic water column (pelagic prey) as well as those in local sediments (benthic prey) (Bulman et al., 2001). As a consequence, the seabed at the shelf-break is productive because it is bathed by upwelling waters that contain high levels of nutrients, particulate organic matter, oceanic pelagic prey, and particular elements of oceanic micronekton at their nearshore limit of distribution (for example, lantern fishes) (Bax and Williams, 1999). Fishing is especially productive in the first few hours of daylight, the time when the feed layer intersects with the bottom. Thus, because fishers and scientists tend to observe the fishery ecosystem at different spatial and temporal scales, their observations are often complementary. Fishers' knowledge may permit scientific observations to be better targeted, and more insightful, while survey data can provide the detail that leads to a more rigorous interpretation of fishers' knowledge.

ROLE OF FISHERS' INFORMATION IN UNDERSTANDING SEASCAPE USE

THE ways in which the seascape of this area is being used and impacted by fishing is the subject of developing interest by fishery mangers, environmental and conservation agencies, the general public and by the industry itself. Management of the seabed is being considered more actively, but whereas spatial management (or zoning) is universally accepted on land, it has only recently been considered as an option, or even necessary, in the ocean (Bohnsack, 1996). Spatial management on the land has benefited from numerous datasets available from visual observation of the landscape – in person, from the air or via satellite. Similar information is not available for the seascape because it cannot be observed directly (except at the shallowest depths).

Increasingly, scientific surveys can be used to provide detailed 'pictures' of the seabed with single-beam acoustics (Kloser et al., 2001a) or multibeam acoustics (Kloser et al., 2001b), but even the most modern techniques are very time consuming and therefore expensive, especially at shallower depths where the acoustic sampling footprint is comparatively small. Only large-scale undersea features such as upwellings of colder water driven by topographic features or sea level rises over submarine ridges can be observed from satellite. What is needed for spatial management, at anything less than the coarsest scale (bioregion and depth), is an information source of sufficient resolution to detect seabed features at the scale where management is possible (less than 1 km for fisheries where satellite transponders are fitted to vessels). Fishers operate below this level of resolution, and we suggest that they have the potential to provide information on the seabed at a scale suitable for spatial management.

In the SEF, the distribution of trawl tows has been used as an index of disturbance (Larcombe et al., 2001). However, interpretation of the resulting maps is limited because fishing is highly targeted at specific seabed features that occur at scales less than the typical three-hour trawl tow. Even unaggregated trawl start (or end) positions are poor representations of tows that are, on average, three hours in duration and therefore up to ~10 nautical miles in length. Analysis based on shot mid-points provides a closer spatial approximation of effort by considering both end-points, but suffers from the introduction of unknown errors because trawl tows do not follow straight lines. They most often follow physical boundaries and may involve several directional changes, for example to navigate through 'broken-ground'; the ~12-nautical-mile 'Snake Track' through the Howe-Gabo Reef complex is one aptly-named example. We conclude that logbook data (start and end positions) enable interpretation of effort distribution at the scale of fishing grounds (tens to hundreds of square kilometres), but provide limited insights into the impacts of seabed use because most significant habitat features occur at a finer spatial scale (tens to thousands of square metres) (Bax and Williams, 2001).

In the SEF, the vulnerability of different seabed types to fishing impacts is highly variable. Fishers have shown us that when areas of low-relief limestone slabs are fished, benthic fauna and some of the actual substratum can be removed. On the other hand, high relief and heavily cemented limestones will never be trawlable and these are regarded as 'natural refuges' by trawl fishers. However, these same 'natural refuges' are often the prime fishing grounds of the non-trawl sector that fishes with static gear such as gillnets, traps, and hook and line. This is a potential source of conflict between industry sectors when spatial management is introduced to the fishery. Habitat features at the scale at which the industry sectors operate will need to be considered if equitable management arrangements are to be introduced, although actual management regulations may operate at a coarser scale. Using the information collected by the fishers themselves is the only feasible way to map the seascape at a resolution similar to that at which fishers operate. However, this information is sometimes highly confidential, being the commercial advantage that one fisher may have over another. In the following section we describe how we set about accessing this confidential information.

INTEGRATION OF FISHERS' KNOWLEDGE IN THE MAPPING PROJECT

'INTEGRATING fishing industry knowledge of fishing grounds with scientific data on habitats for informed spatial management and ESD evaluation in the SEF' – the official title of the mapping project – has the explicit aim of incorporating fishers' knowledge of the seascape into strategic management planning. We have broad support from the industry because the project is viewed as a mechanism through which industry information can be incorporated in decision-making processes for the fishery, and which will help produce better-informed decisions. However, support is not unanimous and this is due, in large part, to many fishers remaining sceptical about whether their information will be used appropriately. Moreover, fishers are not a single cohesive group, and have different views of the system they fish, and short or long-term approaches to sustainability – based, at least in part, on their level of tenure in the fishery. Some fishers are unwilling to share their commercially confidential information with us. Many fear that their information will be used against them, especially to close off valuable fishing areas; they are well aware of the link between areas of high fishery productivity and areas of high biodiversity. In our approach to gathering, storing and releasing industry information, we needed to address these concerns as far as possible; we aimed to maximize support from the industry, while also retaining the option to release aggregated industry knowledge to a broad audience in the form of maps.

We argued the benefits of the project aims to individuals and the peak bodies for several years (including through several failed proposals) before we gained support and funding. Our key argument was that the project would provide a tool to help the industry respond to the raft of upcoming environmental legislation soon to affect the fishery. Legislation includes spatial management of all marine industries under Australia's Oceans Policy – a developing programme of Regional Marine Planning that includes a National Representative System of Marine Protected Areas (Baelde, this volume) as well as fishery specific 'strategic environment impact assessments' that aim to support ecological sustainability. With their information systematically collected and rigorously evaluated, fishers would be positioned to critically evaluate proposed spatial management plans, such as the placement of marine protected areas (MPAs), and could require management agencies to have clearly defined and measurable aims for their proposed management options.

Interestingly, the major industry bodies supported the project, at least in part, because they saw it as a mechanism for the industry to be actively engaged in the process of management planning, rather than just reacting to it. Our hope is that the project, by broadening industry understanding of the seascape they rely on, will encourage proactive thinking and actions from the industry to enhance the sustainability of their fishery. In addition, the project provides the industry with a tool for improving its public image. At present, fishers are concerned about what they see as poorly informed and often misleading media and scientific reporting on interactions between fishing and the environment. This project will provide the industry with some hard facts that they can use to demonstrate their real level of impact on the seascape; the trawl sector, for example, is particularly keen to be able to demonstrate that large areas of the fishery are untrawlable or untrawled.

The project is structured in a very transparent way so as to give fishers a high degree of control over the form in which information is released and the timing of various outputs. We have agreed that habitat maps of the area will be released following review by individual contributors and the relevant associations, and that these maps will include summary detail from commercially confidential information. Higher resolution maps of specific areas of interest, showing precisely the trawled and untrawled areas may also be released but these will require the approval of individual fishers.

In brief, our method of collecting fishers' knowledge in a systematic and rigorous manner centres on the use of vessels' electronic track-plotter data to make maps, with interpretation provided from information gathered via a simple data sheet. This records fishers' impressions of the seabed habitats that make up individual fishing grounds, and is based mainly on fishers' observations of echosounder recordings, wear on fishing gear, and material taken in catches. Information recorded for each ground uses a set of terms commonly used by fishers; it includes what grounds 'are made of' (substratum or bottom type), what 'they look like' (geomorphology), which fishing gears are used, which features contribute to defining the boundaries of grounds (such as physical seabed features, depths or landmarks), and information on our confidence in the data. Confidence levels are scored for each bottom type and boundary on the basis of the general nature of the ground (heterogeneous substratum and complex geomorphology score lower), and whether impressions are independently corroborated by other fishers, or are verified by scientific sampling. Spatially referenced commercial logbook data were overlaid on habitat maps to provide information on the distributions of catch and effort. An overview of the methods, as well as key processes and infrastructure of the project is provided on a project website www.marine.csiro.au/sefmapping.[1]

---

1. Final report on this project added July 2006.

Our approach is adaptive to a degree for two main reasons. First, it is difficult to determine what level of spatial scale and detail is acceptable for map outputs until data are collected and mapped. We have an explicit stepwise protocol for making, reviewing and releasing maps – but there is flexibility to release maps at various resolutions, depending on the specific needs and concerns of the industry and ourselves. Second, the implementation of the new legislation for this fishery is evolving rapidly: the transition from conceptual to operational objectives may make demands on information that we have not anticipated. For that reason we have developed a comprehensive questionnaire, requiring the repeated involvement of active fishers. The resulting data will be available as new management approaches develop, thus allowing the industry to have an input in their development, and managers to access information in a form that best addresses their specific management objectives.

CONCLUSIONS

MANAGEMENT for conservation, multiple-use or fishery goals will benefit from collaboration with the fishing industry because fishers know the seascape considerably better than other stakeholders, and they have a broad understanding of the processes that influence fishery productivity. As concisely stated by Neis (1995), 'fishers deal regularly with a landscape that no-one has seen'. In addition, fishers can provide the means for cost-effective acquisition of mapping data over large areas, and they have an important stake in ensuring that any spatial management of the seabed is based on reliable information interpreted appropriately. Acquiring reliable data requires a structured, verifiable collection process, and methods to reconcile conflicting accounts.

However, collaboration with the industry is not limited to acquiring fishers' data; it requires an ongoing dialogue if the data are to be interpreted judiciously and the industry is to understand the value of any proposed management measures (Neis, 1995). Developing maps of the seabed is one thing, but interpreting them to provide the basis for improved management of the fishery that takes account of the diversity and specialization of fishers' daily activities is another. This is where the ongoing dialogue between the fishing industry and scientists really begins.

ACKNOWLEDGEMENTS

WE gratefully acknowledge the assistance and support provided by many operators in the South East Fishery, and the roles played by the major trawl and non-trawl associations (SETFIA and SENTA). Individual fishers from the catching sectors and fishing cooperatives of Eden in New South Wales and Lakes Entrance in Victoria shared their knowledge of the regional fishery and took us to sea on their fishing vessels during the ecosystem project. We are also indebted to several staff at CSIRO Marine Research, especially Bruce Barker for technical support, Rudy Kloser and the Marine Acoustics group for collaborative efforts in seabed habitat mapping, and the Ocean Engineering Group and workshop, especially Matt Sherlock and Ian Helmond, for considerable technological development and support in past and ongoing field work. Dr Alan Butler and Bruce Barker provided useful comments on an earlier version of the paper. Dr Jeremy Prince played an important role in securing trawl industry support for the mapping project; Dr Pascale Baelde and Crispian Ashby are part of the project team on the mapping project. The funding for both projects was provided jointly by CSIRO Marine Research and the Fisheries Research and Development Corporation.

REFERENCES

ALLISON, G.W.; LICHENS, J.; CARR, M.H. 1998. Marine reserves are necessary but not sufficient for marine conservation. *Ecological Applications*, No. 8, pp. S79–S92.

BAX, N.J.; BURFORD, M.; CLEMENTSON, L.; DAVENPORT, S. 2001. Phytoplankton blooms and productivity on the south east Australian continental shelf. *Marine and Freshwater Research*, No. 52, pp. 451–62.

BAX, N.J.; WILLIAMS, A. 1999. *Habitat and fisheries production in the South East Fishery*. Final report to the Fisheries Research and Development Corporation (FRDC) Project 94/040. Hobart, Tasmania, CSIRO Marine Research, 625 pp.

——. 2001. Seabed habitat on the southeast Australian continental shelf: Context vulnerability and monitoring. *Marine and Freshwater Research*, No. 52, pp. 491–512.

BAX, N.J.; WILLIAMS, A.; DAVENPORT, S.; BULMAN, C. 1999. Managing the ecosystem by leverage points: A model for a multispecies fishery. In: *Ecosystem Approaches for Fisheries Management* (Alaska Sea Grant College Program, AK-SG-99-01). Fairbanks, University of Alaska, pp. 283–303.

BOHNSACK, J.A. 1996. Marine reserves, zoning, and the future of fishery management. *Fisheries*, No. 21, pp. 14–16.

BULMAN, C.M.; HE, X.; BAX, N.J.; WILLIAMS, A. 2001. Diets and trophic guilds of demersal fishes of the south-eastern Australian shelf. *Marine and Freshwater Research*, No. 52, pp. 537–48.

COMMONWEALTH OF AUSTRALIA. 1998. Australia's Oceans Policy, vols. 1 and 2. Canberra, Environment of Australia, 52 pp.

GARCIA-CHARTON, J.A.; PEREZ-RUZAFA, A. 1999. Ecological heterogeneity and the evaluation of the effects of marine reserves. *Fisheries Research*, No. 42, pp. 1–20.

KLOSER, R.J.; BAX, N.J.; RYAN, T.; WILLIAMS, A.; BARKER, B.A. 2001a. Remote sensing of seabed types in the Australian South East Fishery: Development and application of normal incident acoustic techniques and associated 'ground truthing'. *Marine and Freshwater Research*, No. 52, pp. 473–89.

KLOSER, R.K.; WILLIAMS, A.; BUTLER, A.J. 2001b. *Acoustic, biological and physical data for seabed characterisation*. Marine Biological and Resource Surveys of the South East Region, Progress Report No. 2 to the National Oceans Office. Hobart, Tasmania, CSIRO Marine Research, 332 pp.

LANGTON, R.W.; AUSTER, P.J.; SCHNEIDER, D.C. 1995. A spatial and temporal perspective on research and management of groundfish in the northwest Atlantic. *Reviews in Fisheries Science*, No. 3, pp. 201–29.

LARCOMBE, J.W.P.; MCLOUGHLIN, K.J.; TILZEY, R.D.J. 2001. Trawl operations in the South East Fishery, Australia: Spatial distribution and intensity. *Marine and Freshwater Research*, No. 52, pp. 419–30.

MCNEILL, S.E. 1994. The selection and design of marine protected areas: Australia as a case study. *Biodiversity and Conservation*, No. 3, pp. 586–605.

NEIS, B. 1995. Fisher's ecological knowledge and marine protected areas. In: N.L. Shackell and J.H.M. Willison (eds.), *Marine protected areas and sustainable fisheries*. Nova Scotia, Canada, Science and Management of Protected Areas Association, pp. 265–72.

NORSE, E.A. (ed.). 1993. *Global marine biodiversity: A strategy for building conservation into decision making*. Washington, Island Press, 383 pp.

PRINCE, J.; BAELDE, P.; WRIGHT, G. 1998. *Synthesis of industry information on fishing patterns, technological change and the influence of oceanographic effects on SEF fish stocks*. Final report to the Fisheries Research and Development Corporation (FRDC) Project 97/114. Canberra, Fisheries Research and Development Corporation, 103 pp.

SAINSBURY, K.; HAWARD, M.; KRIWOKEN, L.; TSAMENYI, M.; WARD, T. 1997. Multiple use management in the Australian marine environment: principles, definitions and elements. Oceans Planning and Management; Issues Paper 1. Canberra, Environment Australia, 42 pp.

SAKELL, V. 2001. Australia's Oceans Policy: Moving forward. Paper presented to the Global Conference on Oceans and Coasts at Rio+10, 3–7 December 2001. Paris, France, UNESCO.

TILZEY, R.D.J.; ROWLING, K.R. 2001. History of Australia's South East Fishery: A scientist's perspective. *Marine and Freshwater Research*, No. 52, pp. 361–75.

WILLIAMS, A.; BAX, N. 2001. Delineating fish-habitat associations for spatially-based management: an example from the south-eastern Australian continental shelf. *Marine and Freshwater Research*, No. 52, pp. 513–36.

WORM, B.; LOTZE, H.K.; MYERS, R.A. 2003. Predator diversity hotspots in the blue ocean. *Proceedings of the National Academy of Sciences USA*, No. 100, pp. 9884–8.

CHAPTER 19   Using fishers' knowledge goes
beyond filling gaps in scientific knowledge
Analysis of Australian experiences
Pascale Baelde

ABSTRACT

ACCESSING and using fishers' knowledge presents specific challenges in the case of industrial fisheries. In Australia, these fisheries are under increasing pressure from tighter management controls and public demands for greater environmental protection. The principles and practices of fisheries science and management have changed significantly over the last decade with, in particular, greater emphasis on ecosystem-based approaches and implementation of partnership frameworks where fishers share responsibilities and costs for research and management. Three examples of industry–government partnership in Australian fisheries are reviewed. These examples illustrate the evolution in fisheries assessment and management from traditional fish stock assessment approaches to the ecosystem-based implementation of marine protected areas. As fisheries science and management evolve, current perceptions about fishers' role in these areas also need to change. It is clear that today fishers' input extends beyond simply filling gaps in scientific knowledge. The difficulties of integrating fishers' knowledge and expertise into science-driven processes and the influence of some sociocultural factors on the interactions between fishers and scientists/managers are discussed.

INTRODUCTION

THERE is a growing perception, worldwide, that conventional fisheries management is failing. Despite a few recovering stocks, many fish stocks are declining and some fisheries have already collapsed. To help improve the management of fisheries, there is an increasing recognition that more attention should be paid to fishers' knowledge and to the factors that affect fishing behaviour (Hilborn, 1985, 1992; Hilborn and Walters, 1992; Dorn, 1998; Neis et al., 1999a, 1999b; Salmi et al., 1999; Neis and Felt, 2000; and references therein).

Fishers' knowledge, and its communication to scientists, is influenced by the biological, socio-economic and cultural contexts in which fishers operate. Its value and usefulness is most often understood and studied in the case of data-poor fisheries where conventional fisheries research and management methods are not applicable, such as small-scale indigenous fisheries in the tropics (e.g. Johannes, 1998). Management philosophy and problems in these fisheries differ significantly from those in industrial fisheries. Indigenous peoples tend to have long-standing associations with a particular area and environment, whereas in more recently developed industrial fisheries, fishers association with the environment is more transient and is mediated by their tighter integration into technologically, socially and economically dynamic capitalist societies (Neis and Felt, 2000). Also, in industrial fisheries, formal procedures for the assessment and management of fish resources have been in place for some time and usually rely on scientific analysis of fisheries and biological data. This chapter is concerned with fishers' role and the use of their knowledge in Australian industrial fisheries.

Most Australian fisheries are under tight management controls and, since the early 1990s, management systems have rapidly evolved from input-based controls (such as gear control or spatial management) to output-based controls (such as quotas). This is accompanied by the implementation of other mechanisms such as co-management and partnership approaches, allocation of fishing rights, management and research cost recovery mechanisms and, in some fisheries, implementation of industry-driven data collection. In theory, the granting of fishing rights is viewed as a means of providing fishers (and their financial institutions) with greater security of access to resources, thus promoting financial investment and development and long-term stewardship of the resources.

The development of ecosystem-based and precautionary approaches, along with greater and more open recognition of the uncertainty inherent to scientific results (Hilborn, 1992), are also characteristic of ongoing changes in fisheries science and management. Fishers, scientists and managers have had to review and adapt their philosophical beliefs and professional practices to these new approaches. Australian fishers are now more involved in the scientific assessment and management of their fisheries, for which they pay a significant share, or even the entirety, of the costs. However, in a context where fisheries assessment and management remain dominated by science, what is the role of fishers and the value of their knowledge? What are the implications of the participatory approach for scientists, managers and fishers?

In this chapter, the partnership between fishers and scientists/managers in Australia is reviewed through three examples (all relating to fisheries operating off south-east Australia, see Figure 19.1):

- Fishers as information providers: this example relates to an industry survey where fishers provided information on changes in fishing gear and fishing practices.

- Fishers as active collaborators: this example relates to the development of an alternative to conventional fish stock assessment methods that involves collaboration with fishers.
- Fishers as major stakeholders in the development of marine protected areas: this example relates to fishers' role in ecosystem-based management of marine resources and biodiversity conservation.

EXAMPLE 1. FISHERS AS INFORMATION PROVIDERS

THIS example relates to a survey of the Australian south-east trawl fishery (SETF), which was carried out to collect information on, among other aspects, changes in fishing gear and fishing practices. The SETF is a demersal, multi-species fishery in which catches of the sixteen most important species have been controlled by individual transferable quotas (ITQs) since 1992. Fishers also have a long, ongoing,

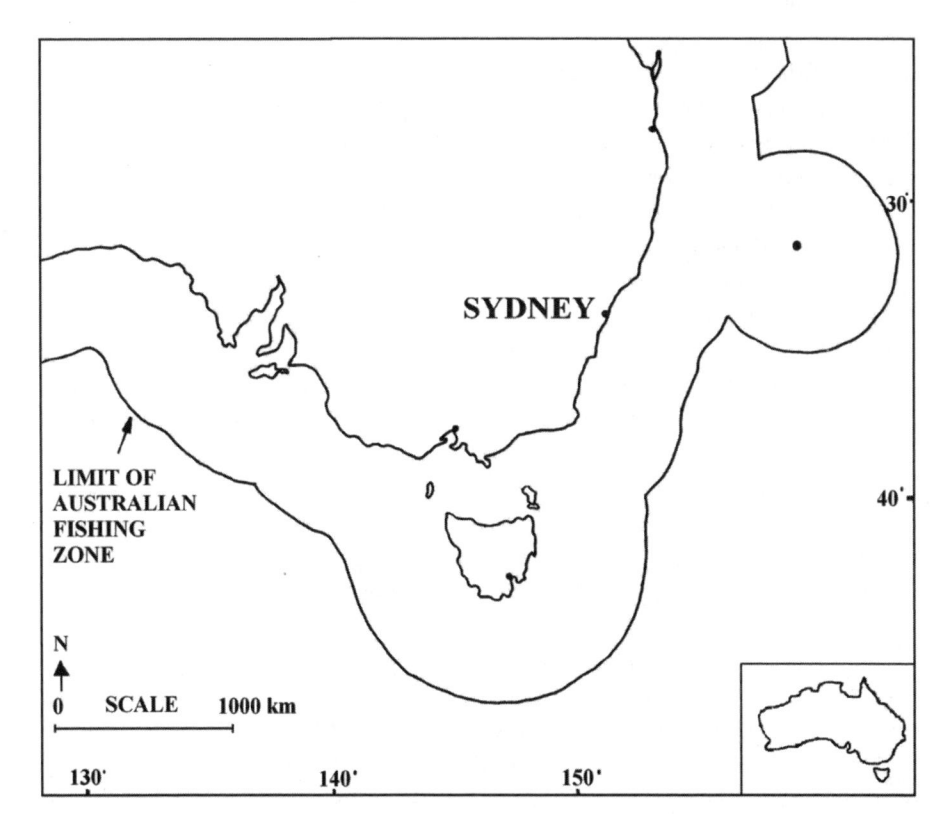

FIGURE 19.1 *SE Australian fishing zone.*

but mostly unappreciated or unacknowledged, history of contributing to research and cooperating with scientists, often on a voluntary basis. In the SETF, their contribution to fisheries assessment and management formally began in 1986 when they started recording catch statistics in compulsory fishing logbooks. They also regularly help with data collection during scientific surveys, and take scientific observers onboard their vessels for routine catch monitoring studies, discarding studies, tagging experiments and fishing gear trials.

Scientific stock assessments are done on a single-species basis and rely for the great majority of species on catch per unit effort (CPUE) analysis using fisheries-dependent catch and fishing-effort data recorded in logbooks (fisheries-independent survey data are available for only two of the sixteen quota species). Both fishers and scientists have long questioned the validity of data recorded in compulsory logbooks, either because of potential misreporting by some fishers (especially since the implementation of the ITQ management system), or because of the influence of changes in fishing gear and fishing practices. Also, the single-species approach to stock assessment in this typically multi-species fishery, and scientists' reliance on CPUE as an index of fish abundance, have become a long-standing point of contention between fishers and scientists. It is well known that using CPUE as an index of fish abundance can lead to misleading results if changes in fishing gear and practices are not taken into account (Megrey, 1989; Hilborn and Walters, 1992; Tilzey, 1999; Baelde, 2001). Over the years, fishers' lack of confidence in scientific methods and advice grew as they repeatedly demanded that scientists integrate changes in fishing technology and the influence of quota management and market demands on fishing practices into their analyses.

Eventually, in 1997, an industry survey was funded to collect this type of information. A questionnaire was used during face-to-face interviews with fishers, which was designed to collect a combination of quantitative information (on vessel and gear description, for example) and qualitative information (such as fishing practice preferences) (see Baelde, 1998, 2001 for more details). Much care was taken to keep the interviews flexible, extending the discussion beyond purely scientific conceptions (Johannes et al., 2000). Besides specific and practical questions, the questionnaire also included open-ended questions to give fishers the opportunity to expand on their answers. The aim of the survey was to provide scientists with information that would help them improve their analysis of logbook data.

A series of pilot interviews was conducted prior to the main survey in order for the interviewer to familiarise herself with the most pertinent issues and with the language/terminology used by fishers. Results from these pilot interviews were used to formulate questions that would be well understood by fishers and relevant to their situation. Also, various validity and reliability checks were built into the survey questionnaire, as well as coding and ranking mechanisms, to assist later quantification and analysis by scientists of the information collected. Following Beed and Stimson

(1985), checking reliability related to checking the consistency of answers given by one fisher to a series of questions which related to common or interconnected topics (for instance, the consistency of answers to questions relating to targeted species and quota availability, to seasonal changes in fishing distribution and seasonal changes in species distribution, to types of net used and types of depths/habitats targeted). Checking validity involved checking the level of individual fishers' knowledge (such as length of experience in the fishery, in targeting particular species, distinction between perceived 'common knowledge' and personal knowledge, or methods by which fishers accumulated their knowledge).

The survey was a great success with fishers; all but two of the forty-seven approached agreed to be interviewed (representing more than half of the skippers actively engaged in the fishery at the time). They provided a large and diverse amount of information including technical details of fishing equipment and description of the influence of environmental, economic and management factors on fishing practices. Their perceptions and beliefs about the status of the fishery and the effectiveness of management were also recorded.

Qualitative analyses of the information collected identified significant changes in fishing practices following the implementation of ITQs (Baelde, 1998, 2001). For example, these changes included fishers' increasing practice of 'running away' from high concentrations of fish (also referred to as 'dodging the fish') to avoid exceeding their allocated quota and to limit dumping unwanted catches (a well-known negative consequence of quota-based management). Fishers said that before quotas were implemented they occasionally needed to 'dodge the fish' when the domestic market was flooded. However, they need to do it more frequently now because of quota restrictions.

In addition, since the mid-1980s, there has been a progressive shift in the fishery from maximizing catch volumes to diversifying the species composition of catches in order to respond better to domestic market demand. ITQs have reinforced this trend, with fishers catching smaller 'mixed-bags' of several species to satisfy both market demand and quota restrictions. In their effort to diversify the species composition of their catches, fishers have modified the original designs of their trawl-nets (where different nets were designed to target different species) to make what they call 'multi-purpose' nets capable of targeting a variety of species over a variety of ground types. New nets and access to the geographical positioning system (GPS) have also allowed fishers to work closer to harder grounds which are more productive. In response to quota restrictions, fishers also tend to focus on non-quota species.

The few fishing practices described above have the potential to selectively drive down the CPUE of some species (for instance, when avoiding large concentration of fish) and increase the CPUE of others (as when accessing more productive grounds). These trends in CPUE would have no or little relation to changes in fish abundance and would not be consistent across the fishery (in other words, would depend on individual fishers' fishing practice preferences) (Baelde, 2001). The survey also showed that, while assumptions about a direct relationship between technological improvement (such as access to GPS) and increase in catches have proved to be correct in some single-species fisheries under input control (Robins et al., 1998), such assumptions are not necessarily justified in the case of multi-species quota-managed fisheries like the SETF (Baelde, 2001).

Another important change in fishing practice worth mentioning here is the increase in communication between fishers, apparently also as a result of ITQs. As observed in other fisheries (Allen and McGlade, 1986; Maurstad, 2000a), communication between fishers influences fishing strategies and the distribution of fishing effort. In the SETF, individual fishers need to know what other fishers are catching (what species and in what quantity) in order to maximize the value of their own restricted catches on the market.

Despite the success of the survey, both in terms of fishers' willingness to participate and volunteer information and in terms of the wealth of information collected, things did not progress much further. Changes in electronic equipment and net design (the details of which are mostly unknown to scientists), and quota and market-driven changes in fishing practices have not been investigated further by scientists. The influences of these changes on CPUE trends are not yet taken into account in stock assessments, despite their potential to seriously undermine the validity of these assessments. In fact, after initially welcoming the results of the survey, scientists appeared to quickly lose interest. It became clear that they had unrealistic expectations and a poor understanding of the nature and content of fishers' knowledge. They failed to appreciate the need for dedicated and specialised work to turn this knowledge into a useful form for science. Institutional inertia quickly overcame their initial interest in favour of established fisheries science practices. Thus, single-species stock assessments and reliance on CPUE still remain a source of contention between fishers and scientists.

EXAMPLE 2. FISHERS AS ACTIVE COLLABORATORS

EXAMPLE 1 described a direct interaction, albeit of limited success in this case, between fishers' information and conventional stock assessments. In Example 2, the blue eye trevalla (*Hyperoglyphe antarctica,* Centrolophidae) fishery, quantitative stock assessment methods are not yet possible because of the limited data available and complex fleet and stock behaviour (Baelde, 1995, 1996, 1999). Thus, another approach is taken which is based on the more holistic harvest and management strategy models that are currently being developed in Australia and elsewhere (Smith et al., 1999; Punt et al., 2001).

Broadly speaking, simulation-based operating models are to be built from hypotheses, or 'what if' scenarios. Hypotheses relate, for example, to the behaviour of the fish (factors driving migration movements, spawning), behaviour of the fishing fleet (factors driving the spatial and temporal distribution of the fleet), consequences of various management regimes (such as varying quota levels) and other factors. These scenarios will be identified using available data and expert opinion from scientists, fishers and managers (see Holm, 2003, for a discussion of using fishers' knowledge as hypotheses and for examples; also Stanley and Rice, this volume). In building the models, management strategies, stock assessment methods, performance indicators and research programmes are simulated and compared (Punt et al., 2001). A working group of scientists, fishers and managers has been created and the process is currently underway. Three major challenges have been identified in using this approach.

The first challenge was to get members of the working group to accept and support a simulation approach. As Smith et al. (1999) pointed out, the development of operating models is an unfamiliar, complex and still experimental approach in fisheries. To go from the principles and concepts of quantitative stock assessment to a simulation approach has proved difficult for everyone involved. For example, without adequate quantitative stock assessments (as in the case of blue eye trevalla), an operating model cannot answer questions regarding the size or status of the fish stock. Thus it cannot provide quantitative advice on quota levels and this is a major setback for fisheries managers working within a quota system.

The second challenge is to get members of the group to commit themselves to the process. The success of the approach depends on genuine participation and collaboration between scientists, fishers and managers. It is important that expertise and interests from all participants are taken into account in developing harvest and management hypotheses. Members must not only share their expertise and interests, but also be able to handle sensitive and/or controversial information in a transparent manner, while respecting confidentiality. Participants also need to openly recognize the uncertainty inherent to their specific knowledge.

The third challenge will be to get members of the group to agree on how to use the results of simulations. Operating models will test the performance of, and risks associated with, various simulated management strategies. For example, 'what if' scenarios could involve proportional splitting of the total allowable catch between fishing methods, or closing particular fishing grounds. On the basis of these tests, the group will then have to decide, and agree upon, a particular set of decision rules that trigger management actions.

In conclusion, in this second example fishers' role is not to inform scientists and fill gaps in scientific knowledge (as in Example 1), but to cooperate with scientists and managers in developing management strategies for the fishery. To develop meaningful simulation models requires effective industry participation and, as noted by Smith et al. (1999), these new trends in research and management fit better with the co-management approach adopted in Australia. However, as a note of caution, Punt et al. (2001) highlighted that hypothesis-based modelling approaches may not resolve contentious issues, but simply move them from being about the validity of data and assumptions in stock assessment methods to being about the plausibility of hypotheses.

EXAMPLE 3. FISHERS AS KEY STAKEHOLDERS IN THE DEVELOPMENT OF MARINE PROTECTED AREAS

There is a growing perception that traditional fisheries management methods are failing, and more and more attention is being paid worldwide to the establishment of marine protected areas (MPAs) to assist fisheries (Attwood et al., 1997; Lauck et al., 1998; Walters, 1998, 2000; Parrish et al., 2000; Pitcher, 2001; Ward et al., 2001). Many problems in fisheries are attributed to a failure by management to adopt a precautionary approach, and the implementation of MPAs (and of no-take areas in particular) is now promoted as the most effective precautionary approach to protect both fisheries resources and biodiversity (Roberts and Hawkins, 2000; Ward et al., 2001). In this fairly recent development in fisheries management philosophy, MPAs are not seen as substitutes for traditional fisheries management methods but as complements.

In Australia, the release of the Oceans Policy in 1998 included acceleration of the implementation of national and regional networks of multiple-use MPAs. This is currently being met with strong resistance from commercial fishers who are directly affected by the establishment of no-take zones within these MPAs. Environmentalists often perceive fishers' resistance as resistance to change and lack of care for the environment. However, in Australia, it is a lack of integration of MPA development with fisheries management which most contributes to fishers' resistance (Baelde et al., 2001). Australian governments clearly state that MPAs are primarily used for biodiversity conservation and not for fisheries management (ANZECC TFMPA, 1998; see also Baelde et al., 2001, for a review of governments' MPA policies). Moreover, fisheries and conservation government agencies show little willingness to cooperate on MPA issues or to accommodate their differing philosophical beliefs and legislative responsibilities.

By relying primarily on spatial management (a form of input control), the development of MPAs tends to conflict with current trends in fisheries management discussed earlier (those based on output controls and allocation of fishing rights). While it is not the purpose of this chapter to discuss the appropriateness or otherwise of Australian fisheries management systems or the value of MPAs, the point here is to highlight the uncertainty caused to fishers by the lack of congruence between the objectives of conservation and of fisheries management.

Governments' MPA policies fail to acknowledge or properly assess the potentially negative impacts of MPAs on commercial fisheries (Baelde et al., 2001). As a consequence, mechanisms to address these impacts (such as more flexibility in designing MPAs, compensation to fishers, or fisheries re-structuring) are not properly investigated. Australian governments are generally reluctant to pay compensation to fishers for loss of access to fishing grounds (and loss of fishing rights), and fishers now tend to use the compensation issue as a bargaining tool in negotiating with governments. Government agencies and MPA advocates fail to recognize that for most fishers, compensation is a last option. They would rather see more compromise between biodiversity conservation needs and use of fish resources in designing MPAs. The current poor integration of conservation and fisheries management, as well as lack of consideration of socio-economic issues, means that the opportunity for using MPAs as tools for re-structuring fisheries (that is, to reduce fishing effort) is being missed (Baelde et al., 2001).

Fishers have to be content with blanket claims that MPAs may benefit their fisheries and provide protection against stock collapse (Roberts and Hawkins, 2000). Ward et al. (2001) have clearly shown that benefits to fisheries from MPAs occur in quite specific circumstances (namely, in the case of overfished and/or unregulated fisheries). Claims of fisheries benefits from MPAs in the Australian context appear largely unsubstantiated and therefore unnecessarily undermine the validity of the conservation message (Baelde et al., 2001).

Another important consequence of the poor integration of fisheries and biodiversity conservation needs is that conservation agencies also fail to recognize and promote the role that fishers could play in the protection of the marine environment. MPAs are selected almost regardless of existing fisheries management systems and with very limited, or inadequate, input from commercial fishers. A recent review of Australian governments' MPA policies and planning processes (Baelde et al., 2001) showed that fishers have little opportunity to input from the start of the planning process for the selection and design of MPAs, and that their concerns and needs are generally overlooked or poorly addressed. This too is in conflict with fisheries co-management and partnership approaches. Whether MPAs are intended solely for biodiversity conservation, for fisheries management or a combination of the two has major implications for their selection and design (size, location, level of protection) and expected benefits and costs for fisheries. This in turn influences fishers' share of MPA management costs (monitoring, compliance and enforcement) and their potential involvement in MPA processes (Baelde et al., 2001).

It is well documented that to achieve effective natural resource management and conservation with minimal conflict and long-term community support requires the involvement of those directly affected by management measures (Fiske, 1992; Neis, 1995; Well and White, 1995; Beaumont, 1997; Crosby, 1997; Johnson and Walker, 2000). However, in Australia, as observed elsewhere (Beaumont, 1997), while government policies and legislation on resource management never fail to mention the importance of stakeholders' participation, they rarely provide practical details and critical accounts of approaches taken (Baelde et al., 2001). Moreover, government policies tend to expect more and more from consulting with fishers. Consultation is expected to resolve many different issues: among others, to provide expert environmental knowledge, provide socio-economic information and assist integrated management by reducing conflict between users. While the stated scope of consultation with fishers continues to expand, there are generally limited resources and expertise, and sometimes limited willingness within government agencies to design effective consultation processes and genuinely engage with fishers' interests and expertise.

Recent events in the Australian state of Victoria are a good, if disappointing, illustration of the situation (see Baelde et al., 2001, for details). After a nine-year investigation, on 17 May 2001 the State Minister for Environment and Conservation proposed to declare twelve MPAs in Victoria's waters (all MPAs were to be highly protected no-take areas where all fishing was to be banned) and tabled a bill in Parliament for their establishment. The hastily drafted bill instantly generated strong opposition from the fishing industry and various political parties because it included a controversial constitutional change. Fishers would have lost their right to seek compensation through the court for loss of property rights, whether or not this loss was related to the creation of no-take areas (the Victorian government later claimed that this was a drafting error). About a month after tabling the proposal, and after stormy street demonstrations, the Victorian government withdrew the bill from parliament on 13 June 2001.

The Victorian government's refusal to pay compensation to fishers has been said to be the major cause of the (temporary[1]) rejection of the MPA bill. However, it more directly reflected a very poor handling of socio-economic issues in the design of MPAs and a lack of proper consultation with the fishing industry. Better protocols to ensure effective fishers' input in the design of MPAs would have helped find a compromise over the bill, and more generally would have helped to mediate the impacts of protected areas on fisheries.

In the Australian South East Fishery discussed in Example 1, fishers are now contributing to spatial management (Williams and Bax, this volume) by providing information on fishing distribution, the types of habitats that exist on fishing grounds, and fishers' operational and socio-economic dependency on these grounds. This is precisely the type of information that was missing in the case of Victoria. It is hoped that such cooperative work between scientists, fishers and conservation agencies will help avoid the difficulties experienced there.

---

1. The MPA bill was finally accepted in 2002 after the government agreed to include compensation mechanisms.

DISCUSSION AND CONCLUSIONS

THERE are an increasing number of studies that describe the detailed knowledge that fishers have of fish stocks, their environment and their exploitation patterns. Most of these studies highlight the usefulness of fishers' knowledge in filling gaps in scientific knowledge. However, as noted by McGoodwin et al. (2000), the integration of scientists' and fishers' types of knowledge remains difficult in practice. By comparison with scientific knowledge, fishers' knowledge is mostly of a qualitative and narrative nature, local and contingent rather than scientifically objective.[2] It reflects not only the biological and the socio-economic contexts within which fishers operate, but also fishers' personal beliefs and values (Baelde, 1998; Neis and Felt, 2000). Various studies have also described techniques to check the validity and reliability of fishers' knowledge (for example, Neis et al., 1999a; Purps et al., 2000).

Studies of fishers' knowledge have generally been concerned with small-scale artisanal fisheries in developing countries. In industrial fisheries, the competitive pursuit of profit and political lobbying partly drive fishers' behaviour and their interaction with scientists and managers (Finlayson, 1994; McGoodwin et al., 2000). This does not mean that fishers' knowledge in industrial fisheries is less useful, but it creates new challenges in accessing and validating it. Also, in industrial fisheries fishers' knowledge and input are often sought only when management is perceived to be ineffective: that is, when these fisheries are already in difficulties. By that time, fishers themselves are under pressure from increasing regulations and may face the ultimate prospect of a ban on fishing (as seen in Example 3). Crisis situations do not facilitate cooperation as scientists' and fishers' information can become political issues in times of conflict over management (Finlayson, 1994; Maurstad and Sundet, 1998).

In Australian fisheries, the partnership framework established by management agencies usually includes the formation of expertise-based (as opposed to representative) scientific and management advisory committees (see Smith et al., 1999, for an analysis of the partnership approach in the case of federally managed fisheries). Membership on these committees comprises scientists, fishers, managers and environmentalists. This framework is, without doubt, a significant step toward promoting fishers' involvement in fisheries assessment and management, and facilitating collaboration between scientists, managers and fishers (see Smith et al., 1999). However, it is only partly effective. Problems are often attributed to fishers' vested interests 'capturing' the process, but Smith et al. (1999) question these perceptions.

---

2. Scientifically objective here means based on observations that are repeatable and independent from the observer, in contrast to fishers' knowledge which is more subjective and based on personal experiences and value judgement. This comparison aims to show the different ways by which scientists and fishers accumulate their respective knowledge, not to make any judgement on the validity of these two types of knowledge.

Other problems are created by the fact that, on the one hand, the partnership framework gives fishers greater access to the assessment and management process, and thus greater opportunity to scrutinize and challenge scientific knowledge with their own knowledge and expertise. But, on the other hand, the partnership framework has not been designed to facilitate the use of the knowledge and expertise that fishers bring into the process. In many fisheries, the scientific fisheries assessment relies largely on conventional quantitative, single-species methods and is not adequately adapted to incorporate the fishers' type of knowledge (as seen in Example 1). Scientists tend to believe that the usefulness of fishers' knowledge is limited because of the difficulties inherent in quantifying it (Holm, 2003). Meanwhile, fishers express growing frustration at scientists' inability to make direct use of industry information and views (Baelde, 1998, 2001; Smith et al., 1999).

Also, the partnership framework based on small expert committees does not facilitate access to and use of broad-based industry knowledge. The communication of information between members of advisory committees and the wider fishing community is not effective, and this generates some tension within the industry. McCay (1999) stated that current partnership practices based on advisory committees tend to create a new type of community, an interest-based community as compared to a place-based community. These 'virtual' communities are defined by their management regimes (by species, area, gear type and the like) and develop new social ties and identities. She suggested that such communities may be the only real hope for a participatory management that encompasses a wide diversity of interest groups. However, experience in Australia shows that they also tend to alienate non-member fishers and may create further divisions within an already divided fishing industry.

It is clear that additional structures that are better adapted to the specific nature of fishers' knowledge must be developed. For example, in the case of the Australian South East Fishery, the management agency funds a team of scientists and managers to conduct annual visits to local fishing ports. The aim is to give grassroots fishers an opportunity to interact with scientists and managers and raise issues about the fishery. However, fishers' low attendance at meetings limits the success of these port visits. Individual fishers tend to be wary of public meetings (especially when there are conflicts about management issues) and one-day-a-year visits to their ports fail to attract their interest: they go fishing instead.

The greatest difficulty with the partnership approach is, possibly, overcoming existing sociocultural barriers that hamper communication and collaboration between fishers and scientists/managers. There is a great sociocultural divide between the moral authority of science (collectively accepted by society and legitimized through rigorous objectivity rules) on the one hand, and the suspicion attached to fishers' information (subjective, non-tested and perceived as biased by vested interests) on the other. The lack of curiosity and interest that scientists showed in the wealth of information that was collected from fishers in Example 1 was surprising at first. However, it quickly became obvious that scientists' attitudes toward fishers' knowledge were influenced by the sociocultural barriers so often described by social scientists (for example, Finlayson, 1994; McCay, 1999; Neis and Felt, 2000; and references therein). In a co-management situation scientists have learned to respect fishers' political power, but they have remained sceptical of the validity of their knowledge. In his analysis of the northern cod fishery, Finlayson (1994) showed that scientists made a clear distinction between fishers' involvement in the scientific process and the incorporation of their knowledge in that process. Even the most sympathetic fisheries scientists are too perplexed by the structure, form and scale of fishers' knowledge, and prefer to retreat into the security and familiarity of established scientific practices (McGoodwin et al., 2000).

Scientists tend to see themselves as possessors of universal knowledge and custodians of the sea (McGoodwin et al., 2000), as defenders of natural resources against an irresponsible fishing industry and an inefficient, or ambivalent, management system (Finlayson, 1994). When asking fishers to share their knowledge, scientists assume that they accept the purpose and methods of science, and that their role is to fill gaps in scientific knowledge. However, this science-driven approach fails to recognize fishers' own values, expectations and methods of gathering knowledge. Besides scientific understanding, other knowledge frameworks and value systems are gaining recognition as products of social, cultural and ecological contexts (McGoodwin et al., 2000). This increasingly challenges the central position of science. We need to explore and test fishers' own understanding and theories about biological processes and market- or management-driven fishing behaviour (Maurstad and Sundet, 1998; Baelde, 2001).

By focusing on the technical difficulties of integrating fishers' knowledge into scientific methods, scientists maintain a narrow and prescriptive view of the nature and value of fishers' knowledge (Baelde, 1998; Maurstad, 2000b). McGoodwin et al. (2000) stressed that it is no longer enough to hire fishers as data-collecting technicians, or even to systematically collect their knowledge in a form that fits with the requirements of existing science. The type of fishers' input that is needed today for assessing and managing industrial fisheries is expanding well beyond simply filling gaps in scientific knowledge. This is because the principles and practices of fisheries research and management are also dramatically and rapidly changing. The three examples described illustrate these changes, from deterministic, quantitative single-species stock assessment (Example 1), to exploratory, hypothesis-based simulation models (Example 2), to a holistic ecosystem approach (Example 3). As a consequence, the role of fishers also diversifies from providing technical knowledge to providing advice and opinions on current and future harvest and management needs.

While the partnership approach is being increasingly adopted and promoted as a tool leading to better resource management, the social and cultural implications and constraints of such an approach are not well understood and appreciated by scientists and managers. They fail to recognize that a truly effective partnership with fishers relies first of all on acknowledging the legitimacy of fishers' knowledge and actively developing ways of overcoming existing technical and sociocultural difficulties (see also Holm, 2003). This would require dedicated research, crossing the boundaries of fisheries and social sciences. Jentoft et al. (1998) also pointed out that, while income is important, the dignity and esteem that come from the occupation of fishing matter a great deal to fishers. Their accumulated knowledge contributes to their pride.

The sweeping changes that are taking place in fisheries assessment and management are partly in recognition of the limitations and uncertainty of traditional fisheries science. As public scrutiny of fisheries issues intensifies, community views and values on the use of common resources play an increasingly important role in fisheries assessment and management. Fisheries management and environmental protection are becoming matters of social debate and negotiations. A balance has to be found between environmental, social and economic values and this cannot be resolved on biological and technical grounds alone. Fishers are (or should be) active players in these social negotiations, contributing not only their knowledge but also their perceptions and values. Jentoft et al. (1998) point out that co-management is a process of social creation through which knowledge is gained, values articulated, culture expressed and community created. Scientists' reluctance to acknowledge, or at least test, the value of fishers' knowledge is anachronistic in today's circumstances. Like fishers, they too are running the risk of being accused of resisting changes in order to protect their own entrenched professional interests.

This contribution is concerned with scientists' and managers' responsibilities in ensuring effective partnership and effective use of fishers' knowledge. However, fishers also have responsibilities toward the community, both as users of common resources and as food providers. They too must realize the extent of societal change with regard to the conservation of marine resources and the consequences for their industry. They cannot operate with the same independence they once did and they must work on developing a more unified and credible voice. The well-known divided nature of the fishing industry is an important factor limiting the use of fishers' knowledge. In the same way that too many scientists tend to retreat behind the comfort and familiarity of established science, too many fishers also tend to retreat behind the belief that resource protection and management is, ultimately, a government responsibility. This too is an anachronistic position, untenable within today's co-management approach.

Fisheries are in crisis and fishers, scientists and managers are under pressure to protect marine resources. Their ability to collaborate and find acceptable and workable solutions to fisheries problems partly depends on their ability to shift from their defensive positions to positions of leadership. They all need to rethink and re-assess their cultural and professional beliefs in order to accommodate and take advantage of each other's complementary expertise.

ACKNOWLEDGEMENTS

I thank all the fishers who have so generously shared their knowledge and expertise with me over the past ten years. This study has been partly funded by the Australian Fisheries Research and Development Corporation.

REFERENCES

ALLEN, P.M.; McGLADE, J.M. 1986. Dynamics of discovery and exploitation: The case of the Scotian Shelf Groundfish Fisheries. *Canadian Journal of Fisheries and Aquatic Sciences,* Vol. 43, pp. 1187–200.

ANZECC TFMPA . 1998. *Guidelines for establishing the national representative system of marine protected areas.* Canberra, Australian and New Zealand Environment and Conservation Council Task Force on Marine Protected Areas, Environment Australia, 15 pp.

ATTWOOD, C.G.; HARRIS, J.M.; WILLIAMS, A.J. 1997. International experience of marine protected areas and their relevance to South Africa. *South African Journal of Marine Science,* Vol. 18, pp. 311–32.

BAELDE, P. 1995. Analysis of the blue eye trevalla (*Hyperoglyphe antarctica*) fishery off Tasmania, Australia. In: *Assessment of the blue eye trevalla fishery and analysis of the impact of mid-water trawling,* Final FRDC Report (Project No. 91/20. Canberra, Fisheries Research and Development Corporation, 155 pp.

——. 1996. Biology and dynamics of the reproduction of blue-eye trevalla, *Hyperoglyphe antarctica* (Centrolophidae), off Tasmania, southern Australia. *Fisheries Bulletin,* Vol. 94, pp. 199–211.

——. 1998. Synthesis of industry information on fishing practices and their effects on catch rate analysis (bottom trawl sector) In: J. Prince, J. P. Baelde and G. Wright (eds.), *Synthesis of industry information on fishing patterns, technological changes and the influence of oceanographic effects on SEF fish stocks.* Canberra, Australia, FRDC Project No 97/114, pp. 1–25.

——. 1999. *Blue eye trevalla: Stock assessment report compiled for the South East Fishery Assessment Group.* Canberra, Australia, Australian Fisheries management Authority, 16 pp.

———. 2001. Fishers' description of changes in fishing gear and fishing practices in the Australian South East Trawl Fishery. *Marine and Freshwater Research*, Vol. 52, No. 4, pp. 411–17.

BAELDE, P.; KEARNEY, R.; MCPHEE, D. 2001. A coordinated commercial fishing industry approach to the use of Marine Protected Areas. *Final FRDC Report* (Project No. 99/163). Canberra, University of Canberra, 190 pp.

BEAUMONT, J. 1997. Community participation in the establishment and management of marine protected areas: A review of selected international experience. *South African Journal of Marine Science*, Vol. 18, pp. 333–40.

BEED, T.W.; STIMSON, R.J. (EDS.). 1985. *Survey interviewing: Theory and techniques.* Sydney, Allen and Unwin, 244 p.

CROSBY, M.P. 1997. Moving towards a new paradigm: Interactions among scientists, managers and the public in the management of marine and coastal protected areas. In: M.P. Crosby, K. Geenen, D. Laffoley, C. Mondor and G. O'Sullivan (eds.), *Proceedings of the Second International Symposium and Workshop on Marine and Coastal Protected Areas: Integrating Science and Management*, 1995. Silver Spring, Md, NOAA, pp. 10–24.

DORN, M.W. 1998. Fine-scale fishing strategies of factory trawlers in a midwater trawl fishery for Pacific hake (*Merlucius productus*). *Canadian Journal of Fisheries and Aquatic Sciences*, Vol. 55, pp. 180–98.

FINLAYSON, A.C. 1994. *Fishing for truth: A sociological analysis of northern cod stock assessments from 1977–1990*. St. John, ISER Books, 176 pp.

FISKE, S.J. 1992. Sociocultural aspects of establishing marine protected areas. *Ocean and Coastal Management*, Vol. 18, pp. 25–46.

HILBORN, R. 1985. Fleet dynamics and individual variation: Why some fishers catch more than others. *Canadian Journal of Fisheries and Aquatic Sciences*, Vol. 42, pp. 2–13.

———. 1992. Current and future trends in fisheries stock assessment and management. *South African Journal of Marine Sciences*, Vol. 12, pp. 975–88.

HILBORN, R.; WALTERS, C.J. 1992. *Quantitative fisheries stock assessment: Choice, dynamics and uncertainty*. New York, Chapman and Hall, 570 p.

HOLM, P. 2003. Crossing the border: On the relationship between science and fishermen's knowledge in a resource management context. *MAST*, Vol. 2, No. 1, pp. 5–33.

JENTOFT, S.; MCCAY, B.J.; WILSON, D.C. 1998. Social theory and fisheries co-management. *Marine Policy*, Vol. 22, No. 4–5, pp. 423–36.

JOHANNES, R.E. 1998. The case for data-less marine resource management: Examples from tropical nearshore fisheries. *Trends in Ecology and Evolution*, Vol. 13, pp. 243–6.

JOHANNES, R.E.; FREEMAN, M.M.R.; HAMILTON, H.J. 2000. Marine biologists miss the boat by ignoring fishers' knowledge. *Fish and Fisheries*, Vol. 1, pp. 257–71.

JOHNSON, A. WALKER, D. 2000. Science, communication and stakeholder participation for integrated natural resource management. *Australian Journal of Environmental Management*, Vol. 7, pp. 82–90.

LAUCK, T.; CLARK, C. W.; MANGEL, M.; MUNRO, G.R. 1998. Implementing the precautionary principle in fisheries management through marine reserves. *Ecological Applications*, Vol. 8, No. 1. Supplement, pp. S72–S78.

MAURSTAD, A. 2000a. To fish or not to fish: Small-scale fishing and changing regulations of the cod fishery in northern Norway. *Human Organisation*, Vol. 59, No. 1, pp. 37–47.

———. 2000b. Trapped in biology: An interdisciplinary attempt to integrate fish harvesters' knowledge into Norwegian fisheries management. In: B. Neis and L. Felt (eds.), *Finding our sea legs: Linking fisheries people and their knowledge with science and management*. St. John's, ISER Books, pp. 135–52.

MAURSTAD, A.; SUNDET, J.H. 1998. The invisible cod: Fishermen's and scientists' knowledge. In: S. Jentoft (ed.), *Commons in a cold climate: Coastal fisheries and reindeer pastoralism in north Norway: the co-management approach*. Man and the Biosphere Series, Volume 22. Paris, UNESCO, pp.167–84.

MCCAY, B.J. 1999. Community-based approaches to the 'fishermen's problem'. Paper presented at the *1999 FishRights Conference*, Fremantle, Australia.

MCGOODWIN, J.R.; NEIS, B.; FELT, L. 2000. Integrating fishery people and their knowledge into fisheries science and managemen: Issues, prospects, and problems. In: B. Neis; L. Felt (eds.), *Finding our sea legs: Linking fisheries people and their knowledge with science and management*. St. John's, ISER Books, pp. 249–64.

MEGREY, B.A. 1989. Review and comparison of age-structured stock assessment models from theoretical and applied points of view. *American Fisheries Society Symposium*, Vol. 6, pp. 8–48.

NEIS, B. 1995. Fishers' ecological knowledge and marine protected areas. In: N.L. Shackell and J.H. Martin Willison (eds.), *Proceedings of the Symposium on Marine Protected Areas and Sustainable Fisheries conducted at the Second International Conference on Science and the Management of Protected Areas*. Halifax, Canada, Dalhousie University, pp. 265–72.

NEIS, B.; FELT, L. (EDS.). 2000. *Finding our sea legs: Linking fisheries people and their knowledge with science and management*. St. John's, ISER Books, 318 pp.

NEIS, B.; FELT L.; HAEDRICH, R.L.; SCHNEIDER, D. C. 1999a. An interdisciplinary methodology for collecting and integrating fishers' ecological knowledge into resource management. In: D. Newell and R. Ommer (eds.), *Fishing places, fishing people: issues and traditions in Canadian small-scale fisheries*. Toronto, University of Toronto Press, pp. 217–38.

Neis, B.; Schneider, D.C.; Felt, L.; Haedrich, R.L.; Fisher, J.; Hutchings, J.A. 1999b. Fisheries assessment: What can be learned from interviewing resource users? *Canadian Journal of Fish and Aquatic Science*, Vol. 56, pp. 1949–63.

Parrish, R.; Seger, J.; Yoklavich, M. 2000. Marine reserves to supplement management of West Coast groundfish resources: phase 1 technical analysis. Pacific Fishery Management Council. Available on http://www.pcouncil.org (last accessed on 26 June 2005).

Pitcher, T.J. 2001. Fisheries managed to rebuild ecosystems? Reconstructing the past to salvage the future. *Ecological Applications*, Vol. 11, No. 2, pp. 601–17.

Punt, A.E.; Smith, A.D.M.; Cui, G. 2001. Review of progress in the introduction of management strategy evaluation (MSE) approaches in Australia's South East Fishery. *Marine and Freshwater Research*, Vol. 52, No. 4, pp. 719–26.

Purps, M.; Damm, U.; Neudecker, T. 2000. Checking the plausibility of data derived from fishing people of the German Wadden Sea. In: B. Neis and L. Felt (eds.), *Finding our sea legs: Linking fisheries people and their knowledge with science and management*. St. John's, ISER Books, pp. 111–23.

Roberts, C.M.; Hawkins, J.P. 2000. *Fully-protected marine reserves: A guide*. Washington, DC, and York, UK, WWF Endangered Seas campaign and Environment Department, University of York, 131 pp.

Robins, C.W.; Wang, Y.G.; Die, D. 1998. The impact of global positioning systems and plotters on fishing power in the northern prawn fishery, Australia. *Canadian Journal of Fisheries and Aquatic Sciences*, Vol. 55, pp. 1645–51.

Salmi, P.; Salmi, J.; Moilanen, P. 1999. Strategies and flexibility in Finnish commercial fisheries. *Boreal Environment Research*, Vol. 3, pp. 347–59.

Smith, A.D.; Sainsbury, K.J.; Stevens, R.A. 1999. Implementing effective fisheries management systems: An Australian partnership approach. *ICES Journal of Marine Research*, Vol. 56, pp. 967–79.

Tilzey, R.D.J. 1999. *The South East Fishery 1998*. Fishery Assessment Report compiled by the South East Fishery Assessment Group. Canberra, Australian Fisheries Management Authority, 199 pp.

Walters, C.J. 1998. Designing fisheries management systems that do not depend on accurate stock assessment. In: T.J. Pitcher, P.J.B. Hart and D. Pauly (eds.), *Reinventing fisheries management*. London, Chapman and Hall, pp. 279–88.

——. 2000. Impacts of dispersal, ecological interactions, and fishing effort dynamics on efficacy of marine protected areas: How large should protected areas be? *Bulletin of Marine Science*, Vol. 66, No. 3, pp. 745–57.

Ward, T.J.; Heinemann, D.; Evans, N. 2001. *The role of marine reserves as fisheries management tools: A review of concepts, evidence and international experience*. Canberra, Bureau of Rural Sciences, 184 pp.

WELL S.; WHITE, A.T. 1995. Involving the community. In: S. Gubbay (ed.), *Marine protected areas: Principles and techniques for management*. London, Chapman and Hall, Conservation Biology Series, pp. 61–84.

CHAPTER 20   Fishers' knowledge?
Why not add their scientific skills while you're at it?
Richard D. Stanley and Jake Rice

ABSTRACT

WE suggest that it is a mistake to focus on fishers simply as data collectors or knowledge sources, ignoring their skills in hypothesis formulation, research design and interpretation. The benefits that accrue from full scientific participation by fishers are demonstrated with two examples from the groundfish fishery in British Columbia, Canada. The first example summarizes a joint acoustic study to estimate the biomass of a shoal of widow rockfish (*Sebastes entomelas*). In addition to providing the essential background information needed to plan and conduct the study, the fishers posed the initial experimental hypothesis, and were full participants in the execution, analysis and documentation. The second example describes the impact of fishers' critique of an age composition-based stock assessment of silvergray rockfish (*S. brevispinis*). They argued that the introduction of individual vessel quotas had caused changes in the spatial distribution of catches and therefore the fishery samples. Thus, the age samples were not comparable over time. In response to their criticism, a preliminary study was jointly conducted and the results supported their concern. These results are now being used to improve the sampling and assessment techniques. The contribution concludes with a summary of the characteristics of the two studies that facilitated the interaction between research partners.

INTRODUCTION

MANY documents suggest that fishery research and management would be more effective if they made more frequent use of fishers as data collectors, and better use of 'local' or 'fisher' or 'traditional' knowledge (see McGoodwin et al., 2000). With respect to the latter attribute, Roepstorff (2000) comments that any catchy first word and 'knowledge' will suffice in this context (see also Agrawal, 1995; Sillitoe, 1998). In fact, it is widely asserted that continued failure to use these assets will lead not only to poorer research but management failure (Dyer and McGoodwin, 1994; Gavaris, 1996; McGoodwin et al., 2000). In support of this premise, many emphasize the wealth of knowledge that fishers possess (Ruddle, 1994; and the exceptional work of Johannes, 1978 et seq.). Some identify specific topics for which fisher knowledge could be most useful, including stock structure, changes in catchability, fish–gear interactions, information on abundance in a closed fishery, and the potential impacts of re-opening (Neis et al., 1999; Fischer, 2000).

Perhaps partly in response to these documents, recognition of this research asset is increasing at the policy and operational levels. Canada's Department of Fisheries and Oceans (DFO) instituted the fisher-based sentinel survey programme in 1994 for East Coast groundfish, in part to 'try to blend the traditional knowledge of fishermen with the objective rigour of scientific data gathering' (public speech by Hon. B. Tobin, Minister of Fisheries and Oceans, September 1994). In the same year, the DFO also made a commitment to bring fisher knowledge into the peer review and advisory process (FRCC, 1994).

The International Council for Exploration of the Seas, the principal marine and fisheries science advisory body for the North Atlantic, is incorporating resource users into its review and advisory processes. The US National Marine Fisheries Service has funded the 'Cooperative Research in the Northeast' programme, as well as numerous cooperative research projects on salmon with fishers, tribal councils and communities in the Pacific Northwest (Office of Management and Budget, 2001). The International Pacific Halibut Commission (IPHC) has established an industry-composed research advisory board that not only reviews but also actively participates in the design and implementation of many research programmes (B. Leaman, personal communication).

We do not question the potential value of this ***-knowledge to fisheries research, nor that it continues to be an under-utilized asset. Clearly government and industry alike recognize its potential. However, we suggest that confining the contributions from fishers to roles as inexpensive data collectors or sources of background knowledge ignores some of the greater potential benefit that can come from truly collaborative work. Publications from agricultural extension work have, for some time, emphasized that these same possessors of knowledge are also effective at hypothesis formulation, experimental design and interpretation (Sajise, 1993; Sillitoe, 1998). To paraphrase Sajise (1993), how could knowledge accrue (as opposed to just being passed on) without someone applying elements of the scientific method? In the words of a long-time salmon troller, 'every time I set my gear, I am conducting an experiment' (Ian Bryce, personal communication).

There is ample evidence now (in the field of agriculture) that local people do their own research; maybe not in the same formal and rigorous way that 'researchers' do it in terms of having statistical designs, replications and analysis, but they do research (Sajise, 1993, p.3). Although these skills are recognized in agricultural research, they are rarely acknowledged in fisheries literature even by those who emphasize that fisher knowledge is under-utilized (for exceptions, see Hutchings, 1996; Neis et al., 1999; and Ames et al., 2000).

The compartmentalization of fishers' contribution to science and management results from relying on the 'data collection' model for linking fisher knowledge (Fischer, 2000) to other sources of information on stock status. It assumes that for fisher knowledge to contribute it must be systematized and stored in much the same way as data from conventional monitoring sources (Ferradás, 1998). Although there is a place for this model, it represents an appending of fishers to conventional scientific research as junior partners. It maintains for researchers, the 'we vs they', and the 'fisher' or 'anecdotal knowledge' vs 'real science' dichotomies (see discussion notes appended to Sillitoe, 1998; Johannes and Neis, this volume). We argue that fishers' experiential knowledge is derived from their skills as experimenters. Fisheries research should move towards adopting the change implicit in the participatory research model long recognized in agriculture (Chambers et al., 1989; Sajise, 1993) but only recently acknowledged in fishery research (McGoodwin et al., 2000; Neis and Felt, 2000).

Fischer (2000) describes participatory research as a joint exercise by a team in which the so-called researcher may be an influential member but does not occupy the top position in the traditional hegemonic framework. In its fullest development, all players can participate in the development of questions, hypotheses, design and execution. This model has well-established precedents in multidisciplinary scientific research wherein fisheries scientists, oceanographers, statisticians and modellers engage in collaborative projects. We draw from the model not necessarily any formal structure, but the attitudinal change wherein each team member acknowledges other team members as peers in planning, conducting and interpreting the science. It should not be considered revolutionary to view partners from the fishing industry in a similar light.

We describe below two studies in which fishers were treated as full participants and provide examples of the scientific elements that fishers can bring to fisheries research. The paper is written in the narrative style of chronicling the studies as they transpired, as opposed to the usual methods–results–discussion sequence of scientific articles. A narrative is more effective in this case, because the message lies in the process as much as in the results. The process, not the concept, develops trust, which is essential for meaningful dialogue.

ACOUSTIC ESTIMATION OF WIDOW ROCKFISH (*SEBASTES ENTOMELAS*)

Background

We first present an acoustic study of a mid-winter shoal of widow rockfish off the central coast of British Columbia (BC), Canada (see Stanley et al., 2000, for details of the acoustics and estimation methodology). Assessment of rockfish (*Sebastes* spp.)

provides an interesting test for collaborative work because of long-standing differences of opinion between government assessment staff and fishers over rockfish biomass estimates and quotas (Leaman and Stanley, 1993). The low productivity of these species, implied by a low natural mortality rate (Archibald et al., 1981), and an early history of overfishing for some species (Archibald et al., 1983), led the DFO to implement restrictive quotas. From the fishers' perspective, which does not include the same corporate memory of proven overfishing, the strong acoustic sign they can observe on their sounders and the high catch rates of these aggregating species lead the fishers to infer large biomasses, inconsistent with the quotas.

Unfortunately, subsequent assessment research has done little to resolve these different viewpoints. Many of these shoaling rockfish species tend to inhabit untrawlable bottom and express high variability in catch rates over time and space. These characteristics combine to reduce the efficacy of swept-area survey methods. Although the shoals of some species can reflect a strong acoustic signal, they stay so close to the bottom it is difficult to separate fish from bottom signal. Nor is it possible to distinguish acoustically between the many cohabiting species. Finally, commercial catch rates (CPUE) are not necessarily comparable over time owing to evolving fishing gear and changing regulations.

In this research context, the senior author, a government biologist, made a trip aboard a commercial trawler in 1996 to discuss stock assessments with the captain, who was also an industry advisor on groundfish management. During the many discussions in the wheelhouse, Captain Mose commented that most fishers felt that widow rockfish quotas and their implied coastwide biomass estimates of 15,000–30,000 tonnes (Stanley and Haist, 1997) were especially conservative. The fishers were aware of one shoal of widow rockfish, which, if estimated, might indicate by itself that coastwide biomass estimates for this species were under-estimated. Furthermore, this shoal, which regularly formed each winter off the central coast of BC, was predominantly widow rockfish, off bottom at dusk, and predictable in its occurrence, thus making it a reasonable candidate for acoustic estimation. Captain Mose also commented that even if the study failed to indicate a large biomass, it would still be helpful since it would be the first directed field research on this species in Canadian waters. It was also noted that the estimation of one rockfish shoal would provide a much-needed quantitative reference point for enhancing dialogue between fishers and biologists about what fishers observed on their sounders. Finally, the principals hoped that the programme would serve as a model for developing closer research collaboration between industry and government staff.

Methods and results

Surveys of the shoal, located at the edge of the continental shelf, were conducted in early February of 1998 and 1999 (Stanley et al., 2000; Wyeth et al., 2000)

(Figures 20.1 and 20.2). Timing and location were based on fishers' knowledge. Two commercial trawlers, the *Frosti* (1998) and the *Viking Storm* (1999), were the catcher vessels and provided the acoustic scouting. During the study, these vessels conducted mid-water trawl hauls to identify the species composition of the shoal. They also sounded the perimeter of the study site for evidence of movement to and from the area. A trawl fishery association, the Canadian Groundfish Research and Conservation Society (CGRCS), paid for the costs of the charter vessels.

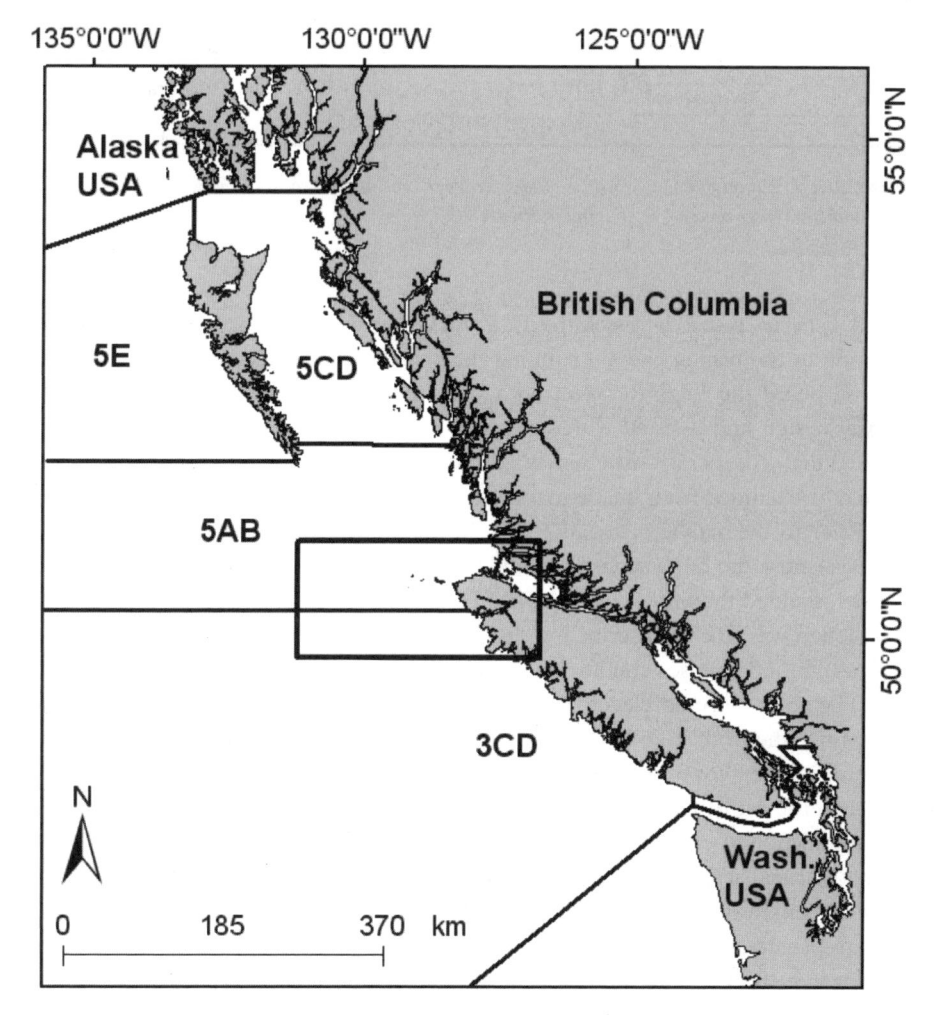

FIGURE 20.1 *Location of widow rockfish study area (inner box) and silvergray rockfish assessment regions.*
Source: From Stanley et al., 2000.

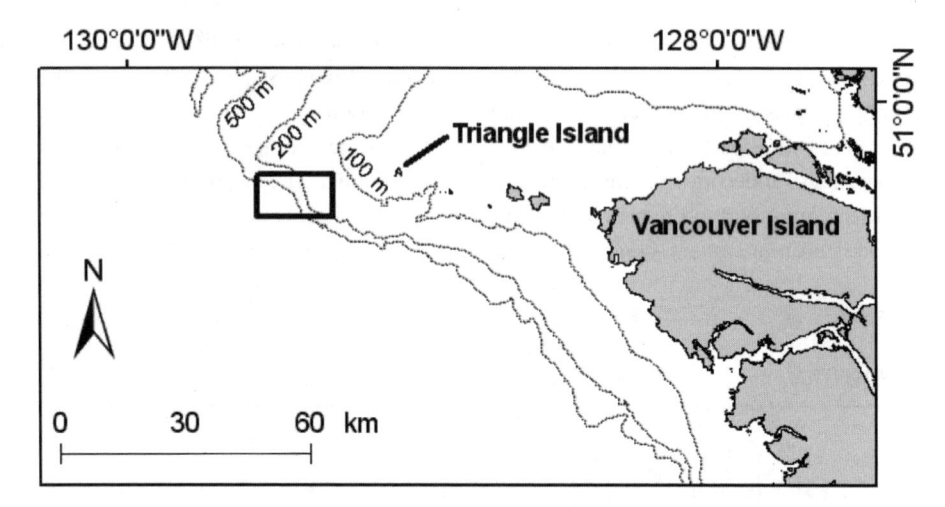

FIGURE 20.2 *Location of widow rockfish study site off the north-west coast of Vancouver Island. Inset indicates inset area corresponding to Figure 20.3.*
Source: From Stanley et al., 2000.

A fisheries research vessel, the *W. E. Ricker*, provided the acoustic platform. In addition to the fishing captain running the charter vessel, a second fishing captain was on board the *W. E. Ricker* in both years. The captains participated equally in survey design and assisted with scrutinizing the acoustic data. For example, they advised that a deeper acoustic sign at 225 m was yellowmouth rockfish (*S. reedi*) and should be excluded from the biomass estimates.

Prior to the arrival of the *W.E. Ricker*, the study team, composed of the chief DFO scientist, the DFO acoustician and two trawl skippers on board the commercial trawler, scouted the site to select the acoustic transects. An important element of the interaction was to circumscribe the shoal to the satisfaction of all participants while still accommodating sea conditions.

The team selected eleven transects in 1998. These were designed to extend across the shelf break and were oriented perpendicular to the longitudinal axis of the shoal. They covered a total area of about 25 km^2 (Figure 20.3). The vessels travelled the set of eleven transects in the same direction and order each time. At the completion of each of twenty replicates of the set, the vessel returned to the start point. During each return trip, the commercial fishing captain piloted the *W.E. Ricker* over the longitudinal axis of the shoal, to re-affirm the general location of the shoal. Each replicate and return trip required two hours. The design was similar in 1999, except that the transects were spread over a broader area in order to encompass additional acoustic sign to the north-west (Wyeth et al., 2000).

One of the most exciting and effective aspects of the study was that the acousticians on the *W.E. Ricker* were able to provide biomass estimates of the shoal during the cruise. Thus, by the completion of the tenth replicate in 1998, the team was aware that the estimates were consistently indicating about 2,000 tonnes of widow rockfish. While disappointing to the team, since these values did not disprove existing quotas, the immediate feedback provided the opportunity to vary the design to see if the shoal was being underestimated.

The captains questioned in particular whether the set of eleven transects were, by chance, consistently missing the denser portions of the shoal. They hypothesized that the biomass estimates could be highly sensitive to transect choice owing to variable density within the shoal. The team had planned to repeat the same transects to study the daily vertical movements by the widow rockfish. However, since the credibility of the estimates was at stake, the team chose to vary the design. Starting with replicate set No. 11, the entire set of transects was shifted to the north-west by approximately 180 m, thereby generating a new set of 11 transects, slightly offset from the original. Again with replicate No. 12, the new set of transects in No. 11, was moved a further 180 m to the north-west (Stanley et al., 2000).

The new sets of transect still indicated about 2,000 tonnes, but the modification was useful in testing the sensitivity of the estimates to the choice of transects. It also confirmed to the fishing captains that they were equal participants in the study. The

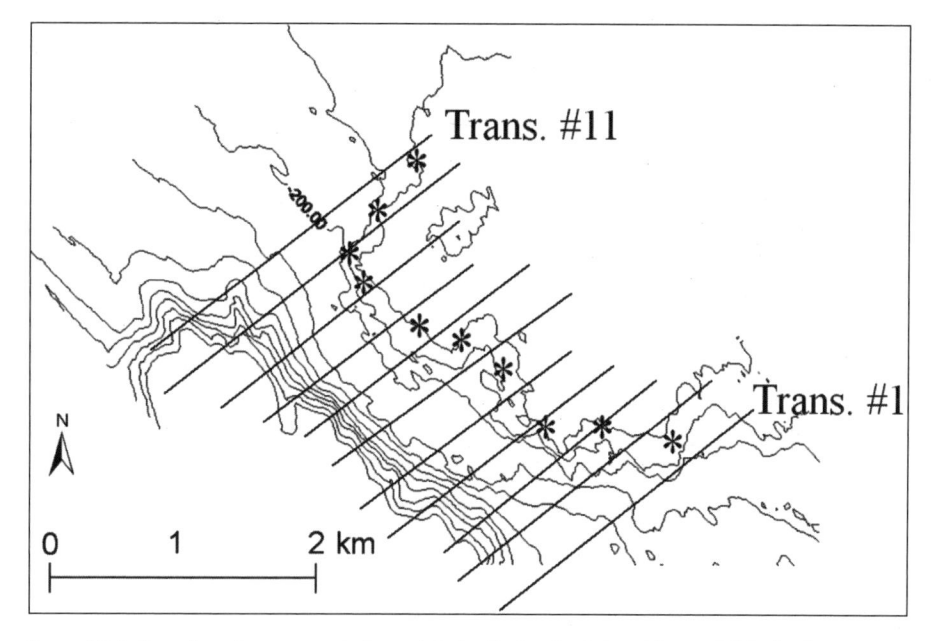

FIGURE 20.3 *Location of transects relative to longitudinal axis of the widow rockfish shoal.*
* = approximate shoal location.
Source: From Stanley et al., 2000.

fishers' concerns about the representativeness of the first set of eleven transects was viewed as a valid scientific question by the DFO scientists and one that should be addressed at the expense of other questions.

Although changing the transects helped reduce scepticism about the estimation process, the fishing captains still remained concerned that the survey spent too little time over the shoal and too much time where there were no fish. This, they surmised, could lead to an underestimate. The team accommodated this concern by extrapolating an independent biomass estimate from each return trip that ran over the longitudinal axis of the shoal. Fish density estimates for these transects were extrapolated under the assumption that the shoal was 0.5 km wide, the approximate average width of the shoal on the echogram. The team found in both years that although these single transects concentrated on the shoal and provided a consistent display on the monitor, the extrapolations did not indicate any more biomass than the standard design.

Finally, the captains questioned whether a 2,000 tonnes estimate was consistent with the fact that they could catch 50–100 tonnes from this shoal in a few minutes. The team therefore convened a small meeting on board the *W.E. Ricker* during the 1998 cruise. The observed acoustic density estimates were converted to potential catch rates based on the net specifications and simplified assumptions of catchability. These estimates were found congruent with the catch rates the fishers had observed. The importance of this interaction was that the referential ground-truths of all participants were given their due. Government researchers might have responded that commercial catch rates were simply not relevant. They had used an acoustic methodology that was documented, peer reviewed, and scientifically sound; however, what comes up in a net is the real point of contact between those who fish and what is in the sea. The fishing captains were correct in suggesting that the acoustic estimates of densities and their maximum catch rates had to be congruent, or there was a mistake somewhere.

Encouraged by the success with one shoal in 1998, the senior author was ready to expand the approach and attempt a coastwide biomass estimate in 1999; however, the fishing fleet commented that while widow rockfish were caught elsewhere on the coast, they were so unpredictable in time and space that a large-scale survey would likely be unproductive. Thus, within the course of this project, the fishers not only identified a fruitful direction of research, but also prevented a wasteful one.

Instead of an expanded survey, the team re-examined the same study site in 1999. The two main objectives were to ensure that the 1998 estimates were not anomalous and to obtain estimates during days with stronger tides. In reviewing a draft report of the 1998 study, Captain Mose commented that the shoal had been estimated during days of the weak neap tides. Fishers typically observed more acoustic sign and higher catch rates on days when the tides were strengthening, just prior to the new or old moon. However, the 1999 results indicated a similar biomass to that in 1998, and provided no evidence of a significantly larger biomass during days with stronger tides. At the conclusion of the study, there was a consensus that the project had exhaustively addressed the initial question of estimating the biomass of the shoal.

The team's success in 1998 also led them to examine the potential for conducting acoustic biomass estimates directly from commercial vessels. The practicability of adapting commercial fishing vessels to acoustic research platforms had already been demonstrated for herring on the Canadian East Coast (Melvin et al., 1998). The 1999 field trip successfully connected digitizing equipment to the sounder on the *Viking Storm*. The calibration was successful and inter-calibration with the *W.E. Ricker* system indicated that the acoustic output was comparable (Wyeth et al., 2000). This confirmed that future rockfish shoal estimation could be conducted directly from commercial vessels.

This study also provided benefits beyond the stated objectives. Fishers involved in the project were introduced to research quality acoustics and to the methodologies and assumptions required to convert backscatter measurements (the returning echo) to biomass. They not only became educated about the strengths and weaknesses of acoustic biomass assessment, but also learned that output from split-beam sounders can provide information on fish size frequencies. Fishers are now using this equipment to reduce bycatch during midwater trawling (B. Mose, personal communication). In turn, fishers educated acoustic staff about the extent to which side-lobe acoustic interference over high-relief bottom can generate false fish sign. Although this phenomenon is well known, the actual examples surprised acousticians, leading to changes in how research echograms are scrutinized following surveys of near-bottom species (R. Kieser, personal communication).

SILVERGRAY ROCKFISH (*SEBASTES BREVISPINIS*)

Background

The second example chronicles the events related to an assessment review of silvergray rockfish (Stanley and Kronlund, 2000). This species is a minor element in the BC bottom-trawl fishery with an annual coastwide harvest of approximately 1,000 tonnes from four areas (Figure 20.1). Although the harvest is small, the size of the quota is critical to the fishery. Each vessel requires a sufficient individual vessel quota (IVQ) of silvergray rockfish to accommodate the bycatch of silvergray rockfish that accumulates as they target other species. If an increase in silvergray rockfish abundance is not matched by a higher quota, the species becomes an increasing nuisance in that other species cannot be fished without the vessels exceeding their IVQ*s* for silvergray rockfish. Given 100 per cent observer coverage, vessels may have to stop fishing completely when they reach their area-specific silvergray rockfish IVQ*s*, losing the opportunity to catch their remaining IVQ*s* for other species.

Assessment information on silvergray rockfish is limited. Fishery-dependent trends in commercial CPUE cannot be assumed to reflect trends in abundance. Fishery independent surveys are not available (Stanley and Kronlund, 2000). This lack of a credible abundance index forced the assessment to rely on age composition data. These analyses indicated that the fisheries on three of the four stocks were relying on a strong recruitment mode centred on the 1981 year-class, and, although difficult to distinguish from increasing recruitment, the analysis also indicated a modest fishing down of older age classes. The reduction in the proportion of older fish was interpreted as indicating that exploitation has been equal to, or above, a sustainable rate, at least prior to the current recruitment pulse. Hence, although the stock is currently benefiting from the presence of a large incoming year-class, the long-term trend in age composition indicated that harvests should not be increased.

The trawl fishers commented that silvergray rockfish were becoming harder to avoid, and therefore the biomass was increasing and quotas should be raised. They suggested it was incorrect to assume comparability in the age composition over time, because the fishing locations, and thus the locations of the samples, had changed. Most of the samples had been collected from commercial landings; thus the fishery determined the sample locations. With the introduction of IVQ*s* in 1997, the relatively small IVQ*s* for silvergray rockfish had forced the fleet to move away from areas of high CPUE. They reasoned that the age composition might differ in the new locations. Therefore, it was incorrect to infer population dynamics from trends in the age composition. The senior author responded that a brief review of the spatial distribution of the samples had not revealed gross changes, and, because there was no demonstrable bias in age composition, the stock assessment advice was accepted as the basis for the management plan. The trawl fishers requested that the next silvergray rockfish assessment examine more closely the spatial effects on age composition.

Methods and results

Following the review process, the senior author conducted an observer trip on another commercial trawler, the *E.J. Safarik*, in February 2001 with another industry associate, Captain Reg Richards. Captain Richards was one of the principal trawl fishers operating in Area 5E, the north-west coast of the Queen Charlotte Islands (Figures 20.1 and 20.4). He had been an advisor during work on the assessment, and was critical of the resulting quota. The objective of the trip was to provide the senior author with an opportunity to discuss the assessment as well as providing Captain Richards the opportunity to demonstrate how IVQ*s* might have changed the sampling of silvergray rockfish in Area 5E.

FISHERS' KNOWLEDGE AND SCIENTIFIC SKILLS

Captain Richards explained that the fishery for silvergray rockfish had traditionally concentrated on the 'Frederick Spit' grounds (Figure 20.4). With

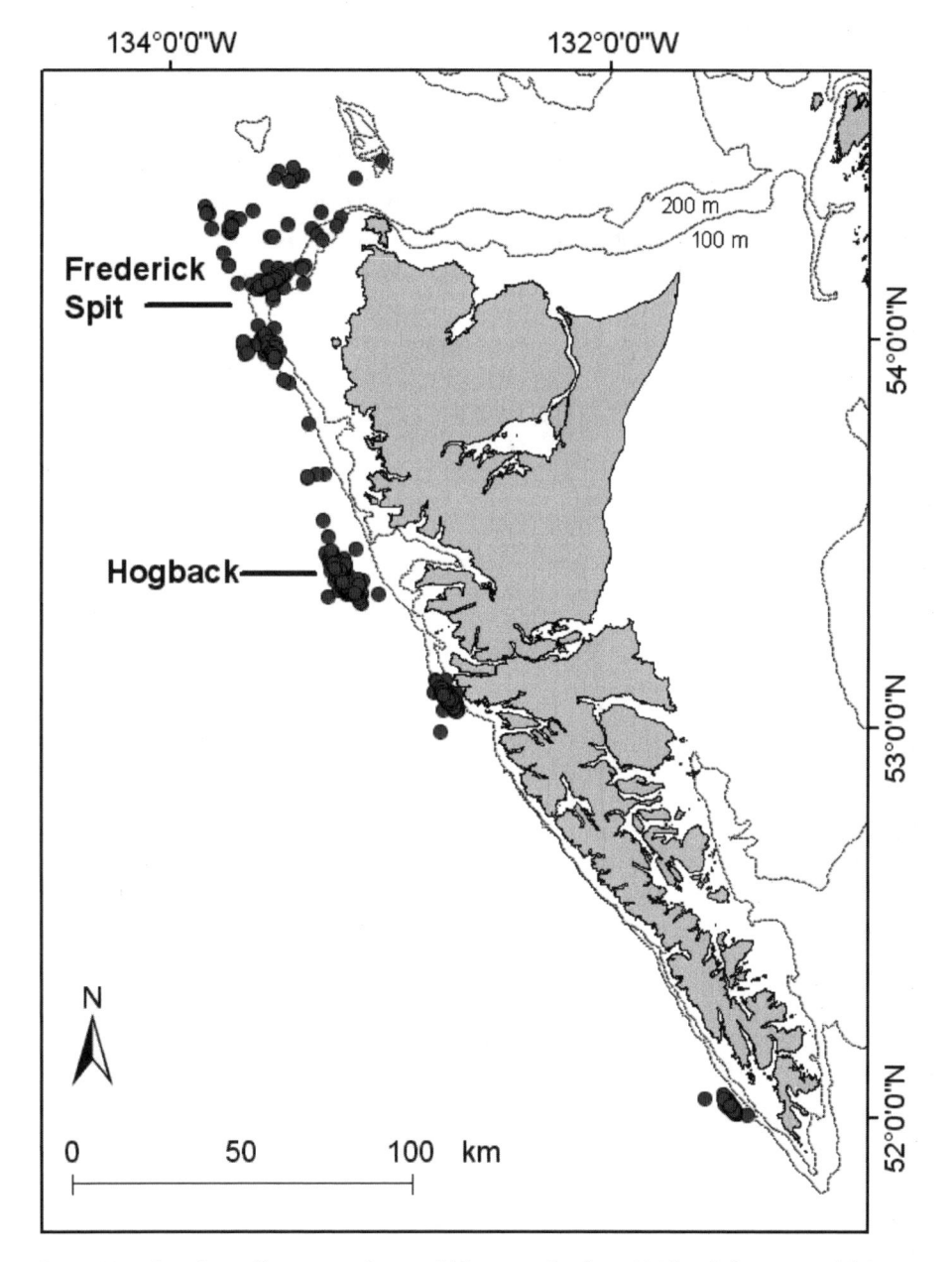

FIGURE 20.4 *Locations of bottom-trawl tows which captured at least 200 kg of silvergray rockfish.*
Note: Data are from 1996 to September, 2001.

COMMERCIAL FISHERIES

introduction of IVQs in 1997, the fishers had switched to the 'Hogback' grounds to avoid high catch rates of silvergray rockfish. There they targeted redstripe rockfish (*S. proriger*) while slowly accumulating their IVQ of silvergray rockfish. He questioned whether the relative absence of older fish from recent samples might have resulted from shifting the source of the samples from the Frederick Spit to the Hogback fishing grounds. He offered to conduct tows on both spots so that the senior author could obtain a comparison of the age composition.

When the three samples were collected and analysed, they indicated a significant difference in age composition (Figure 20.5). The Hogback sample indicated the typical 1981 recruitment mode, whereas the two samples from Frederick Spit were much older. This led the senior author to look more closely at the spatial distribution of the samples used in the stock assessment. These indicated that through 1998, the samples were representative of the entire area and consistent in age composition over time and space (Figure 20.6). Thus, the assessment, based on data through 1998, was not biased in that respect; but, from 1999 onward, the samples were tending

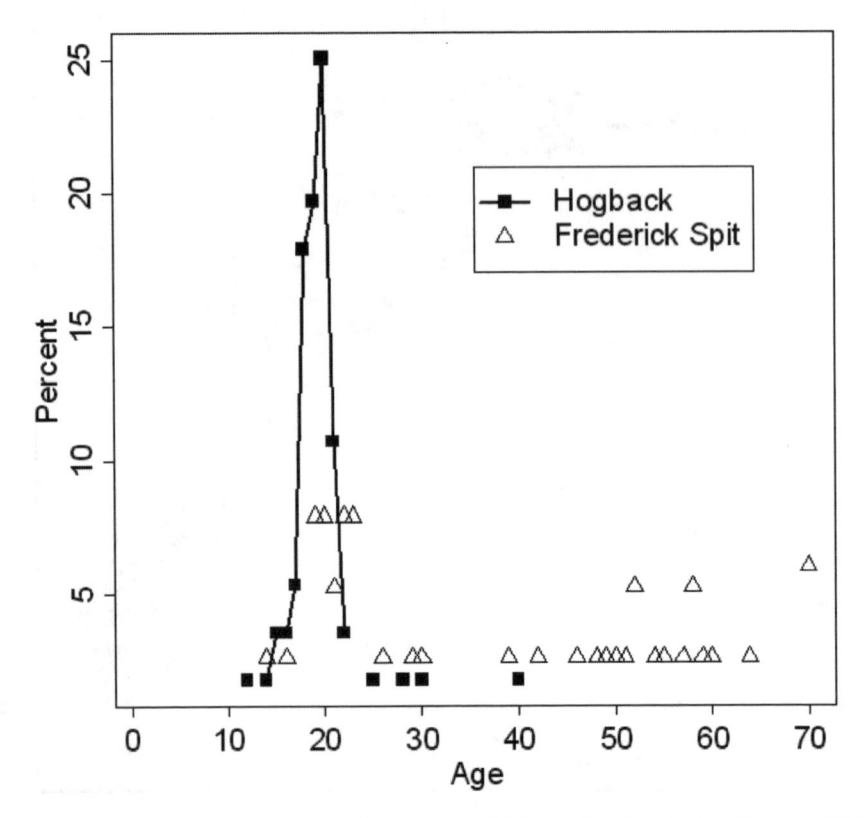

FIGURE 20.5 *Percent composition by age of silvergray rockfish samples taken during February 2001 observer trip.*

to come from the Hogback. Thus, the fishing captains' concerns had revealed that the fortuitous representativeness of the commercial sampling was deteriorating and future assessments would be compromised.

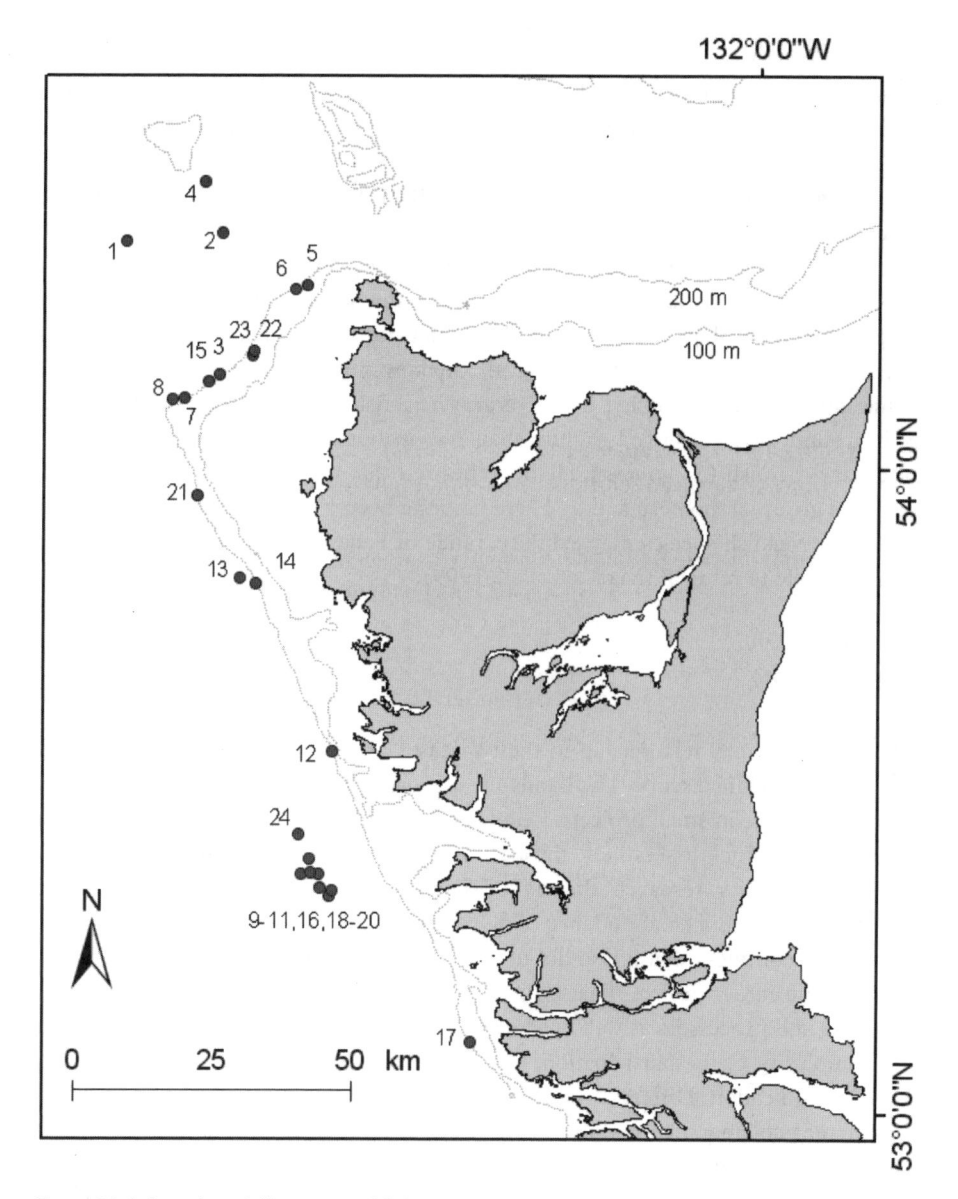

FIGURE 20.6 *Location of silvergray rockfish samples used in 2000 silvergray rockfish assessment for area 5E.*
Source: Stanley and Kronlund, 2000.

The concern that age composition might vary with changing fishing patterns led to discussions with Captain Richards over how to obtain representative samples from the whole area. The senior author could not envision a survey that could provide the samples without capturing most of the 5E quota of about 200 tonnes. Captain Richards proposed a solution wherein he would trawl a set of specified tow locations but avoid excessive catches by cutting a hole in the forward part of the codend. In addition to this attempt to modify sample collection, the senior author noted that future assessments would pay more attention to the fine-scale spatial distribution of the samples.

Further analysis indicated that the two 2001 Frederick Spit samples had a high percentage of older fish, and thus differed not only from the 2001 Hogback samples but from all previous Frederick Spit samples. They differed although they were collected within a few kilometres, in the same months, and only a few meters shallower than previous samples. When informed of the results, Captain Richards commented that these slightly shallower locations were rarely fished. While attempting to provide the silvergray rockfish samples from the Frederick Spit ground, he had moved slightly shallower in hopes of also obtaining samples of canary rockfish (*S. pinniger*). He hypothesized that the older fish represented an unfished group of 'homesteaders'. There has long been a suspicion among biologists that some rockfish species may exhibit a range of behavioural modes, ranging from highly mobile to refugial (MacCall et al., 1999).

DISCUSSION

THE participatory relationship in the two cases was facilitated by their small and narrowly defined scope. The hands-on approach of the two studies will obviously be more feasible in small projects. Leaman and Stanley (1993) describe an attempt to improve a stock assessment through a combination of adaptive management and participatory research that failed because of a lack of preparation and communication, and an abundance of naïvete. With the additional hindsight of these studies, we comment that it also did not work as well as hoped because of the larger scope of the earlier project. That project ended up involving a much larger number of fishers and processors as well as a local community. As the scope expands, so does the complexity of the participation, leading to an exponential increase in the need for communication. Different and more complex contexts may require a blending of different collaborative research methods, many of which are discussed in other papers in this volume (see also Mackinson and Nøttestad, 1998).

The participation was also facilitated by computer technology that helped to provide biomass estimates during the survey and the use of three-dimensional graphics to present the finished results (Stanley et al., 2001). They illustrate the benefits in participatory research of rapid feedback (Zwanenburg et al., 2000) and the value of mutually understandable graphical images (Walters et al., 1998).

A strategic issue that contributed to these studies was the growing importance of industry-funded research. It not only increases the pool of resources but, by decentralizing the control, it leads research in new directions (Chambers, 1989). This partnership also educates industry groups in the cost of science, just as joint authorship of a primary paper from this work (Stanley et al., 2000) conveyed to fishers the commitment required to communicate research results.

Industry-based research organizations, such as the CGRCS, the Herring Research and Conservation Society or the Canadian Sablefish Association on Canada's Pacific coast, also provide venues for fishers to discuss scientific ideas, directions and hypotheses away from the tense atmosphere of stock assessment meetings (B. Turris, personal communication). It has been conjectured that an essential step in maximizing the value of resource users as research partners is to support mechanisms that encourage the users to seek excellence and test ideas together, on their own terms, and in their own language (Rice, 1998). These meetings are thus industry analogues to scientific conferences and workshops. Assisting the blending of ideas is the tendency for these organizations to fund science–industry liaison positions and hire fisheries research analysts. The liaison activity works to keep communication lines open, and the discussions with their own analysts provide a less intimidating forum for fishers to question and learn technical issues.

Making fishers full partners in research ultimately requires a strategy to enhance communication and build trust. An example of the cost of not communicating is provided by a retrospective look at the earlier days of groundfish stock assessment in BC, from 1980 to the mid–1990s. During those years, fishers were excluded from assessment meetings because it was felt that their presence would inhibit debate among the scientists. It was assumed that fishers would equate uncertainty with a lack of knowledge (Preikshot, 1998), which would further erode credibility in the assessment advice. However, excluding fishers from healthy debate acted to reinforce their belief that researchers overestimated the accuracy of their stock assessments.

Once fishers observed and participated in the debate, they were reassured that researchers understood the limitations of their data and analyses. Fishers were already aware of how hard it must be to estimate abundance; what worried them was the possibility that research staff were not. Fishers may become more sceptical of the science the more they know, but scepticism is a good thing when it prompts constructive follow-up (McGoodwin et al., 2000). While still evolving, the process for BC groundfish research has now progressed to where fishers and other interested groups are present at a series of meetings that include work-plan prioritization and pre-assessment meetings in which authors outline the data sources and methods that will be used. This trend is present throughout Canada, as stakeholders have taken an increasing role in assessment meetings. Documents from the Canadian Stock Assessment Secretariat Proceedings provide growing evidence of their interpretative skills; and summary documents frequently include a section on 'Industry perspective'.[1]

Fishers were also excluded because it was felt that the economic pressures of commercial fishing would make it difficult for them to participate objectively. This risk cannot be ignored, but it is well documented that scientists do not have a flawless record of objectivity. For example, instances of confirmatory bias are very common in science (Nicholls, 1999). Even when scientists specify the error rates that are the basis for traditional hypothesis testing (or the probabilities associated with Bayesian decision support), the estimates derived from fisheries data are highly uncertain (McAllister and Kirkwood, 1998), as are the likelihood profiles, which attempt to characterize the probability of a given estimate being correct. Reliance on the formal use of probability-based methods can be more a matter of form than reality (Patterson et al., 2001). Often an assessment meeting must focus more on the justification for assuming that alternative information sources and interpretations are reliable and credible, rather than on statistical nuances that have weak empirical foundations.

Increased participation by fishers can thus be both a means and an end (Sajise, 1993) if it builds trust. It appears to be a means for coping with the 'conflicting dogma of the omniscience' that researchers know better because of their formal education, and fishers know better because of their experiential background.

While we extol the potential benefits from a more hands-on and participatory approach, we acknowledge its risk and costs. Producing better research is not as simple as parachuting biologists onto fishing boats or dragging fishers to stock assessment meetings. All participants need to learn how to critique each other's hypotheses and information, jeopardizing neither rigour nor respect. Even after individual fishers and scientists have learned to respect and value each others' creative hypotheses, criticisms and sources of new information, the relationship can be strained by the challenging function of peer review (Sillitoe, 1998).

It is fine to argue that 'one cannot communicate too much' and endorse the idea of paid liaison positions, but these resources come from a finite pool. Time spent by government biologists on commercial fishing boats is time away from detailed likelihood profiling and ecosystem modelling, and industry advisors are now complaining of meeting fatigue. Strategic planning has to cope with these conflicts but our underlying belief is that any initiative that brings more research assets into the process must ultimately be cost effective. As learning to fish can be thought of by Icelandic fishers as a journey, so might we perceive truly collaborative or participatory research (Pálsson, 2000). It is perhaps a long process of small steps wherein harvesting and research become the same thing (J. Prince, personal communication).

1. See Stock Status Reports on http://www.dfo-mpo.gc.ca/csas.

The studies described in this chapter are only two of many examples currently underway within the fisheries on Canada's Pacific coast. In fact, joint research with industry has a long history on Canada's Pacific coast, as we are sure it has had elsewhere. Although the '***-knowledge' keywords cannot be identified in the publication titles, we know many fishery researchers who have accrued extensive commercial fishery time, and fishers, much meeting time, in the process of conducting joint research.

ACKNOWLEDGEMENTS

THE present paper benefited greatly from discussions and fishing trips with trawl fishing captains, including Captains Kelly Anderson, Brian Dickens, Brian Mose, Reg Richards, and John Roche. Review comments from Bruce Turris, Norm Olsen, Bruce Leaman, Barbara Neis and two anonymous reviewers greatly improved the manuscript. We noted above the long history of joint research in the Pacific fisheries of Canada. In particular, we acknowledge the tradition established by individuals such as Neil Bourne, Dan Quayle, Terry Butler, Keith Ketchen, and Bruce Leaman.

REFERENCES

AGRAWAL, A. 1995. Dismantling the divide between indigenous and scientific knowledge. *Development and Change*, Vol. 26, pp. 413–39.

AMES, E.; WATSON, S.; WILSON, J. 2000. Rethinking overfishing: Insights from oral histories of retired groundfishermen. In: B. Neis; L. Felt (eds.), *Finding our sea legs: Linking fishery people and their knowledge with science and management*. St. John's, Newfoundland, ISER books, pp. 153–64 pp.

ARCHIBALD, C.P.; SHAW, W.; LEAMAN, B.M. 1981. Growth and mortality estimates of rockfishes (*Scorpaenidae*) from B.C. waters, 1977–1979. *Canadian Technical Report of Fisheries and Aquatic Sciences*, 1048. Ottawa. Minister of Supply and Services Canada. 57 pp.

ARCHIBALD, C.P.; FOURNIER, D.; LEAMAN, B.M. 1983. Reconstruction of stock history and development of rehabilitation strategies for Pacific ocean perch in Queen Charlotte Sound, Canada. *North American Journal Fisheries Management*, Vol. 3, pp. 283–294.

CHAMBERS, R. 1989. Reversal, institutions and change. In: R. Chambers, A. Pacey and L.A. Thrupp (eds.), *Farmer first: Farmer innovation and agricultural research*. Exeter, Intermediate Technology Publications, Short Run Press, pp. 181–195.

CHAMBERS, R.; PACEY, A.; THRUPP, L.A. (eds.). 1989. *Farmer first: Farmer innovation and agricultural research*. Exeter, Intermediate Technology Publications, Short Run Press, 218 pp.

DYER, C.L.; McGOODWIN, J.R. (eds.). 1994. *Folk management in the world's fisheries: Lessons for modern fisheries management*. Boulder, University Press of Colorado, 347 pp.

FERRADÁS, C. 1998. In: Appended discussion notes of Sillitoe, P. 1998. The development of indigenous knowledge. p. 239. *Current Anthropology*, Vol. 39, No. 2, pp. 223–52.

FISCHER, J. 2000. Participatory research in ecological fieldwork: A Nicaraguan framework. In: B. Neis; L. Felt (eds.), *Finding our sea legs: Linking fishery people and their knowledge with science and management*. St. John's, Newfoundland, ISER books, pp. 41–54.

FRCC. 1994. *Conservation: Stay the course*. Report to the Minister of Fisheries and Oceans. 1995 Conservation measures for Atlantic Groundfish. Ottawa, Fisheries Resource Conservation Council, 145 p. Available online at: http://www.frcc.ca/scanned%20reports/FRCC94R4.pdf (last accessed on 26 April 2006).

GAVARIS, S. 1996. Population stewardship rights: Decentralised management through explicit accounting of the value of uncaught fish. *Canadian Journal of Fisheries and Aquatic Sciences*, Vol. 53, pp. 1683–91.

HUTCHINGS, J.A. 1996. Spatial and temporal variation in the density of northern cod and a review of the hypotheses for the stock's collapse. *Canadian Journal of Fisheries and Aquatic Sciences*, Vol. 53, pp. 943–62.

JOHANNES, R.E. 1978. *Words of the lagoon: Fishing and marine law in the Palau District of Micronesia*. Berkeley, Calif., University of California Press, 245 pp.

LEAMAN, B.M.; STANLEY, R.D. 1993. Experimental management programs for two rockfish stocks off British Columbia, Canada. In: S.J. Smith; J.J. Hunt; D. Rivard (eds.), *Risk evaluation and biological reference points for fisheries management*. Canadian Special Publications Fisheries Aquatic Sciences 120. Ottawa, National Research Council of Canada, pp. 403–18.

McALLISTER, M.K.; KIRKWOOD, G.P. 1998. Using Bayesian decision analysis to help achieve a precautionary approach for managing developing fisheries. *Canadian Journal of Fisheries and Aquatic Sciences*, Vol. 12, 2642–61.

MACCALL, A.D.; RALSTON, S.; PEARSON, D.; WILLIAMS, E. 1999. Status of bocaccio off California in 1999 and outlook for the next millenium. Appendix to: *Status of the Pacific coast groundfish fishery through 1999 and recommended acceptable biological catches for 2000*. Oregon, US, Pacific Fisheries Management Council. 45pp. http://www.pfmc.org. Last accessed 26 June 2005.

MACKINSON, S.; NØTTESTAD, L. 1998. Combining local and scientific knowledge. *Reviews in Fish Biology and Fisheries*, Vol. 8, pp. 481–90.

McGoodwin, J.R.; Neis, B.; Felt, L. 2000. Integrating fishery people and their knowledge into fisheries science and resource management. Issues, prospects, and problems. In: B. Neis; L. Felt (eds.), *Finding our sea legs: Linking fishery people and their knowledge with science and management.* St. John's, Newfoundland, ISER books, Memorial University, pp. 249–64.

Melvin, G.; Li, Y.; Mayer, L.; Clay, A. 1998. The development of an automated sounder/sonar acoustic logging system for deployment on commercial fishing vessels. *ICES CM, 1998/S,* No. 14, 23 pp.

Neis, B.; Felt, L. (eds.). 2000. *Finding our sea legs: Linking fishery people and their knowledge with science and management.* St John's, Newfoundland, ISER books, 318 pp.

Neis, B.; Schneider, D.C.; Felt, L.; Haedrich, R.; Fischer, J. 1999. Fisheries assessment: What can be learned from interviewing resource users? *Canadian Journal of Fisheries and Aquatic Sciences,* Vol. 56, No. 10, pp. 1949–63.

Nicholls, N. 1999. Cognitive illusions, heuristics, and climate prediction. *Bulletin of the American Meteorological Society,* Vol. 80, No. 7, pp. 1385–97.

Office of management and budget. 2001. NOAA FY 2002 Budget Summary. Washington, D.C., Office of Management and Budget, United States Government.

Pálsson, G. 2000. Learning, the process of enskilment, and integrating fishers and their knowledge into fisheries science and management. In: B. Neis; L. Felt (eds.), *Finding our sea legs: Linking fishery people and their knowledge with science and management.* St. John's, Newfoundland, ISER books, Memorial University, pp. 26–40.

Patterson, K.; Cook, R.; Darby, C.; Gavaris, S.; Kell, L.; Lewy, P.; Mesnil, B.; Punt, A.; Restrepo, V.; Skagen, D.W.; Stefannsson, G. 2001. Estimating uncertainty in fish stock assessment and forecasting. *Fish and Fisheries,* Vol. 2, No. 2, pp. 125–57.

Preikshot, D.B. 1998. Reinventing the formulation of policy in future fisheries. In: T.J. Pitcher; P.J.B. Hart; D. Pauly (eds.), *Reinventing fisheries management.* London, Kluwer Academic, pp.114–23.

Rice, J. 1998. Fostering sustainable development and research by encouraging the right kind of institutions. In: T.J. Pitcher; P.J.B. Hart; D. Pauly (eds.), *Reinventing fisheries management.* London, Kluwer Academic, pp. 195–200.

Roepstorff, A. 2000. The double interface of environmental knowledge: Fishing for Greenland halibut. In: B. Neis; L. Felt (eds.), *Finding our sea legs: linking fishery people and their knowledge with science and management.* St. John's, Newfoundland, ISER books, Memorial University, pp. 165–88.

Ruddle, K. 1994. Local knowledge in the folk management of fisheries and coastal marine environments. In: C.L. Dyer; J.R. McGoodwin (eds.), *Folk management in the world's fisheries: Lessons for modern fisheries management.* 1994. Boulder, University Press of Colorado, pp.161–206.

SAJISE, P.E. 1993. Participation in research or research for participation: Its relevance to sustainable development. *Out of the Shell,* Vol. 3, No. 2, pp. 1–5.

SILLITOE, P. 1998. The development of indigenous knowledge. *Current Anthropology,* Vol. 39, No. 2, pp. 223–52.

STANLEY, R.D.; KIESER, R.; COOKE, K.; SURRY, A. M; MOSE, B. 2000. Estimation of a widow rockfish (*Sebastes entomelas*) shoal off British Columbia, Canada as a joint exercise between assessment staff and the fishing industry. *ICES Journal of Marine Science,* Vol. 57, pp. 1035–49.

STANLEY, R.D.; HAIST, V. 1997. *Shelf rockfish assessment for 1997 and recommended yield options for 1998.* Canadian Stock Assessment Secretariat Research Document 97/132. Ottawa, Fisheries and Oceans Canada, 76 pp.

STANLEY, R.D.; KRONLUND, A.R. 2000. *Silvergray rockfish (Sebastes brevispinis) assessment for 2000 and recommended yield options for 2001/2002.* Canadian Stock Assessment Secretariat Research Document 2000/173. Ottawa, Fisheries and Oceans Canada, 116 pp.

STANLEY, R.D.; KIESER, R.; HAJIRAKAR, M. 2001. Three-dimensional visualization of a widow rockfish (*Sebastes entomelas*) shoal over interpolated bathymetry. *ICES Journal of Marine Science,* Vol. 59, pp. 151–5.

WALTERS, C. J.; PRESCOTT, J.H.; McGARVEY, R.; PRINCE, J. 1998. Management options for the South Australian rock lobster (*Jasus edwardsii*) fishery: A case study of co-operative assessment and policy design by fishers and biologists. In: G.S. Jamieson; A. Campbell (eds.), *Proceedings of the North Pacific Symposium on Invertebrate Stock Assessment and Management.* Canadian Special Publication Fisheries Aquatic Science No. 125. Ottawa, NRC Research Press, pp. 377–383.

WYETH, M.R.; STANLEY, R.D.; KIESER, R.; COOKE, K. 2000. *Use and calibration of a quantitative acoustic system on a commercial fishing vessel.* Canadian Technical Report of Fisheries and Aquatic Sciences, 2324. Ottawa, Fisheries and Oceans Canada, 46 pp.

ZWANENBURG, K.; KING, P.; FANNING, P. 2000. Fishermen and scientists research society: A model for incorporating fishermen and their knowledge into stock assessment. In: B. Neis; L. Felt (eds.), *Finding our sea legs: Linking fishery people and their knowledge with science and management.* St. John's, Newfoundland, ISER books, pp. 124–32.

CHAPTER 21 The changing face of fisheries science and management

Nigel Haggan and Barbara Neis

'Great complaints are made against the use of the net called "wondyrchoun" [beam trawl] which drags from the bottom of the sea all the bait that used to be the food of great fish. ... [It] runs so heavily and hardly over the ground when fishing that it destroys the flowers of the land below the water and also the spat of oysters, mussels, and other fish upon which the great fish are nourished. ... Through means of this instrument fishermen catch "such great plenty of small fish that they do not know what to do with them, but fatten their pigs with them".'

UK Rolls of Parliament (1376/77)

Today, 99 per cent of the world's 51 million fishers are small-scale, producing over half of the global foodfish catch of 98 million tonnes. One billion people rely on aquatic resources as their main source of dietary protein (Berkes et al., 2001, and references therein). Globally, many fish stocks are depleted. Overall, our capacity to harvest fish continues to outpace our capacity to monitor the effects of fishing, let alone design, implement and enforce effective conservation measures. Fish populations once deemed inexhaustible (Huxley, 1883), have been reduced to a fraction of their past abundance (Hilborn et al., 2003). High-level predators in the North Atlantic hover round 10 per cent of their 1900 levels (Christensen et al., 2003; Myers and Worm, 2003). Some sharks have suffered declines of over 50 per cent since the mid-1980s (Schindler et al., 2002; Baum et al., 2003). Other species as diverse as marine turtles (Hays et al., 2003) and many species of whales hover at very low levels (Roman and Palumbi, 2003). In too many cases, stocks are so depleted that conserving what is left would amount to sharing the present misery (Pitcher, 2001). In these cases, the only meaningful option is recovery but we generally know even less about how to achieve recovery than we do about conservation.

'Fisheries science' and 'management', as currently practised, are relatively new phenomena. However, knowledge about marine and freshwater ecosystems and social institutions mediating human relationships with those ecosystems is ancient, being a necessity of survival as well as the product of natural human interest in the surrounding world. Together, these have led, throughout the world, to acute observation, experimentation, the formulation and testing of hypotheses, and the development of theories and practices as well as social institutions to regulate resource use and transmit knowledge from generation to generation (Berkes, 1999).

This book has brought together many case studies from different parts of the world where the knowledge of fishers and their institutions is being actively integrated into fisheries science and management. The chapters represent different points on a number of continua; between contexts where mutual respect, cooperation and reciprocity (Stanley and Rice, this volume) are just evolving and those where formal co-management arrangements operate (Baird et al, this volume); between Indigenous management, state management, and state management mitigated by the re-emergence of elements of traditional management and values (Hickey, this volume); and between documenting the richness and scope of the indigenous knowledge still in use in some fisheries (Nsiku, this volume) and clear cut applications of fishers' knowledge (FK) and science in the struggle to understand and conserve fish stocks (Baird, this volume). In this concluding chapter, we underscore the urgent need for approaches to fisheries science and management that promote the collection, critical examination and synthesis of all potential types of fisheries knowledge and all effective mechanisms for promoting sharing of information and of the responsibility and struggle to protect, and ideally restore, the world's endangered wild fish stocks.

'One man stood before the microphone, his face grey with fatigue and anxiety, and said in a breaking voice: "Let's face it: we've caught them all".'

(Storey, 1993, cited in Ommer, 1994).

This quote from a participant in a post-mortem after the collapse of the Atlantic cod (*Gadus morhua*) captures, like nothing else, the dawn of awareness that people now have the capacity to destroy not only local stocks and ecosystems but great resources distributed over thousands of cubic miles of ocean. Frank et al. (2005) suggest that changes induced by overfishing are so profound as to make recovery of the Atlantic cod unlikely.

Despite enormous investments in 'science' and 'management', marine and riverine ecosystem structures and responses to stress are poorly understood. There is mounting evidence that some key, taken-for-granted assumptions about the behaviour of fish and fishers associated with fisheries science and management are incorrect. For example, perversely, that the increasing power of fishing technology now appears to have been reinforced by the catchability effects of shoaling behaviour found in Atlantic cod, herring (*Clupea harengus*), capelin (*Mallotus villosus*) and many other important food species. Pitcher (1997) notes that under adverse conditions, shoaling fish concentrate in areas of prime habitat rather than getting 'thin on the ground' over their entire previous range. Fish-finding equipment, mobile vessels and dense shoals help to account for situations where catch rates remain high until stocks collapse.

Similarly, for much of the history of industrial fishing, we have assumed that overfishing would lead to a reduction in the density of fish populations, declining catch rates and, eventually, to cessation of fishing for economic reasons. We now know that industrial fisheries and contemporary artisanal fisheries tend to respond to declining catches by increasing effort and shifting to other, frequently lower trophic level species (Pauly et al., 1998). Patterns of intensification and expansion with spatial, temporal and ecological dimensions have been documented (Neis and Kean, 2003). Fish do eventually become scarce, but fishers often keep on fishing, sometimes ploughing their own resources into bigger boats, more horsepower and more sophisticated gear, a form of self-subsidization. Alternatively, they are displaced by large, mobile, corporate-owned fisheries. Big and small industry players are often kept in business through annual subsidies estimated at US$ 20 billion worldwide (Milazzo, 1998).

A third common but problematic assumption in fisheries science and management is that marine fish stocks are 'panmictic': in other words, that individuals are unrestrictedly capable of interbreeding. Cury (1994) observes that there can be a high degree of variability within stocks of small pelagic fish, including Peruvian anchoveta (*Engraulis ringens*) and Northwest Atlantic herring. The implication is that marine fish species may well be composed of many different stocks, some small and a few large, a characteristic previously associated only with salmon. The information that Ames (this volume) collected from retired fish harvesters confirms the existence of many now 'extinct' cod and haddock (*Melanogrammus aeglefinus*) spawning areas in the Gulf of Maine, supplementing other cod-related findings that suggest population structures are more complex than formerly thought. Hutchinson et al. (2003) used DNA recovered from archived otoliths of North Sea cod (*G. morhua*) to show a significant decline in genetic diversity between 1954 and 1998, with implications for the ability of this species to withstand fishing pressure and environmental change. Prince (2003) makes a case for abalone (*Haliotid*) populations being composed of numerous 'micro stocks' that are 'myriad and complex to study, monitor, assess and manage'.

The Hauser et al. (2002) study of a New Zealand snapper (*Pagurus auratus*) showed a reduction in genetic diversity comparable to that described by Hutchinson for North Sea cod. They also observed that, unlike terrestrial species, where a small number of survivors can regenerate a population, marine fish populations tend to have a relatively small percentage of highly effective breeders. In the case of the New Zealand snapper, only 10,000 out of the residual population of three million were found to be effective breeders (Hauser et al., 2002). Similar research on cod and lobster points to the importance of protecting larger, experienced and more fecund spawners (Corson, 2004; Trippel, 1998). Berkeley et al. (2004) report that the eggs of older spawners have a higher oil content leading to greatly enhanced larval survival.

All the above scientific findings may come as no surprise to indigenous and artisanal fishers; indeed the Haida, Heiltsuk and Nuu-chah-nulth First Nations[1] of BC have all taken action to close herring fisheries approved by Canada's Department of Fisheries and Oceans under the assumption that stocks could withstand a 20 per cent harvest rate, an assumption the aboriginal people felt to be untenable (Jones this volume; Lucas, this volume; Ross Wilson, Chief Councilor, Heiltsuk Nation, personal communication). Many inshore fishers in Newfoundland expressed concern that large mesh gill-nets and dragger fisheries that targeted large, spawning cod were destroying the 'mother fish' and thereby affecting recruitment to local populations (Neis et al., 1999).

These and other recent findings on the state of the world's fisheries and the vulnerability of fishery communities, as well as on the ways interactions between fish, fisheries, science and management have contributed to the precariousness of fish and fisheries, all point to the need for new approaches. Science and management need to take place at finer temporal and spatial scales with rapid feedback between fishers, scientists and managers. We need longer time series so that short-term fluctuations can be separated from long-term trends, and so that current and former habitat and stock remnants can be identified, protected, and ideally enhanced. The dynamism of fisheries means that scientists and managers tend to follow fisheries around (Neis and Kean, 2003), often discovering after the fact what used to be there and why it is now gone.

We need to learn more about former and current practices that have conserved and in the best cases enhanced fish populations and the larger ecosystems upon which they depend, and to substantially expand the tool kit of cultural, scientific and management practices that have been shown to promote recovery and preservation and encourage effective stewardship by all stakeholders. To illustrate, Haggan et al. (2004) suggest that the 'inexhaustible' fish, forests and wildlife seen by early explorers of the Pacific North-west of Canada were, to a large extent, the result of active enhancement and stewardship of a very wide range of aquatic and terrestrial resources. While more research is needed, it is apparent that the Pacific Northwest was far from a 'wilderness' when Europeans arrived in the mid-1700s. What is indisputable is that, compared to pre-contact times, the Pacific Northwest is now a wasteland, and that the resources of government agencies are not adequate for the task of conserving and managing what remains for us, to say nothing of the reinvestment in natural capital needed to restore some level of historic abundance, a point made eloquently throughout this volume.

The failure of centralized management to avert major stock collapse, along with the 'vicious cycle' where depleted resources require more study and more

1. West-coast and other aboriginal peoples living in what is now Canada describe themselves as 'First Nations', a term that serves the dual purpose of affirming both their presence long before the 'discovery' of Canada and that they must be treated on a nation-to-nation basis.

management, but produce fewer benefits or negative returns to government, is prompting two divergent courses of action. The first is to transform commercial fishing licences into transferable quotas that have most of the characteristics of property rights. This enables those with deep pockets to purchase a sufficient quota to make harvesting efficient. A standard proviso is that the quota holders pay the research and management costs for 'their' fishery. Canada's west coast fisheries for halibut (*Hippoglossus stenolepis*), blackcod (*Anoplopoma fimbria*), geoduck clam (*Panope generosa*), sea urchin and sea cucumber have been effectively privatized, raising quota values to levels beyond the reach of small-scale fishers and creating a problem for the return of resources to First Nations under the modern-day treaty-making process in BC.[2] A proposal by McRae and Pearse (2004) to privatize the Pacific salmon fishery is even more problematic (Haggan et al. 2004; Jones et al., 2004). It is clear that this approach is having serious social and economic consequences for many fishers and their communities.

The second trend is the effort to 'bale out' or beef up fisheries science and management, by 'harnessing', 'capturing', 'incorporating', or sometimes 'integrating' FK. This, as one would properly expect, raises no end of practical, ethical and epistemological problems, most of which result from trying to fit FK into the current framework of resource management and ownership: in other words, to the benefit of those who control the lion's share of the resource and to the preservation of existing institutions (see Holm, 2003 and responses from Neis and Pinkerton in the same issue). FK does not come out of thin air. It is 'situated' knowledge in that it is tied to place (Newell and Ommer, 1999) and to particular social, ecological and historical contexts, and in some cases may be based on hundreds or thousands of years of interdependence with particular environments, resources and social institutions. It is often expressed in ways that are hard for those from other sociocultural circumstances to understand, even if they speak the same language. FK is generally collective and individual, a socio-ecological product mediated by ecology, technology, divisions of labour, culture, knowledge sharing and transmission. To some degree the same is true of scientific knowledge and the knowledge of managers (see, for example, Finlayson, 1994; Neis and Morris, 2002; Hutchings et al., 2002). People of other cultures have often developed similar, if somewhat less rigid, ways to transmit their knowledge and practices about fish, fisheries and ecosystems (many chapters, this volume).

FK is generally partially passed down through oral traditions, sometimes by people trained to prodigious feats of memory. In the Haida Nation on the west coast of Canada, students learning from elders were required to repeat lessons verbatim. It was not acceptable to stumble, recall and continue; each lesson had to be repeated

---

2. BC is unlike the rest of Canada in that few treaties were negotiated, leaving a vacuum. In 1990 the First Nations Summit, a group representing many BC First Nations, entered into an agreement with the governments of Canada and BC to negotiate treaties. The process has been slow, and currently a number of First Nations have dropped out of the treaty process and are pursuing their claims through the courts.

from the beginning (Russ Jones, Technical Director of Haida Fisheries Program, personal communication). Science is generally written, but is also influenced by dominant paradigms. Informal cultures and knowledge systems can also be found in science. Intergenerational transfer of FK and science (Narcisse and others, this volume) maintains continuity, and ensures a shared knowledge of past events and interpretations of those events, as well as lessons learned and institutions that reflect those lessons. However, both FK and science may have minority points of view that can be valid and important but tend to get marginalized and forgotten.

Local knowledge and management systems, like formal science and management, are also dynamic in that they expand and change in response to changes in policies and practice, species distribution, species targeted, the movement of people from one area to another, fishery and climate-induced ecosystem shifts, market preference and other factors. Often marginalized in the process of colonization through displacement, mortality, indoctrination and resource degradation, FK can contract in scope and complexity under unfavourable conditions, but is also resilient and can re-emerge as conditions change (Hickey; Satria and others, this volume). When resurrected and given support, indigenous and local knowledge and management systems can contribute to the development of innovative approaches to research, conservation and management of species as diverse as turtles (Küyük et al., this volume), serranid spawning aggregations (Phelan, this volume) and trout (Spens, this volume).

The 'First Salmon Ceremony' practised in various forms by First Nations throughout the Pacific Northwest (Swezey and Heizer, 1993) is an example of the symbolic importance of rituals within which knowledge is contextually nested. The ceremony gives thanks for the gift of salmon and shows proper respect to ensure their return in future years. It includes a feast where salmon are eaten and the bones and intestinal organs are returned to the river, while the flesh passes, sooner or later, back to the land.

Recent scientific research on the pivotal importance of marine nitrogen and phosphorous from salmon carcasses to the forest as well as the freshwater ecosystem is a longer way of saying the same thing, but important in that it increases scientific and public awareness of connections not obvious to most people. The huge runs of Pacific salmon transport thousands of tonnes of nutrients from the ocean to inland watersheds (Stockner, 2003).[3] Core samples taken from riverside trees indicate the strength of past salmon runs in the differential ring growth (Reimchen, 2001). Watkinson (2001), a member of the Tsimshian Nation on the north coast of British Columbia, used ecosystem modelling to quantify the importance of these nutrients to the coastal ecosystem and identified around forty species, ranging from bears

---

3. Commercial salmon catch on the west coast of Canada varies between 60,000 and 80,000 tonnes, a substantial amount of which would have been returned to the watersheds prior to European contact and development of the commercial salmon fishery.

to insects, that depend on salmon for food and that transport nutrients up to 1 km, or even more, from the riverbank. This is an excellent example of traditional knowledge and science working effectively together. Legend, theory and practical implications for the 'resource economy' converge in our growing awareness of salmon as an ecological keystone species (Paine, 1969; Power et al., 1996). The salmon is also a 'cultural keystone species' (*sensu* Garibaldi and Turner, 2004), in that it is central to the spiritual well being of aboriginal peoples and a key focal point for the intergenerational transfer of knowledge (Narcisse and others, this volume).

KNOWLEDGE GAINS POWER WHEN IT IS SHARED

THIS saying of the Stó:lõ Nation, whose traditional territory is on the Lower Fraser River in BC, identifies both the reward and the price of knowledge integration (Haig-Brown and Archibald, 1996). The nature and application of FK cannot be fully understood outside the culture, belief and value systems in which it is embedded. Such knowledge is the product of long interaction with its cultural as well as its geographical and biological context. Some elements transplant readily to neighbouring or even distant communities, but caution is indicated (Sultana and Thompson, this volume). It is also the product of a belief system or worldview, and differs from the 'fisheries science' of the last 200 years in the particular values that it embraces (Hickey; Poepoe et al., this volume). The local science of more recent fishing communities, such as the small fishing ports of Canada's east and west coasts, is also 'situated' knowledge and commonly incorporates important values such as continuity and respect for the environment and species that are the reason for the community's existence.

Industrial fishers who are part of the same fleet or fish for the same species are also 'communities' that often share a common corporate culture, shaped by the societies from which they originate, industry associations and interactions with scientists and managers over the life of their fishery. Industrial fishers have detailed knowledge of their target and associated species; indeed, it is only through the sheer amount of time spent on, and indeed under, the water (Meeuwig et al., this volume) that we can hope to be able to comprehend and map the fine-scale detail shown in the fisher maps of Ames (this volume); and Williams and Bax (this volume), and bring their knowledge to bear in the design and analysis of information and new management processes (Baelde; Mulrennan; Stanley and Rice, this volume; see also Neis and Felt (2000) for other examples of efforts to combine local ecological knowledge and science related to commercial fisheries).

Common to all the cases presented in this book is the fact that some form of cooperative management is often required to maximize the utility of the knowledge.

Contributions from Baird, Kalikoski and Vasconcellos, Mulrennan, Satria, Sultana and Thompson and others (this volume) emphasize that legal recognition of local management and guaranteed access to resources is a precondition of conservation and potential sustainability. This can only be achieved when we expand our concept of 'science' to include the wealth of knowledge at local levels and the analytic power of stock assessment and modelling tools, and commit to a genuine sharing of the benefits that come from the application of our collective wit and wisdom.

In summary, the knowledge of those who live, move and have their being on or beside the water contains much biological and ecological information, sometimes explicitly expressed in ways that scientists can easily comprehend, and sometimes opaquely entwined in rituals and other sociocultural aspects. Viewed from the point of view of fishers, there are similar problems with science. FK has value to the holders for management, food security, wealth (however defined), spirituality, and ultimately for peace, order and good government. It is of interest and value to scientists and educators, and it has inherent value among the holders of the knowledge, regardless of whether outsiders understand it, accept it, are able to validate it, make use of it, or not. Knowledge of fisheries science and the principles that guide formal management are also of potential value to fishers. Above all, FK is not a fossil, but dynamic and capable of change and adaptation.

More than twenty years ago, in *Words of the lagoon*, the late Bob Johannes (1981) drew attention to the serious neglect of the FK of 'native fishermen' by social and natural scientists. He commented that the type of research he had done for this book:

> 'offers a shortcut to some of the basic natural history data we need in order to understand these vast and valuable resources. ... I gained more new (to marine science) information during sixteen months of fieldwork ... than I had during the previous fifteen years using more conventional research techniques. This is because of my access to a store of unrecorded knowledge gathered by highly motivated observers over a period of centuries. This book, then, is really the work of uncounted individuals carried out over many generations.'

Johannes was seriously concerned that without more attention to FK, a vast storehouse of primarily oral information would be lost. Twenty years later, some of his comments about the relationship between FK and science seem naïve and even ethnocentric, particularly to social scientists. We have also heard natural scientists complain that he should have done more to verify the information he learned from the Palauan fishing elders. Whatever the validity of these criticisms, Johannes cannot be faulted for working persistently, in his dedicated but humble way, to promote attention to FK and to the fishers who carried it. Twenty years later, with the world's supplies of freshwater and marine species much diminished, and much FK lost with the passing of more generations of elders, and with rapid changes in fisheries and the erection of serious barriers to intergenerational transmission of FK, we can only wish that we had all reacted faster and more fully to his call for more attention to FK. Meaningful and respectful exchanges of knowledge between fishers, scientists and managers remain the exception rather than the norm, even in fisheries like the lobster fishery in Maine where enormous resources have been dedicated to one fishery and one species, and where differences of class, ethnicity and background between fishers and scientists are relatively small (Corson, 2004).

While it is heartening to know that there are more cases where fishers and their knowledge are being integrated into science and management than it would take to fill one book, it is disturbing to note that there is still no international institution dedicated to the ethical collection, preservation and dissemination of FK. Bob wanted us to change this and it is about time that we did.

REFERENCES

BAUM, J.K.; MYERS, R.A.; KEHLER, D.; WORM, B.; HARLEY, S.J.; DOHERTY, P.A. 2003. Collapse and conservation of shark populations in the Northwest Atlantic. *Science*, Vol. 299, pp. 389–92

BERKELEY, S.A.; CHAPMAN, C.; SOGARD, S.M. 2004. Maternal age as a determinant of larval growth and survival in a marine fish, *Sebastes melanops*. *Ecology*, Vol. 85, No. 5, pp. 1258–64.

BERKES, F. 1999. *Sacred ecology: Traditional ecological knowledge and resource management*. Philadelphia, Taylor Francis, 209 pp.

BERKES, F.; MAHON, R. MCCONNEY, P.; POLLNAC, R.; POMEROY, R. 2001. *Managing small-scale fisheries: Alternative directions and methods*. Ottawa, International Development Research Center, 320 p. Available online at: http://www.idrc.ca/en/ev-9328-201-1-DO_TOPIC.html#begining (accessed on 12 APRIL 2006).

CHRISTENSEN, V.; GUÉNETTE, S.; HEYMANS, J.J.; WALTERS, C.J.; WATSON, R.; ZELLER, D.; PAULY, D. 2003. Hundred-year decline of North Atlantic predatory fishes. *Fish and Fisheries*, Vol. 4, No. 1, p. 24.

CORSON, T. 2004. *The secret life of lobsters*. New York, Harper Collins, 298 pp.

CURY, P. 1994. Obstinate nature: An ecology of individuals. Thoughts on reproductive behavior and biodiversity. *Canadian Journal of Fisheries and Aquatic Sciences*, Vol. 51, 1664–73.

FINLAYSON, A.C. 1994. *Fishing for truth: A sociological analysis of northern cod stock assessments from 1977–1990*. St John's, ISER Books, 176 pp.

FRANK, K.T.; PETRIE, B.; CHOI, J.S.; LEGGETT, W.C. 2005. Trophic cascades in a formerly cod-dominated ecosystem. *Science*, Vol. 308, pp. 1621–3.

GARIBALDI, A.; TURNER, N. 2004. Cultural keystone species: Implications for ecological conservation and restoration. E*cology and Society*, Vol. 9, No. 3, p 1. Available online at: Http://www.ecologyandsociety.org/vol9/iss3/art1 (last accessed on 27 June 2005).

HAGGAN, N.; TURNER, N.J.; CARPENTER, J.; JONES, J.T.; MENZIES, C.; MACKIE, Q. 2004. 12,000+ years of change: Linking traditional and modern ecosystem science in the Pacific Northwest. In R. Seaton (ed.), *Proceedings of the Sixteenth Society for Ecological Restoration Conference*, Victoria, BC.

HAIG-BROWN, C.; ARCHIBALD, J. 1996. Transforming First Nations research with respect and power. *Qualitative Studies in Education*, Vol. 9, No. 2, pp. 245–67.

HAUSER, L.; ADCOCK, G.J.; SMITH, P.J.; BERNAL RAMÍREZ, J.H.; CARVALHO, G.R. 2002. Loss of microsatellite diversity and low effective population size in an overexploited population of New Zealand snapper (*Pagrus auratus*). *Proceedings of the National Academy of Science USA*, Vol. 99, No. 18, pp. 11742–7.

HAYS, G.C.; BRODERICK, A.C.; GODLEY, B.J.; LUSCHI, P.; NICHOLS, W.J. 2003. Satellite telemetry suggests high levels of fishing-induced mortality in marine turtles. *Marine Ecological Progress Series*, Vol. 262, pp. 305–9.

HILBORN, R.; BRANCH, T.A.; ERNST, B.; MAGNUSSON, A.; MINTE-VERA, C.V.; SCHEUERELL, M.D.; VALERO, J.L. 2003. State of the world's fisheries. *Annual Review of Environment and Resources*, Vol. 28, pp. 359–99.

HOLM, P. 2003. Crossing the border: On the relationship between science and fishermen's knowledge in a resource management context. *MAST*, Vol. 2, No. 1, pp. 5–33.

HUTCHINGS, J.A.; NEIS, B.; RIPLEY, P. 2002. The nature of cod: Perceptions of stock structure and cod behaviour by fishermen, 'experts' and scientists from the nineteenth century to present. In: R. Ommer (ed.), *The resilient outport: Ecology, economy and society in rural Newfoundland*. St John's, ISER Books, pp. 140–85.

HUTCHINSON W.F.; VAN OOSTERHOUT, C.; ROGERS, S.I.; CARVALHO, G.R. 2003. Temporal analysis of archived samples indicates marked genetic changes in declining North Sea cod (*Gadus morhua*). Proceedings of the Royal Society London. *Biological Sciences*, Vol. 270, pp. 2125–32.

HUXLEY, T.H. 1883. Inaugural address. *The Fisheries Exhibition Literature*. London, International Fisheries Exhibition, Vol. 4, pp. 1–22.

JOHANNES, R.E. 1981. *Words of the lagoon: fishing and marine lore in the Palau district of Micronesia*. Berkeley, University of California Press, 245 pp.

JONES, R.R., SHEPHERD, M.; STERRITT, N.J. 2004. *Our place at the table: First Nations and the BC Fishery*. Vancouver, BC Aboriginal Fisheries Commission, 85 pp. Available online at: http://www.bcafc.org/documents/FNFishPanelReport0604.pdf (last accessed 21 April 2006).

MCRAE, D.M.; PEARSE, P.H. 2004. *Treaties and transition: Towards a sustainable fishery on Canada's Pacific coast*. Vancouver, Hemlock Printers. 58 pp. Available online at: http://www.gov.bc.ca/tno/down/pearse_mcrae_report-joint_fish_task_group.pdf (last accessed 27 June 2005).

MILAZZO, M. 1998. *Subsidies in world fisheries: A re-examination*. World Bank Technical Paper No. 406. Washington, World Bank. 130 pp.

MYERS, R.A.; WORM, B. 2003. Rapid worldwide depletion of predatory fish communities. *Nature*, Vol. 423, pp. 280–3.

NEIS, B.; FELT, L. (eds.). 2000. *Finding our sea legs: Linking fishery people and their knowledge with science and management*. St John's, Newfoundland, ISER books, 318 pp.

NEIS, B.; KEAN, R. 2003. Why fish stocks collapse: An interdisciplinary approach to understanding the dynamics of 'fishing up'. In: R. Byron (ed.), *Retrenchment and regeneration in rural Newfoundland*. Toronto, University of Toronto Press, pp. 65–102.

NEIS, B.; MORRIS, M. 2002. Fishers' ecological knowledge and fisheries science. In: R. Ommer (ed.), *The resilient outport: Ecology, economy, and society in rural Newfoundland*. St John's, ISER Books, pp. 205–40.

NEIS, B.; SCHNEIDER, D.; FELT, L.; HAEDRICH, R.; HUTCHINGS, J.; FISCHER, J. 1999. Northern cod stock assessment: What can be learned from interviewing resource users? *Canadian Journal of Fisheries and Aquatic Sciences*, Vol. 56, pp. 1949–63.

NEWELL, D.; OMMER, R.E. (EDS). 1999. *Fishing places, fishing people: Traditions and issues in Canadian small-scale fisheries*. Toronto, University of Toronto Press, 374 pp.

OMMER, R. 1994. One hundred years of fishery crisis in Newfoundland. *Acadiensis*, Vol. 23, No. 2, pp. 5–20.

PAINE, R.T. 1969. A note on trophic complexity and community stability. *The American Naturalist*, Vol. 103, pp. 91–93.

PAULY, D.; CHRISTENSEN, V.; DALSGAARD, J.; FROESE, R.; TORRES JR., F.C. 1998. Fishing down marine food webs. *Science*, Vol. 279, pp. 860–3.

PITCHER, T.J. 1997. Fish shoaling behaviour as a key factor in the resilience of fisheries: Shoaling behaviour alone can generate range collapse in fisheries. In: D.A. Hancock, D.C. Smith, A. Grant and T.P. Beumer (eds.), *The development and sustainability of world fisheries*. Collingwood, Australia, CSIRO, pp. 143–8.

——. 2001. Fisheries managed to rebuild ecosystems? Reconstructing the past to salvage the future. *Ecological Applications*, Vol. 11, 601–17.

POWER M.E.; TILMAN, D.; ESTES, J.A.; MENGE, B.A.; BOND, W.J.; MILLS, L.S.; DAILY, G.; CASTILLA, J.C.; LUBCHENCO, J.; PAINE, R.T. 1996. Challenges in the quest for keystones. *Bioscience*, Vol. 46, No. 8, pp. 609–20.

PRINCE, J.D. 2003. The barefoot ecologist goes fishing. *Fish and Fisheries*, Vol. 4, No. 4, pp. 359–71.

REIMCHEN, T. E. 2001. Salmon nutrients, nitrogen isotopes and coastal forests. *Ecoforestry*, Vol. 16, pp. 13–17.

ROMAN, J.; PALUMBI, S.R. 2003. Whales before whaling in the North Atlantic. *Science*, Vol. 301, pp. 508–10.

SCHINDLER, D.E.; ESSINGTON, T.E.; KITCHELL, J.F.; BOGGS, C.; HILBORN, R. 2002. Sharks and tunas: Fisheries impacts on predators with contrasting life histories. *Ecological Applications*, Vol. 12, No. 3, pp. 735–48.

STOCKNER, J.G. (ed.) 2003. *Nutrients in salmon ecosystems: Sustaining production and biodiversity*. Bethesda, Maryland, American Fisheries Society, Symposium 34, 285 pp.

STOREY, K. 1993. Defining the Reality. *Proceedings of the conference on The Newfoundland Groundfish Fisheries: Defining the Reality*. St John's, ISER Press, 246 pp.

SWEZEY, S.L.; HEIZER, R.F. 1993. Ritual Management of Salmonid Fish Resources. In: T.C. Blackburn and K. Anderson (eds.), *Before the Wilderness: Environmental Management by Native Californians*. Menlo Park, Calif, Ballena Press, pp. 299–327.

TRIPPEL, E.A. 1998. Egg size and viability and seasonal offspring production of young Atlantic cod. *Transactions of the American Fisheries Society*, Vol. 127, pp. 339–59.

UK ROLLS OF PARLIAMENT. 1376/77. Vol. 2, p. 369. 50 Edward III, (99H3).

WATKINSON, S. 2001. *The importance of salmon carcasses to watershed function*. Vancouver, Canada, University of British Columbia Fisheries Centre, MSc thesis, 111 pp.

CHAPTER 22   The last anecdote
Ian G. Baird

This last anecdote is dedicated to Bob Johannes, the driving force behind the conference that led to this book. Bob's major contribution was revealing the value of local ecological knowledge (LEK) to natural scientists and fisheries managers. Like him, we see the 'value of anecdote'. Thus, we close with one last anecdote about Bob, LEK and fisheries in the Mekong River basin in mainland South-East Asia, where I have worked for many years.

In recent years, I have been at the centre of a vigorous debate on the value of LEK for the management of highly migratory species in the Mekong region. The debate relates to the validity of villager efforts to establish fish conservation zones (FCZs) to ban or seriously restrict fishing in deep-water pool areas in the mainstream Mekong River in southern Laos (for more information about FCZs see Baird, Chapter 12 this volume; Baird, 2001; Baird and Flaherty, 2005; Baird, 2006).

The Lao Community Fisheries and Dolphin Protection Project (LCFDPP or 'the project') was established in 1993 as a small non-governmental organization (NGO) effort to support community-based fisheries co-management in Khong district, Champasak province. I was the senior advisor, working with a group of capable Lao colleagues. Apart from small-scale lowland rice cultivation, fishing is the most important livelihood activity (Baird, 2001). Local people were reporting that wild-capture fisheries were in decline (Roberts, 1993; Roberts and Baird, 1995); so improving fisheries management was high on the local agenda. The project thus came into being at an opportune time.

The project started from the premise that fishers with a long history in the area have considerable LEK and that given the right circumstances they could make a significant contribution to improving local fisheries management. We had no preconceptions of how exactly this could be done, and were in no particular rush to advocate a specific strategy. Instead, we decided to spend plenty of time living with the fishers and learning from them. It was during the early days of the project that I first read Bob's now classic book, *Words of the lagoon* (Johannes, 1981). The book was immediately inspiring and it greatly influenced my thinking and approach to fisheries work over the coming years.

After months of daily interactions with many villagers, I was approached by an elderly fisher who said, in Lao, 'One important way of protecting fish would be to protect the deep-water pools in the river, because that is where all the large fish concentrate, especially during the dry season.' This statement was apparently the result of my many conversations with him, and a series of related community discussions about the value of deep-water pools as FCZs ensued. The first government-recognized FCZ in the Mekong River was eventually established in December 1993 – almost a year after the project began.

The villagers know that water levels in the Mekong River decline up to thirty-fold during the dry season, more than in any other major river in the world (Cunningham, 1998). The result is massive changes and reductions in fish habitat, and water temperatures that can be unbearable for many fishes during the heat of the day during low-water periods. Local people know that at the height of the dry season fish concentrate in deep-water areas ranging from just a few metres to between 20 and 50 metres or more in depth (Baird, 2001; Baird and Flaherty, 2005; Baird, 2006). They also believe that protecting vulnerable large fish, including brood stock, in these dry season refuges makes sense as a way of ensuring that there will be more fish available in the future.

The project's role was limited to facilitating communications within and between communities and with local government, leading to the establishment of village-level co-management fisheries regulations. Before we would get involved, villagers had to ask us, in writing, to participate. The project had no say over what regulations would be established, but we did facilitate discussions within and between communities on the social and ecological appropriateness of local regulations. In the end, each village established different regulations, but most decided that setting up at least one FCZ would be an important part of their aquatic resource management strategies. Villages with no deepwater areas near them were the exception, as locals believe that it would be pointless to establish FCZs anywhere but in deep-water pool areas (see Baird, 2001; Baird and Flaherty, 2005; Baird, 2006; Baird, Chapter 12 this volume). Using their LEK, which the project helped disseminate from fisher to fisher and from village to village, over sixty communities in Khong District established more than seventy deepwater pool FCZs by the end of the 1990s when the project finally ended (Baird, 2001; Baird and Flaherty, 2005; Baird, 2006; Baird, Chapter 12 this volume).

While the local government in Khong was quick to endorse local ideas, FCZs were strongly opposed by outside experts working on fisheries in the Mekong River basin, who at that time did not appear to value LEK highly and believed in promoting a Western model of fisheries management. They were annoyed that local people were coming up with their own ideas and that we were supporting what they saw as 'untested LEK' as opposed to 'sound fisheries science'. But untested by whom? Fishers had certainly had ample opportunity to test their ideas on the water, over decades or hundreds of years.

The opponents of FCZs accused the project of 'providing unfounded and unscientific information to local people about the value of FCZs', and formally asked the provincial government in Champasak to disallow the establishment of new FCZs. Fortunately, the official they spoke to was from Khong, and knew from personal experience that these fishers knew what they were talking about.

He therefore defended their right to implement measures they believed could help protect fisheries resources.

At the height of the controversy in the late 1990s, Bob Johannes visited the Mekong region at the invitation of the fisheries component of the newly established Mekong River Commission (MRC). As a long admirer of his important work in Micronesia, I was eager to read an interview with him that appeared in the MRC's newsletter, *Catch and Culture*, soon after his short trip to the region (Jensen, 1999). I was shocked to find that instead of acknowledging the LEK of fishers and endorsing their decisions to establish FCZs in deepwater pool areas, he seemed to endorse a different vision that essentially ruled out local management efforts, including FCZs for Mekong fish. It appears that he based this opinion on the views of outside experts that most or all of the fish in the Mekong River are highly migratory. In which case, local management efforts would have little effect.

I finally got to meet Bob at the August 2001 Putting Fishers' Knowledge to Work Conference. We went for a walk, during which time I broached the subject of his troubling views on Mekong River fisheries management. He listened patiently and attentively as I explained why I believed that the LEK of local people in Laos was sound, and why villager decisions to establish FCZs were justified. I explained that there may be 1,000 or more fish species in the Mekong River basin, giving it one of the most diverse freshwater fish faunas in the world, and certainly the most biodiverse in Asia (Roberts, 1993; Rainboth, 1996).

Based on this enormous biodiversity, I suggested that while some species certainly are highly migratory (Baird et al., 2001a; Baird et al., 2003; Hogan et al., 2004), others undertake only local seasonal movements, while some could be considered almost sedentary (Baird et al., 1999; Baird et al., 2001b). Moreover, FCZs have the potential to protect resources at larger spatial scales when networks of FCZs are situated along the course of a single river, leading to both synergistic and cumulative positive impacts (Baird and Flaherty, 2005; Baird, 2006). I explained why local people believe that protecting fish during the dry season in deepwater pools is very important to the lifecycles of many Mekong species, including those of many highly migratory fishes, which frequent the areas during critical times in their lifecycles. While acknowledging that there was still much to be learned, I expressed my belief that the LEK of fishers in Khong was sophisticated enough to provide a sound basis for management decisions. I also told him that it would be very difficult to improve fisheries management only through transboundary or long-distance management strategies, as there are six nation states in the Mekong River basin. It therefore makes sense to at least try to strengthen local fisheries management, as villagers have been doing in Khong. Finally, I indicated that independent evaluations had already confirmed that local people firmly believe that FCZs have benefited fish stocks and that they have often resulted in improved fish catches (Meusch, 1997; Chomchanta et al., 2000).

Since I spoke with Bob in 2001, hydro-acoustic studies of deep-water pools conducted in the mainstream Mekong River in southern Laos and north-east Cambodia have confirmed what the villagers had always told us – that deepwater pools are critical keystone habitats for many migratory and non-migratory fish species during the dry season, and that they are full of fish (Kolding, 2002; Phounsavath et al., 2004). This research supports our conclusion that, so long as locals support the idea and feel that such protection is appropriate for their area, FCZs can make an important contribution to management. Over ten years after villagers first suggested establishing FCZs in deep-water areas in the Mekong River, the wisdom of their decisions have finally been verified by 'science'!

What struck me most about my conversation with Bob that day was that he was able to listen to me with an open mind and reassess his original position on the value of LEK in that particular situation. Based on what I told him, he was willing to set aside his ego, admit his mistake and even apologize for making it! Unfortunately, this is not a common characteristic amongst highly revered scientists and I could immediately see that it was this quality – Bob's ability to remain open to information from different sources and to understand the importance of local knowledge to fisheries science and management – that led him to do so much important research over his career. Bob was a wise and determined scientist. This is why he was able to inspire and excite so many, not only with his writings, but also through the personal connections that he was able to establish with fishers, scientists and others over his career. Bob was the father of 'ethno-ichthyology' and his energy and leadership in the field will be greatly missed.

REFERENCES

BAIRD, I.G. 2001. Towards sustainable co-management of Mekong River aquatic resources: The experience in Siphandone wetlands. In: G. Daconto (ed.), *Siphandone Wetlands*. Bergamo, Italy, CESVI, pp. 89–111.

BAIRD, I.G. 2006. Strength in diversity: Fish sanctuaries and deep-water pools in Laos. *Fisheries Management and Ecology*, Vol. 13, pp. 1–8.

BAIRD, I.G.; INTHAPHAISY, V.; KISOUVANNALATH, P.; PHYLAVANH, B.; MOUNSOUPHOM, B. 1999. *The fishes of southern Lao* (In Lao). Pakse, Lao PDR, Lao Community Fisheries and Dolphin Protection Project, Ministry of Agriculture of Forestry, 162 pp.

BAIRD, I.G.; HOGAN, Z.; PHYLAVANH, B.; MOYLE, P. 2001a. A communal fishery for the migratory catfish *Pangasius macronema* in the Mekong River. *Asian Fisheries Science*, Vol. 14, pp. 25–41.

BAIRD, I.G.; PHYLAVANH, B.; VONGSENESOUK, B.; XAIYAMANIVONG, K. 2001b. The ecology and conservation of the smallscale croaker *Boesemania microlepis*

(Bleeker 1858–59) in the mainstream Mekong River, southern Laos. *Natural History Bulletin of the Siam Society,* Vol. 49, pp. 161–76.

BAIRD, I.G.; FLAHERTY, M.S.; PHYLAVANH, B. 2003. Rhythms of the river: Lunar phrases and migrations of small carps (*Cyprinidae*) in the Mekong River. *Natural History Bulletin of the Siam Society,* Vol. 51, No. 1, pp. 5–36.

BAIRD, I.G.; FLAHERTY, M.S. 2005. Mekong River Fish conservation zones in southern Laos: Assessing effectiveness using local ecological knowledge. *Environmental Management,* Vol. 36, No. 3, pp. 439–54.

CHOMCHANTA, P.; VONGPHASOUK, P.; SOUKHASEUM, V.; SOULIGNAVONG, C.; SAADSY, B.; WARREN, T. 2000. *A preliminary assessment of Mekong fishery conservation zones in the Siphandone area of southern Lao PDR, and recommendations for further evaluation and monitoring.* LARReC Technical Paper No. 0001. Vientiane, Lao PDR, The Living Aquatic Resources and Research Centre, 13 pp.

CUNNINGHAM, P. 1998. Extending a co-management network to save the Mekong's giants. *Catch and Culture,* Vol. 3, No. 3, pp. 6–7.

HOGAN, Z.S.; MOYLE, P.B.; MAY, B.; VANDER ZANDEN, M.J.; BAIRD, I.G. 2004. The imperilled giants of the Mekong: Ecologists struggle to understand – and protect – Southeast Asia's large migratory catfish. *American Scientist,* Vol. 92, pp. 228–37.

JENSEN, J.G. 1999. Co-management and traditional fishing rights: Bob Johannes visits the Mekong. *Catch and Culture,* Vol. 4, No. 3, pp. 1–3.

JOHANNES, R.E. 1981. *Words of the lagoon: Fishing and marine lore in the Palau District of Micronesia.* Berkeley, Calif, University of California Press, 245 pp.

KOLDING, J. 2002. *The use of hydro-acoustic surveys for the monitoring of fish abundance in the deep pools and fish conservation zones in the Mekong River, Siphandone area, Champassak Province, Lao PDR.* LARReC technical report. Vientiane, Lao PDR, 32 pp.

MEUSCH, E. 1997. Participatory Evaluation workshop of the LCFDPP. Savannakhet, Lao PDR, Asian Institute of Technology Aqua Outreach Program report, 7 pp.

PHOUNSAVATH, S.; PHOTITAY, C.; VALBO-JORGENSEN, J. 2004. *Deep pools survey in Stung Treng and Siphandone area 2003–2004.* LARReC Progress Report. Vientiane, Lao PDR, The Living Aquatic Resources and Research Centre, 13 pp.

RAINBOTH, W.J. 1996. *Field guide to fishes of the Cambodian Mekong.* Rome, Food and Agriculture Organization of the United Nations, 265 pp.

ROBERTS, T.R. 1993. Artisinal fisheries and fish ecology below the great waterfalls of the Mekong River in southern Laos. *Natural History Bulletin of the Siam Society,* Vol. 41, No. 1, pp. 31–62.

ROBERTS, T.R.; BAIRD, I.G. 1995. Traditional fisheries and fish ecology on the Mekong River at Khone Waterfalls in Southern Laos. *Natural History Bulletin of the Siam Society,* Vol. 43, pp. 219–62.

Achevé d'imprimer sur les presses de l'Imprimerie BARNÉOUD
B.P. 44 - 53960 BONCHAMP-LÈS-LAVAL
Dépôt légal : Février 2007 - N° d'imprimeur : 612015
Imprimé en France